ENZYME BIOCATALYSIS

ENZYME BIOCATALYSIS
PRINCIPLES AND APPLICATIONS

Driss Cherqaoui
Diana Avetyan
Jordan Karubian

Kruger Brentt
Publishers

2024

Kruger Brentt Publishers UK. LTD.
Company Number 9728962

Regd. Office: 68 St Margarets Road, Edgware, Middlesex HA8 9UU

ISBN: 978-1-78715-257-1

For information on all our publications visit our website at http://krugerbrentt.com/

PREFACE

Enzymes, the molecular architects of life, stand as intricate catalysts orchestrating the myriad biochemical reactions that sustain living organisms. These remarkable proteins act as nature's catalysts, accelerating chemical transformations with unparalleled precision and efficiency. In the intricate dance of cellular processes, enzymes emerge as the conductors, ensuring that reactions vital for life occur at rates compatible with the dynamic equilibrium of living systems. The study of enzymes and catalysis unveils the secrets behind the astonishing speed and specificity with which these biological catalysts operate, offering profound insights into fundamental biological processes and paving the way for revolutionary applications in fields as diverse as medicine, biotechnology, and environmental science. This exploration into the realm of enzymes and catalysis is a journey into the heart of life's molecular machinery, where nature's catalysts govern the symphony of biochemistry.

Enzymes, the molecular workhorses, play a crucial role in governing the rate and specificity of biochemical reactions. Their ability to accelerate reactions under mild conditions makes them invaluable tools in various fields, ranging from biotechnology and medicine to environmental science and industrial processes. This book delves into the latest advancements and groundbreaking research, providing a thorough understanding of the principles, applications, and emerging trends in the field of enzymes and biocatalysis. Dive into the forefront of enzymology with a collection of chapters featuring the latest research findings. From novel halogenases in bacterial consortia to innovative approaches in enzyme production and purification, each chapter presents a unique perspective on the evolving landscape of biocatalysis.

The importance of enzymes and catalysis resonates across the spectrum of life sciences, industry, and environmental sustainability. Enzymes are the unsung heroes of biological systems, orchestrating the precise and timely execution of countless chemical reactions that underpin the complexity of life. Their catalytic prowess is pivotal in maintaining cellular function, regulating metabolic pathways, and ensuring energy efficiency within organisms. Beyond the realms of biology, enzymes play a transformative role in industry, facilitating greener and more sustainable processes by replacing traditional chemical methods. Their specificity and efficiency make them indispensable tools in biotechnology, from the production of pharmaceuticals to the synthesis of biofuels. Moreover, enzymes contribute to environmental stewardship, participating in the remediation of pollutants and the sustainable management of natural resources. In essence, the study and application of enzymes and catalysis not only deepen our understanding of the fundamental processes of life but also hold the key to addressing global challenges with innovation, efficiency, and environmental responsibility.

The book explores the practical applications of enzymes in various industries. From the use of exogenous enzymes in animal feed for zootechnical benefits to the employment of lipases in eco-friendly processes, the book provides insights into how biocatalysis is shaping the future of sustainable and efficient technologies. Uncover the potential of metagenomic approaches in discovering novel biocatalysts from natural resources like soil and water. The exploration of microbial diversity opens new avenues for biocatalyst discovery, showcasing the wealth of untapped enzymatic potential in our environment. Witness the advancements in protein engineering with chapters dedicated to the rational design of enzymes. The development of trypsin variants independent of calcium highlights the innovative strategies employed to tailor enzymes for specific applications. This book serves as a valuable resource for researchers, students, and professionals in the fields of biochemistry, biotechnology, and environmental science, offering a comprehensive overview of the current state of enzymes and biocatalysis.

We are grateful to all those persons as well as various books, manuals, periodicals, newsletters, magazines, journals etc. that helped in the preparation of this book. In spite of the best efforts, it is possible that some errors may have occurred into the compilation and editing of the book. Further queries, constructive suggestions and criticisms for the further improvement of the book are always welcome and shall be thankfully acknowledged.

Driss Cherqaoui

Diana Avetyan

Jordan Karubian

CONTENTS

CHAPTER-1

INTRODUCTION TO ENZYMES AND BIOCATALYSIS

1. INTRODUCTION: AN OVERVIEW

Enzymes, also known as biocatalysts, are proteins produced by living cells and found in a wide range of species, including animals, plants, and microorganisms. Due to their specificity, enzymes are often widely used to catalyze various chemical reactions as proficient biocatalysts in many applications. Since an enzyme is a protein, it easily becomes denatured and loses its activity under unfavourable conditions of the surrounding environment. Therefore, in order to prolong the activity and increase enzyme stability, immobilization technology is commonly used for improving the overall efficiency of enzyme catalysis. Immobilized enzymes can maintain high efficiency, specificity, and mild reaction characteristics to overcome the shortcomings of free enzymes [1]. Immobilized enzymes have many advantages, such as exhibiting high storage stability [2], easy separation and recovery [3], reusability [4], continuous operation in a bioreactor [5,6], and high stability. The commonly used methods for immobilizing enzymes on carriers are adsorption, cross-linking, covalent binding, and entrapment [7]. Among these methods, adsorption is the simplest and most common way to immobilize enzymes. The adsorption method mainly works through van der Waals interactions, hydrophobic interactions, or electrostatic interactions between the carrier and the enzyme. Therefore, adsorption can maintain the maximum activity of enzymes since there is no or significantly less configurational change in the enzyme's structure.

Remonatto et al. [8] investigated the immobilization process of lipases by physical adsorption using clay supports, including diatomite, vermiculite, montmorillonite KSF (MKSF), and kaolinite. The results showed the lipase immobilized on MKSF support had an improvement in the catalytic performance that presented a 69.47% immobilization yield and higher hydrolytic activity (270.7 U g^{-1}). The V_{max} value of immobilized lipase was 13 times greater than the free one, indicating that the lipase activity was improved. Lipases are versatile enzymes that have several industrial applications [9]. The lipases immobilized on MKSF demonstrate high temperature and pH stability, making them suitable for industrial applications. More specifically, inorganic clays are low-cost supports with high adsorption capacity, environmentally friendly properties, and renewable abundance.

2. PROCESS OF ENZYMES AND BIOCATALYSIS

2.1 .Biocementation of Soil

The biocementation of soil has received significant attention as an environmentally friendly alternative to chemical stabilization methods. An enzyme called urease is required to catalyze the chemical reaction that generates calcium carbonate precipitates from urea hydrolysis [10,11]. Pinto Vilar and Ikuma [12] reported the adsorption of a specific target protein, urease, as part of a complex, natural protein extract on soils through different surface chemistries with emphasis on the retained activity of the adsorbed urease. The adsorption and retention of urease activity is a critical first step for successful biocementation, which is a catalytic method used for soil stabilization. The results showed that the mass of proteins adsorbed was similar across soils with different surface chemistries (i.e., sand only, iron-coated sand, and hydrophobic sand). Urease was preferentially adsorbed in soils with hydrophobic contents greater than 20% (*w/w*) compared to total proteins contained in crude bacterial protein extracts. A comparison of the urease adsorption into silica sand and soil mixtures, as well as iron-coated sand, showed that it was much lower than the total protein adsorption. There was preferential adsorption of urease within the crude protein extract onto hydrophobic surfaces, which in turn yielded higher urease activity. These results suggest that soil surface manipulations may significantly enhance enzymatic activity, leading to improved outcomes for biocementation.

2.2. Packed-Bed Bioreactor

The packed-bed bioreactor is a simple piece of equipment that includes a pump and a column packed with immobilized enzymes and is commonly used for continuous production [13,14]. In the industry, the packed-bed bioreactor is commonly used to maximize the efficiency of enzyme reactions since it is energy-efficient, reduces reaction volumes, and is convenient to operate continuously [15]. Ethyl esters of omega 3 fatty acids, especially docosahexaenoic acid (DHA) and eicosapentaenoic acid (EPA) ethyl esters, are active pharmaceutical ingredients used to reduce triglyceride levels in the treatment of hyperlipidemia [16]. The first time DHA and EPA ethyl esters were produced continuously from DHA + EPA concentrate and ethyl acetate in an ultrasonic packed-bed bioreactor was reported by Kuo et al.

[17]. The mass transfer kinetic model was used to evaluate the efficiency of the ultrasound. The results showed that the ultrasonic packed-bed bioreactor had a higher external mass transfer coefficient compared to the packed-bed bioreactor. With ultrasonication, the highest conversion of 99% was obtained at a flow rate of 1 mL min^{-1} and substrate concentration of 100 mM. When increasing the substrate concentration and flow rate to 500 mM and 5 mL min^{-1}, the conversion remained at 93%. The effect of ultrasonication was also evaluated by the kinetic model using an ordered mechanism in the batch reaction. There was an 8.9 times greater V'_{max}/K_2 value in the ultrasonic bath, indicating that ultrasonication greatly enhanced lipase-catalyzed transesterification efficiency. For long-term operation, the ultrasonic packed-bed bioreactor that performed under the highest conversion conditions showed that the enzyme remained stable for at least 5 days and maintained a conversion of 98%. The ultrasonic packed-bed bioreactor is a continuously operating system that, in terms of enzyme usage and production cost, is superior to batch production. Additionally, it has a good chance of being used in industrial mass production or production capacity in the future.

2.3 Enzymes from microorganisms

Enzymes from microorganisms play a crucial role as metabolic catalysts, and they can be used in numerous industries and applications. Recent developments in the discovery of stable enzymes have expanded their uses to encompass organic synthesis and the production of specialized chemicals, pharmaceutical intermediates, and agrochemicals [18]. In order to improve enzyme performance, microbial screening or gene cloning is generally used [19,20,21]. It is also common for enzyme studies to use a particular medium for microbiological fermentation in order to produce enzymes.

Dakhmouche Djekrif et al. [22] demonstrated the ability of a yeast strain *Clavispora lusitaniae* ABS7, isolated from wheat grains that produced significant amounts of amylopullulanase on by-products of milk manufacturing (whey). The chromatographic profile on Sephacryl S-200 revealed two α-amylase and pullulanase-specific activities eluted together with the protein peak. The elution on DEAE-cellulose confirmed the presence of both activities in the same fraction. The α-amylase and pullulanase were purified with purification rates of 50.45 and 44.59 and yields of 23.88% and 21.11%, respectively. The purified enzyme showed a single band on the SDS-PAGE gel with an estimated molecular weight of 75 KDa. The coexistence of the two pullulanase and amylase activities was analyzed from the band, which suggested that the *C. lusitaniae* ABS7 strain had a bifunctional amylolytic enzyme with two active sites: one for α-amylase and the other for pullulanase, thus allowing simultaneous hydrolysis of α-1,4 and α-1,6 glycosidic bonds. The thin layer chromatography also confirmed that the enzyme digested starch to maltose and glucose and pullulan to maltotriose, maltose, and glucose. These two activities are probably localized in two distinct active sites of a type II amylopullulanase with saccharifying power. The α-amylase and pullulanase had pH optima at 9 and temperature optima at 75°C and 80°C, respectively. After heat treatments for 3 h at 75°C, 2 h at 100°C, and 3 h at 100°C, α-amylase retained 88%, 51.76%, and 38.6% activity, respectively. The pullulanase maintained 91% and 42% activity after incubations at 80°C and 100°C for 3 h, respectively. The results showed the excellent thermostability of *C. lusitaniae* ABS7 amylopullulanase type II. The effect of various metal ions and chemical reagents on the α-amylase and pullulanase activities were also studied. Because of its excellent stability and compatibility with commercial laundry detergents, the alkalothermophilic amylopullulanase of *C. lusitaniae* ABS7 is suitable for industrial use, especially in detergents.

2.3 Sequential fermentation (SF)

Sequential fermentation (SF) is usually used in the fermentation process of rice wine [23,24], which combines solid-state (SSF) and submerged (SmF) fermentation. De Oliveira et al. [25] developed an SF method for producing protease from *Aspergillus tamarii* URM4634 using wheat bran as a substrate. A 2^3 full factorial design was used to examine the effects of glucose concentration, medium volume, and inoculum size on protease production. Moreover, glucose concentration and medium volume were optimized using a 2^2 central composite rotational design to achieve a maximum protease activity of 180.17 U mL^{-1}. The protease activity produced by the SF method was increased approximately 9-fold over the conventional SmF method. A temperature of 50 °C and pH of 7 were found to be optimal for *A. tamarii* URM4634 protease. In the kinetic and thermodynamic study, the enzyme exhibited values of the Michaelis constant (Km), maximum rate (V_{max}), and turnover number (kcat) of 16.26 mg mL^{-1}, 147.06 mg mL^{-1} min^{-1}, and 195.37 s^{-1}, respectively. The activation energy (E*a) of 40.38 kJ mol^{-1} was estimated for azocasein hydrolysis. Protease thermo stability was performed in the temperature range of 50 to 80 °C. The results indicated that the protease was thermostable at 50°C, which is a commonly used temperature in many industrial processes, as evidenced by a half-life of 231.05 min and a decimal-reduction time of 767.53 min.

2.4 Nitrile hydratase

Nitrile hydratase is a biocatalytic enzyme that is commonly used to synthesize acrylamide from acrylonitrile [26,27]. Grill et al. [28] investigated a new nitrile hydratase, *Gordonia hydrophobica* nitrile hydratase (GhNHase), which was produced in *Escherichia coli* and applied as a cell-free extract (CFE). An enzymatic dynamic kinetic resolution with GhNHase was employed to prepare the precursor of API levetiracetam, 2-(pyrrolidine-1-yl) butanamide, via enantioselective. The GhNHase was used to convert (RS)-2-(pyrrolidine-1-yl) butanenitrile to (S)-2-(pyrrolidine-1-yl) butanamide. The effect of GhNHase, substrate concentration, and reaction temperature on the product titer and enantiomeric excess was explored. The results showed that the substrate concentration, reaction pH, and amount of biocatalyst were critically important factors for achieving (S)-configured amide conversions. Among APIs with amide groups, nitrile is often used as the precursor to introduce the amide group. GhNHase showed full conversion of the nitriles, including benzonitrile, 3-cyanopyridine, and pyrazine-2-carbonitrile, indicating that substrates with cyano groups directly attached to aromatic systems might be preferred.

2.5 Cyanide hydratases

Cyanide hydratases (CynHs) (EC 4.2.1.66) belong to the nitrilase family and catalyze the hydration of cyanide to formamide. Sedova et al. [29] used a new CynH from *Exidia glandulosa* (protein KZV92691.1, namely, NitEg), which was overproduced in *Escherichia coli*. The cell-free extract had a specific activity of 280 U mg protein^{-1}, which was increased to 697 U mg protein^{-1} after purification via cobalt ion affinity chromatography. The purified NitEg (4.0 μg protein mL^{-1}) could convert 25 mM free cyanide (fCN) to formamide in 60 min. The NitEg exhibited values of the Michaelis constant (Km), maximum rate (V_{max}), and turnover number (k_{cat}) of 22.2 mM, 1335 U mg^{-1}, and 927 s^{-1}, respectively. Enzyme performance is often represented by the V_{max}/K_m parameter [30]. The V_{max}/K_m ratio (U mg^{-1} mM^{-1}) was 60 in NitEg, which was similar to *Aspergillus niger* CynH (62) but higher than *Gloeocercospora sorghi* CynH (49), indicating the high performance of NitEg. The optimal activity of NitEg occurred at 40–45 °C and a pH range of approximately 6–9. A simulated electroplating effluent was composed of 100 mM fCN and 1 mM $AgNO_3$ or 1 mM $CuSO_4$. The results showed that the NitEg at a concentration of 14 μg enzyme mL^{-1} was sufficient to remove more than 97% fCN in the simulated electroplating effluent after 1 h. The NitEg demonstrated excellent performance in fCN solutions of up to 100 mM concentrations under alkaline conditions [29].

2.6 Phenylalanine ammonia lyase (PAL)

Phenylalanine ammonia lyase (PAL) is an important enzyme involved in the phenylpropanoid pathway, which catalyzes the biosynthesis of polyphenolic compounds such as phenylpropanoids, flavonoids, and lignin in plants. Hsieh et al. [31] investigated the expression of BoPAL4, a Bamboo PAL protein, in *Escherichia coli* Top10. BoPAL4 contained a 2106 bp open-reading frame and encoded a 701- amino acids polypeptide. L-Phe, L-Tyr, and L-3,4-dihydroxy phenylalanine (L-DOPA) were used as substrates to examine PAL, tyrosine ammonia-lyase (TAL), and L-DOPA ammonia-lyase (DAL) activities of BoPAL4. The optimal reaction pH and temperature for BoPAL4 on three substrates were at 9.0, 8.5, 9.0, and 50, 60, 40°C, respectively. In addition, the k_{cat} value of 1.87 s^{-1} and K_m value of 640 μM indicated that the Phe-123 to His mutation of BoPAL4 had a higher L-Phe binding affinity than wild-type BoPAL4. BoPAL4 can catalyze L-Phe, L-Tyr, and L-DOPA to yield trans-cinnamic acid, p-coumaric acid, and caffeic acid, respectively. Trans-cinnamic acid derivatives have been shown to have beneficial effects on human health and may therefore enhance the utility of BoPAL4.

2.7 Trypsin

Trypsin is a serine protease that hydrolyzes peptides and proteins on the carboxyl side of lysines and arginines. The robustness and high enzymatic activity of trypsin make it to be use in diverse products and many biotechnological applications [32,33]. The robustness of trypsin is based on three intrinsic disulfide bridges and is also influenced by the direct presence of Ca^{2+}-ions in the calcium-binding loop. Simon et al. [34] inserted an additional fourth disulfide bridge by substitution of Glu70 and Glu80 to Cys70 and Cys80, named aTn. The results found that disulfide bonds eliminated the enzyme's dependence on Ca^{2+}-ions. The disulfide bond in aTn prevents autolysis in absence of Ca^{2+}-ions situation. In addition, the aTn did not seem to denature completely even at 90 °C indicated the disulfide bridge also contributed to increase thermal stability. The non-structural antigen protein 3 of the hepatitis C virus (HCV NS3), commonly used for ELISA diagnosis of HCV, has protease and helicase activities. Huang et al. [35] developed a clone with a special design to produce a truncated NS3 recombinant protein (without protease domain) and over expressed in the *E. coli* expression system. As the temperature shifted from 37 to 25 °C, the yield of the soluble fraction of HCV NS3 was increased from 4.15 to 11.1 mg L^{-1}. In terms of solubility, purity, antigenic efficacy, and stability, low-temperature (25 °C) protein expression exhibits superior performance than high-temperature expression at 37 °C. The comparison of recombinant HCV ELISA with two commercial kits, Abbott HCV EIA 2.0 and Ortho HCV EIA 3.0, was also carried out and thoroughly discussed. It was demonstrated that truncated NS3 produced at 25 °C showed a better discriminating ability than the proteins produced at 37 °C and was competitive with commercial kits, suggesting that it might have potential for HCV ELISA diagnosis.

Velázquez-De Lucio et al. [36] discussed exogenous enzymes as zootechnical additives in animal feed. Exogenous enzymes have been shown to increase the bioavailability and digestibility of nutrients as well as eliminate some anti-nutritional factors from agro-industrial and agro-forestry wastes used as animal feed. A general classification of enzymes depends on the substrate on which they act; commercially, in animal nutrition, enzymes are divided into three categories based on their purpose: carbohydrases, proteases, and phytase. The enzymes used as zootechnical additives in animal feed, such as those for poultry, swine, ruminant, fish, and dogs, were also discussed. Additionally, the production methods of various enzymes for animal feeding are also introduced. A description of enzymes used for animal nutrition, their mode of action, production, and new sources of production are presented in this review, along with studies on different animal models.

Neubauer et al. [37] described the application of a Hidden Markov Model (HMM) for the identification of novel flavin-dependent halogenases in metagenomes. They chose metagenomic data since they contain many genomes, even from uncultivable organisms, which expands the scope for identification. With this HMM, 254 complete and partial putative flavin-dependent halogenase genes in 11 metagenomic data sets were identified. In another study, the HMM strategy was used to screen the bacterial associates of the *Botryococcus braunii* consortia (PRJEB21978) [38] and identify several putative flavin-dependent halogenase genes. Two of these novel halogenase genes were found in one gene cluster of the *Botryococcus braunii* symbiont *Sphingomonas* sp. In vitro activity tests revealed that both heterologously expressed enzymes are active flavin-dependent halogenases able to halogenate indole, indole derivatives, and phenol derivatives while preferring bromination over chlorination.

2.8 Metagenomics

Metagenomics is the study of directly obtaining all the genetic material in the environment. Metagenomics can be used as an innovative strategy to study these unculturable microorganisms by using DNA extracted from environmental samples. Sousa et al. [39] discussed the most relevant information reported in the last decade about the application of metagenomics for the discovery of promising biocatalysts in soil and water. Metagenomic studies are classified into the categories of raw resources, human-manipulated resources, and unspecified resources. A compilation of metagenomic data obtained from different environments was provided. The metagenomic studies conducted for enzyme discovery demonstrated that a considerable number of promising environments are yet to be explored. The authors analyzed and discussed the research on metagenomics from a global perspective. The data and discussion offered in this review will be fascinating for a large community of researchers working on metagenomic studies to examine microbial functionality.

In conclusion, this Special Issue shows that enzymes and biocatalysis have found numerous applications in various fields. These applications include the production of APIs, wastewater treatment, transformation of health ingredients, laundry detergent applications, fortification of feed, and biocementation. Moreover, there are studies focusing on the immobilization and production procedures of enzymes.

Furthermore, an extensive review of metagenomics studies related to water and soil is included.

REFERENCES

1. Li, G.; Zhu, F.; Gu, F.; Yin, X.; Xu, Q.; Ma, M.; Zhu, L.; Lu, B.; Chen, N. Enzymatic preparation of L-malate in a reaction system with product separation and enzyme recycling. *Catalysts* **2022**, *12*, 587. [**Google Scholar**]
2. Kuo, C.-H.; Liu, Y.-C.; Chang, C.-M.J.; Chen, J.-H.; Chang, C.; Shieh, C.-J. Optimum conditions for lipase immobilization on chitosan-coated Fe_3O_4 nanoparticles. *Carbohydr. Polym.* **2012**, *87*, 2538–2545. [**Google Scholar**]
3. Xie, W.; Huang, M. Enzymatic production of biodiesel using immobilized lipase on core-shell structured Fe_3O_4@ MIL-100 (Fe) composites. *Catalysts* **2019**, *9*, 850. [**Google Scholar**]
4. Kuo, C.-H.; Chen, G.-J.; Twu, Y.-K.; Liu, Y.-C.; Shieh, C.-J. Optimum lipase immobilized on diamine-grafted PVDF membrane and its characterization. *Ind. Eng. Chem. Res.* **2012**, *51*, 5141–5147. [**Google Scholar**]
5. Huang, S.-M.; Huang, H.-Y.; Chen, Y.-M.; Kuo, C.-H.; Shieh, C.-J. Continuous production of 2-phenylethyl acetate in a solvent-free system using a packed-bed reactor with Novozym® 435. *Catalysts* **2020**, *10*, 714. [**Google Scholar**]
6. Alvarez-Gonzalez, C.; Santos, V.E.; Ladero, M.; Bolivar, J.M. Immobilization-Stabilization of β-Glucosidase for Implementation of Intensified Hydrolysis of Cellobiose in Continuous Flow Reactors. *Catalysts* **2022**, *12*, 80. [**Google Scholar**]
7. Lyu, X.; Gonzalez, R.; Horton, A.; Li, T. Immobilization of enzymes by polymeric materials. *Catalysts* **2021**, *11*, 1211. [**Google Scholar**]
8. Remonatto, D.; Ferrari, B.R.; Bassan, J.C.; Mussagy, C.U.; de Carvalho Santos-Ebinuma, V.; Veloso de Paula, A. Utilization of clay materials as support for Aspergillus japonicus lipase: An eco-friendly approach. *Catalysts* **2021**, *11*, 1173. [**Google Scholar**]
9. López-Fernández, J.; Benaiges, M.D.; Valero, F. Rhizopus oryzae lipase, a promising industrial enzyme: Biochemical characteristics, production and biocatalytic applications. *Catalysts* **2020**, *10*, 1277. [**Google Scholar**]
10. Leeprasert, L.; Chonudomkul, D.; Boonmak, C. Biocalcifying Potential of Ureolytic Bacteria Isolated from Soil for Biocementation and Material Crack Repair. *Microorganisms* **2022**, *10*, 963. [**Google Scholar**] [**CrossRef**]
11. Murugan, R.; Suraishkumar, G.; Mukherjee, A.; Dhami, N.K. Influence of native ureolytic microbial community on biocementation potential of Sporosarcina pasteurii. *Sci. Rep.* **2021**, *11*, 20856. [**Google Scholar**]

12. Pinto Vilar, R.; Ikuma, K. Effects of Soil Surface Chemistry on Adsorption and Activity of Urease from a Crude Protein Extract: Implications for Biocementation Applications. *Catalysts* **2022**, *12*, 230. [Google Scholar]

13. Rodriguez-Colinas, B.; Fernandez-Arrojo, L.; Santos-Moriano, P.; Ballesteros, A.O.; Plou, F.J. Continuous packed bed reactor with immobilized β-galactosidase for production of galactooligosaccharides (GOS). *Catalysts* **2016**, *6*, 189. [Google Scholar]

14. Hollenbach, R.; Muller, D.; Delavault, A.; Syldatk, C. Continuous Flow Glycolipid Synthesis Using a Packed Bed Reactor. *Catalysts* **2022**, *12*, 551. [Google Scholar]

15. Sen, P.; Nath, A.; Bhattacharjee, C. Packed-bed bioreactor and its application in dairy, food, and beverage industry. In *Current Developments in Biotechnology and Bioengineering*; Elsevier: Amsterdam, The Netherlands, 2017; pp. 235–277. [Google Scholar]

16. Watanabe, Y.; Tatsuno, I. Omega-3 polyunsaturated fatty acids focusing on eicosapentaenoic acid and docosahexaenoic acid in the prevention of cardiovascular diseases: A review of the state-of-the-art. *Expert Rev. Clin. Pharmacol.* **2021**, *14*, 79–93. [Google Scholar]

17. Kuo, C.-H.; Tsai, M.-L.; Wang, H.-M.D.; Liu, Y.-C.; Hsieh, C.; Tsai, Y.-H.; Dong, C.-D.; Huang, C.-Y.; Shieh, C.-J. Continuous Production of DHA and EPA Ethyl Esters via Lipase-Catalyzed Transesterification in an Ultrasonic Packed-Bed Bioreactor. *Catalysts* **2022**, *12*, 404. [Google Scholar]

18. Demirjian, D.C.; Moís-Varas, F.; Cassidy, C.S. Enzymes from extremophiles. *Curr. Opin. Chem. Biol.* **2001**, *5*, 144–151. [Google Scholar]

19. Nigam, P.S. Microbial enzymes with special characteristics for biotechnological applications. *Biomolecules* **2013**, *3*, 597–611. [Google Scholar]

20. Wang, Y.-T.; Wu, P.-L. Gene cloning, characterization, and molecular simulations of a novel recombinant chitinase from Chitinibacter Tainanensis CT01 appropriate for chitin enzymatic hydrolysis. *Polymers* **2020**, *12*, 1648. [Google Scholar]

21. Guo, B.; Li, P.-Y.; Yue, Y.-S.; Zhao, H.-L.; Dong, S.; Song, X.-Y.; Sun, C.-Y.; Zhang, W.-X.; Chen, X.-L.; Zhang, X.-Y. Gene cloning, expression and characterization of a novel xylanase from the marine bacterium, Glaciecola mesophila KMM241. *Mar. Drugs* **2013**, *11*, 1173–1187. [Google Scholar]

22. Dakhmouche Djekrif, S.; Bennamoun, L.; Labbani, F.Z.K.; Ait Kaki, A.; Nouadri, T.; Pauss, A.; Meraihi, Z.; Gillmann, L. An Alkalothermophilic Amylopullulanase from the Yeast Clavispora lusitaniae ABS7: Purification, Characterization and Potential Application in Laundry Detergent. *Catalysts* **2021**, *11*, 1438. [Google Scholar] [CrossRef]

23. Yang, K.-R.; Yu, H.-C.; Huang, C.-Y.; Kuo, J.-M.; Chang, C.; Shieh, C.-J.; Kuo, C.-H. Bioprocessed production of resveratrol-enriched rice wine: Simultaneous rice wine fermentation, extraction, and transformation of piceid to resveratrol from Polygonum cuspidatum roots. *Foods* **2019**, *8*, 258. [Google Scholar] [CrossRef] [PubMed]

24. Zou, J.; Ge, Y.; Zhang, Y.; Ding, M.; Li, K.; Lin, Y.; Chang, X.; Cao, F.; Qian, Y. Changes in Flavor-and Aroma-Related Fermentation Metabolites and Antioxidant Activity of Glutinous Rice Wine Supplemented with Chinese Chestnut (Castanea mollissima Blume). *Fermentation* **2022**, *8*, 266. [Google Scholar] [CrossRef]

25. de Oliveira, R.L.; de Souza Claudino, E.; Converti, A.; Porto, T.S. Use of a Sequential Fermentation Method for the Production of Aspergillus tamarii URM4634 Protease and a Kinetic/Thermodynamic Study of the Enzyme. *Catalysts* **2021**, *11*, 963. [Google Scholar] [CrossRef]

26. Mashweu, A.R.; Chhiba-Govindjee, V.P.; Bode, M.L.; Brady, D. Substrate profiling of the cobalt nitrile hydratase from Rhodococcus rhodochrous ATCC BAA 870. *Molecules* **2020**, *25*, 238. [Google Scholar] [CrossRef]

27. Ogutu, I.R.; St. Maurice, M.; Bennett, B.; Holz, R.C. Examination of the Catalytic Role of the Axial Cystine Ligand in the Co-Type Nitrile Hydratase from Pseudonocardia thermophila JCM 3095. *Catalysts* **2021**, *11*, 1381. [Google Scholar] [CrossRef]

28. Grill, B.; Horvat, M.; Schwab, H.; Gross, R.; Donsbach, K.; Winkler, M. Gordonia hydrophobica Nitrile Hydratase for Amide preparation from nitriles. *Catalysts* **2021**, *11*, 1287. [Google Scholar] [CrossRef]

29. Sedova, A.; Rucká, L.; Bojarová, P.; Glozlová, M.; Novotný, P.; Křístková, B.; Pátek, M.; Martínková, L. Application potential of cyanide hydratase from Exidia glandulosa: Free cyanide removal from simulated industrial effluents. *Catalysts* **2021**, *11*, 1410. [Google Scholar] [CrossRef]

30. Kuo, C.-H.; Hsiao, F.-W.; Chen, J.-H.; Hsieh, C.-W.; Liu, Y.-C.; Shieh, C.-J. Kinetic aspects of ultrasound-accelerated lipase catalyzed acetylation and optimal synthesis of 4′-acetoxyresveratrol. *Ultrason. Sonochem.* **2013**, *20*, 546–552. [Google Scholar] [CrossRef]

31. Hsieh, C.-Y.; Huang, Y.-H.; Yeh, H.-H.; Hong, P.-Y.; Hsiao, C.-J.; Hsieh, L.-S. Phenylalanine, Tyrosine, and DOPA Are bona fide Substrates for Bambusa oldhamii BoPAL4. *Catalysts* **2021**, *11*, 1263. [Google Scholar] [CrossRef]

32. Wang, R.; Han, Z.; Ji, R.; Xiao, Y.; Si, R.; Guo, F.; He, J.; Hai, L.; Ming, L.; Yi, L. Antibacterial activity of trypsin-hydrolyzed camel and cow whey and their fractions. *Animals* **2020**, *10*, 337. [Google Scholar] [CrossRef] [PubMed]

33. Wang, Y.-H.; Kuo, C.-H.; Lee, C.-L.; Kuo, W.-C.; Tsai, M.-L.; Sun, P.-P. Enzyme-assisted aqueous extraction of cobia liver oil and protein hydrolysates with antioxidant activity. *Catalysts* **2020,** *10*, 1323. [**Google Scholar**] [**CrossRef**]

34. Simon, A.H.; Liebscher, S.; Kattner, A.; Kattner, C.; Bordusa, F. Rational design of a calcium-independent trypsin variant. *Catalysts* **2022,** *9*, 990. [**Google Scholar**] [**CrossRef**]

35. Huang, C.-J.; Peng, H.-L.; Patel, A.K.; Singhania, R.R.; Dong, C.-D.; Cheng, C.-Y. Effects of lower temperature on expression and biochemical characteristics of hcv ns3 antigen recombinant protein. *Catalysts* **2021,** *11*, 1297. [**Google Scholar**] [**CrossRef**]

36. Velázquez-De Lucio, B.S.; Hernández-Domínguez, E.M.; Villa-Garcia, M.; Diaz-Godinez, G.; Mandujano-Gonzalez, V.; Mendoza-Mendoza, B.; Alvarez-Cervantes, J. Exogenous enzymes as zootechnical additives in animal feed: A review. *Catalysts* **2021,** *11*, 851. [**Google Scholar**] [**CrossRef**]

37. Neubauer, P.R.; Blifernez-Klassen, O.; Pfaff, L.; Ismail, M.; Kruse, O.; Sewald, N. Two Novel, Flavin-Dependent Halogenases from the Bacterial Consortia of Botryococcus braunii Catalyze Mono-and Dibromination. *Catalysts* **2021,** *11*, 485. [**Google Scholar**] [**CrossRef**]

38. Neubauer, P.R.; Widmann, C.; Wibberg, D.; Schröder, L.; Frese, M.; Kottke, T.; Kalinowski, J.; Niemann, H.H.; Sewald, N. A flavin-dependent halogenase from metagenomic analysis prefers bromination over chlorination. *PLoS ONE* **2018,** *13*, e0196797. [**Google Scholar**] [**CrossRef**]

39. Sousa, J.; Silvério, S.C.; Costa, A.M.; Rodrigues, L.R. Metagenomic approaches as a tool to unravel promising biocatalysts from natural resources: Soil and water. *Catalysts* **2022,** *12*, 385. [**Google Scholar**] [**CrossRef**]

CHAPTER-2

TWO NOVEL, FLAVIN-DEPENDENT HALOGENASES FROM THE BACTERIAL CONSORTIA OF *BOTRYOCOCCUS BRAUNII* CATALYZE MONO- AND DIBROMINATION

1. INTRODUCTION: AN OVERVIEW

Halogenated organic compounds often show higher biological activities than the non-halogenated correlates. Therefore, chemical halogenation is an important methodology, but often lacks regioselectivity. Moreover, it requires relatively harsh reaction conditions and Lewis acid catalysts, and is usually performed in an organic solvent [1]. Hence, a regioselective and facile halogenation method would provide a more environmentally friendly alternative. Nature has evolved enzymatic halogenation, making use of different cofactors [2]. The enzymatic halogenation by flavin-dependent halogenases may overcome the drawbacks of chemical halogenation, since it often is highly regioselective and only requires a halide salt as a halogen source, water, oxygen, and reduced flavin–adenosine–dinucleotide ($FADH_2$) as a cofactor [1,3,4,5,6]. In general, flavin-dependent halogenases belong to the super family of flavin-dependent monooxygenases, which are able to activate molecular oxygen by using reduced flavin ($FADH_2$) [7], thus allowing diverse reactions, such as hydroxylation, epoxidation, and Baeyer–Villiger oxidation [8].

Tryptophan (Trp) halogenases are flavin-dependent halogenases, categorized according to their specific position of halogenating tryptophan. Trp halogenases have already been employed for halogenation on a gram scale, making use of enzyme immobilisation as cross-linked enzyme aggregates (CLEAs) [1]. The enzymatic halogenation approach has been combined with Suzuki–Miyaura cross-coupling [4,9,10,11,12], Sonogashira–Hagihara cross-coupling [13], and Mizoroki–Heck cross-coupling [14,15] in reaction cascades [4,10,11]. The tryptophan 7-halogenases PrnA, RebH, and KtzQ halogenate tryptophan regioselectively in the C7 position; the tryptophan 6-halogenases Thal, SttH, and Th-Hal prefer the C6 position, while the tryptophan 5-halogenases PyrH and ClaH modify the C5 position [16,17,18,19,20,21,22,23,24]. It has also been shown that it is possible to switch regioselectivity by exchanging amino acid residues. Following this strategy, Moritzer et al. redirected the regioselectivity of the tryptophan 6-halogenase Thal to the C7 position [25]. A combination of directed evolution, rational design, as well as site-saturation mutagenesis was also implemented for the design of a thermostable Thal variant with stronger elevated activity [26].

All known flavin-dependent halogenases require $FADH_2$ as a cofactor, whereas other flavin derivatives like riboflavin and flavin-mononucleotide (FMN) are not accepted [27]. $FADH_2$ is bound in the flavin binding site (motif GxGxxG) of the enzymes and reacts with oxygen to give a flavin hydroperoxide (FAD–OOH). This is followed by a nucleophilic attack of a halide forming hypohalous acid (HOX). The substrate binding site is positioned far away from the flavin binding site, but both sites are connected by a 10 Å long tunnel [28,29,30]. HOX passes through this tunnel, as verified by molecular dynamics calculations [31]. Different amino acid residues in the substrate binding site, as well in the tunnel, either interact with HOX or the substrate, while others are responsible for positioning the substrate to effect regioselective halogenation. Yeh et al. [32] elucidated the important role of a conserved lysine residue, e.g., K79 (PrnA) [28], K79 (RebH) [32], K83 (BrvH) [33], K74 (RadH) [34] that supposedly interacts with HOX. The potential formation of a long-lived chloramine (Lys–N^{ϵ}H–Cl) is still under debate [32]. Alternatively, the conserved lysine residue has been postulated to stabilize and position HOX by a hydrogen bond [35]. In both cases, the electrophilic HOX species reacts with the substrate bound to the substrate binding site. This leads to a Wheland intermediate in the course of electrophilic aromatic substitution (S_EAr), which is stabilized by highly conserved glutamic acid (E346 (PrnA) [28]; E357 (RebH) [36]). The carboxylate group of the conserved glutamic acid residue interacts with HOX or the chloramine, which increases electrophilicity and aligns it for regioselective halogenation. Flavin-dependent halogenases with substrates other than tryptophan do not possess the glutamic acid residue [37]. Other compounds, such as phenols or pyrroles, might be more susceptible for halogenation by HOX or the chloramine species [35]. The conserved amino acid motif WxWxIP is located at the flavin binding domain, and is believed to block binding of the substrate near the flavin cofactor, which suppresses monooxygenase activity of the enzymes [28,29,38].

Over the last years, several new flavin-dependent halogenases have been identified and heterologously expressed in different host organisms, leading to integration of halogenated substrates in the biosynthetic production of host compounds [39,40,41,42]. PrnA was the first flavin-dependent halogenase to be identified in 1997, within the pyrrolnitrin gene cluster [42]. The tryptophan 7-halogenase RebH described in 2002 is responsible for the chlorination of the secondary metabolite rebeccamycin [18]. By employing degenerative primers for highly conserved motifs, Zehner et al. identified PyrH [23], and Fujimori et al. identified KtzR and KtzQ [19] in different bacterial genomes, while Smith et al. retrieved KrmI from a marine sponge metagenome [43]. With the advent of next-generation sequencing methods, many genomes and associated metagenomes have been sequenced, and the obtained data led to the identification of many more halogenases, elucidated based on these conserved amino acid regions [22,24,44,45,46,47,48,49,50,51]. Neubauer et al. created in 2018 a profile hidden Markov model (pHMM), which led to the identification of several putative, flavin-dependent halogenases in different metagenomic data sets [33]. One of them, BrvH, was characterized with respect to halide and substrate preference in vitro, leading to the conclusion that it halogenates indole and indole derivatives, as well as phenol derivatives, and prefers bromination over chlorination. BrvH was crystallized and revealed a structure similar to Trp halogenases, in addition to an open substrate binding site, which might lead to the acceptance of larger substrates, such as peptides [33,52]. BrvH was the first reported flavin-dependent halogenase that accepts chloride and bromide with a preference for bromination over chlorination. Only the phenol halogenases Bmp2 and Bmp5 had been reported as brominases that do not accept chloride [46]. Ismail et al. identified three novel, flavin-dependent halogenases from *Xanthomonas campestris* that exclusively brominate the substrates indole, 7-azaindole, 5-hydroxytryptophan, tryptophol, and other heterocyclic derivatives, even in the presence of an excess of chloride [51]. Like for BrvH, the crystal structure of Xcc4156 from *Xanthomonas campestris* showed an open substrate binding site. However, crystallization with FAD and bromide resulted in disruption of the crystal, which leads to the conclusion that this binding leads to positive cooperativity and conformational change [53].

Within the present work, the metagenomes of *Botryococcus braunii* communities were screened for the presence of putative, flavin-dependent halogenases by applying the generated profile hidden Markov model (pHMM) [33]. Two of the identified novel halogenases were heterologously expressed in *E. coli* and analysed in detail.

2. RESULTS AND DISCUSSION

2.1. Identifying and analyzing newly discovered halogenases that rely on flavin as a cofactor from the Botryococcus braunii consortia

Since the utilization of hydrocarbons by microorganisms relies on a set of different monooxygenases [54], we considered the microbial consortia accompanying the hydrocarbon-secreting microalga *Botryococcus braunii* as a potential source for novel monooxygenases [55] and potential halogenases. *Botryococcus braunii* is a colony-forming green microalga belonging to the class *Treboxiophyceae*, which can be sub-divided into distinct races depending on the type of hydrocarbon synthesized [56]. *Botryococcus* readily releases large amounts of organic carbon into the extracellular medium [57], creating a phycosphere that naturally attracts many microorganisms, including various taxa known to utilize hydrocarbons [58].

We screened metagenomic data sets of *Botryococcus braunii* consortia [59] for novel, flavin-dependent halogenases by applying our pHMM strategy [33], which is based on the PFAM database (**http://pfam.xfam.org/**, accessed on 8 April 2021) tryptophan–halogenase model (Trp_halogenase, PF04820), to increase the scope of possible halogenases. The metagenomes of four *Botryococcus braunii* consortia were analysed; they contained at least 33 distinct bacterial species, as indicated by 16 S rDNA amplicon and metagenome sequencing analyses [59]. For the analysis of the metagenomes, according to Neubauer et al., the pHMM was employed, which was based on Trp halogenases and was optimised in a two-step approach [33]. The analysis led to the identification of 18 complete and seven partial putative flavin-dependent halogenase genes, which possess the known conserved amino acid regions of Trp halogenases. With one exception, all halogenases found were encoded in the genomes of alphaproteobacteria, while 16 belonged to the sphingomonads genera, such as *Sphingomonas* and *Sphingopyxis* (**Table 1**).

Table 1. Newly found putative, flavin-dependent halogenases with the conserved amino acid motif of Trp halogenases and their taxonomic assignment, based on BLAST analyses (https://blast.ncbi.nlm.nih.gov, accessed on 8 April 2021; Basic Local Alignment Search Tool).

No.	Length (bp)	Complete	Contig Location	Phyla	Putative Origin
1	225	no	contig-391000007_2	*Alphaproteobacteria*	*Porphyrobacter*
2	394	no	contig-1913000010_2	*Alphaproteobacteria*	*Sphingopyxis*
3	348	no	contig-565000012_1	*Alphaproteobacteria*	*Caulobacter*

No.	Length (bp)	Complete	Contig Location	Phyla	Putative Origin
4	272	no	contig-1532000014_2	*Alphaproteobacteria*	*Sphingopyxis*
5	503	yes	contig-1000028_60	*Alphaproteobacteria*	*Sphingomonas*
6	502	yes	contig-1029000036_1	*Alphaproteobacteria*	*Sphingopyxis*
7	501	yes	contig-3201000046_41	*Gammaproteobacteria*	*Stenotrophomonas*
8	511	yes	contig-2000047_22	*Alphaproteobacteria*	*Sphingomonas*
9	501	yes	contig-2000047_33	*Alphaproteobacteria*	*Sphingomonas*
10	533	yes	contig-2000047_110	*Alphaproteobacteria*	*Sphingomonas*
11	505	yes	contig-2000047_112	*Alphaproteobacteria*	*Sphingomonas*
12	501	yes	contig-2000047_113	*Alphaproteobacteria*	*Sphingomonas*
13	418	no	contig-1419000070_1	*Alphaproteobacteria*	*Caulobacter*
14	359	no	contig-832000073_1	*Alphaproteobacteria*	*Porphyrobacter*
15	501	yes	contig-252000084_3	*Alphaproteobacteria*	*Sphingopyxis*
16	501	yes	contig-2000086_90	*Alphaproteobacteria*	*Brevundimonas*
17	443	no	contig-2264000090_1	*Alphaproteobacteria*	*Sphingopyxis*
18	521	yes	contig-16000092_39	*Alphaproteobacteria*	*Sphingomonas*
19	502	yes	contig-867000143_2	*Alphaproteobacteria*	*Caulobacter*
20	513	yes	contig-3212000146_163	*Alphaproteobacteria*	*Sphingomonas*
21	512	yes	contig-3212000146_164	*Alphaproteobacteria*	*Sphingomonas*
22	553	yes	contig-3212000146_184	*Alphaproteobacteria*	*Sphingomonas*
23	493	yes	contig-1000154_7	*Alphaproteobacteria*	*Sphingomonas*
24	522	yes	contig-981000159_36	*Alphaproteobacteria*	*Brevundimonas*
25	516	yes	contig-981000159_37	*Alphaproteobacteria*	Brevundimonas

A phylogenetic tree was constructed based on amino acid sequence to further analyse the identified hits and to compare these to known flavin-dependent halogenases (**Figure 1**). The phylogenetic analysis was conducted in MEGAX [60] by using the neighbour-joining (NJ) method and a bootstrap of 1000 [61,62]. The phylogenetic tree is divided into five groups of flavin-dependent halogenases. The phenol, pyrrol, and Trp halogenases build their own clades, respectively.

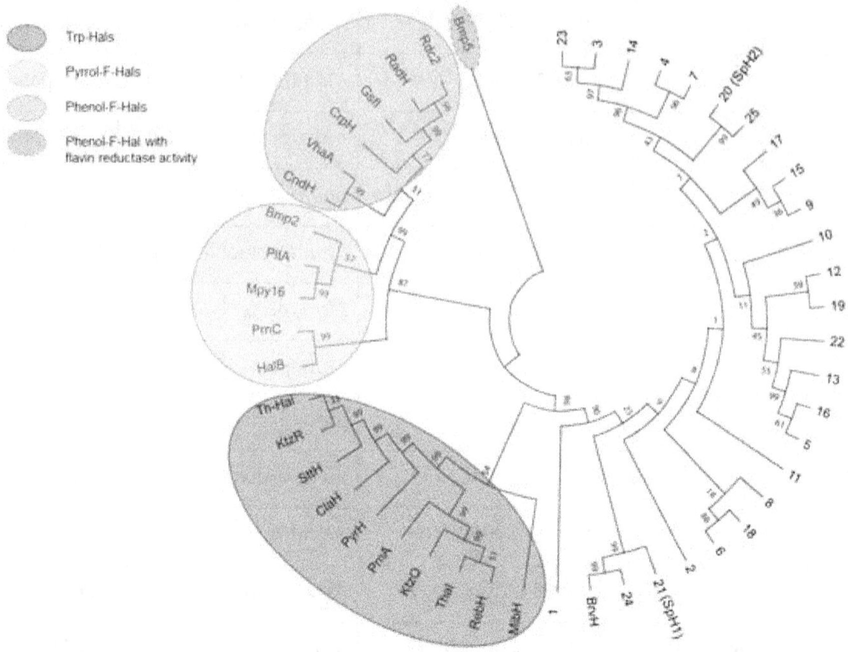

Figure 1. Phylogenetic tree of the hits from the *Botryococcus braunii* consortia, as well as known Trp halogenases, phenol halogenases, and pyrrol halogenases. The optimal tree, with the sum of branch lengths = 17.80254621, is shown. The newly identified hits all cluster with the flavin-dependent halogenase BrvH. We decided to further analyse hit 21 (SpH1) and hit 20 (SpH2).

Noteworthy, Bmp5 is outside the phenol halogenase clade, although it is a phenol halogenase. However, it also possesses flavin reductase activity, which might be the reason for building its own clade [46]. The newly identified hits from the *B. braunii* bacterial consortia cluster close to Trp halogenases, but build their own clade together with BrvH. BrvH was found to be very similar to hits 24 and 21 (pairwise identity of 95% and 64%, respectively), which represent flavin-dependent halogenases encoded in the metagenome-assembled genomes (MAG 10 and MAG 21, respectively [59]) of *Brevundimonas* and *Sphingomonas*. Thus, these enzymes might represent their own group within the flavin-dependent halogenases with likely similar substrate preferences. According to antiSMASH analyses (antibiotics & Secondary Metabolite Analysis Shell [63]), both MAGs encode several metabolic gene clusters for bacteriocins and antibiotic compounds [59,64]. In addition, the MAG similar to *Sphingomonas* contains a gene cluster potentially responsible for siderophore synthesis.

We decided to investigate the identified gene 21 (J) from *Sphingomonas*. However, during the gene cluster analysis, the presence of another flavin-dependent halogenase gene (20, I) located upstream to 21 (J) was detected (**Figure 2**). Some of the known flavin-dependent halogenases have been reported to originate from the same gene cluster, e.g., PrnA and PrnC that play a role in pyrrolnitrin biosynthesis [**42**]. Bmp2 and Bmp5 likewise halogenate different building blocks in the biosynthesis of natural, polybrominated marine products [**41**]. KtzR and KtzQ act together in the biosynthesis of kutzneride. In this case, KtzQ is responsible for the chlorination of tryptophan at C7, while KtzR catalyses the second halogenation to give 6,7-dichlorotryptophan. KtzR also catalyses chlorination of tryptophan but possesses a higher affinity to 7-chlorotryptophan [**19**].

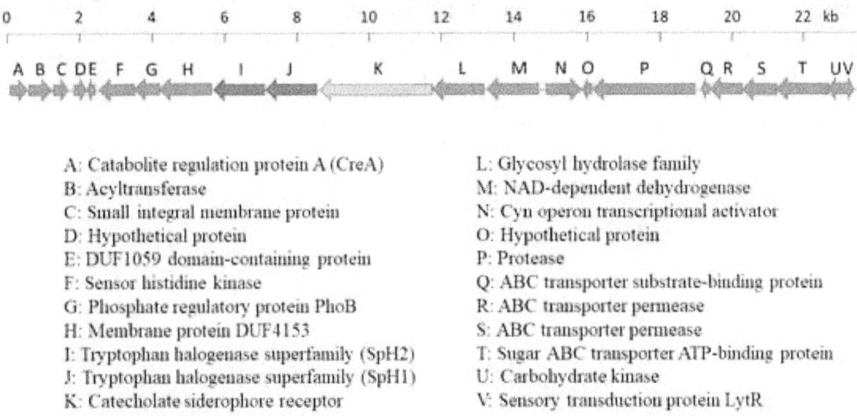

Figure 2. Gene cluster in the vicinity of hit 21 (J; SpH1) in the metagenome-assembled genome (MAG) 21 of the *B. braunii* bacterial consortia. Another flavin-dependent halogenase gene, *sph2* (J), was identified upstream to *sph1* (I). The genes I, J, K, and L were found to be encoded on the same contig in the metagenomics dataset.

Since many flavin-dependent halogenases originating from the same gene cluster were reported to act in concert [**17**,**19**,**46**], both putative, flavin-dependent halogenases 21 (J) and 20 (I)—in the following named SpH1 (21) and SpH2 (20)—were heterologously expressed in *E. coli* and subjected to further analyses. Both enzymes possess the typical conserved amino acid regions of flavin-dependent halogenases like the flavin binding domain and the conserved lysine residues. However, both show 42% pairwise identity to each other, while the pairwise identity to BrvH was 64% for SpH1 and 42% for SpH2.

2.2. Conducting laboratory experiments using the flavin-dependent halogenases SpH1 and SpH2 in vitro.

2.2.1. Determination of Halogenation Activity and Substrate Scope

In order to confirm that the identified genes are active halogenating enzymes, experiments with the novel enzymes were conducted in vitro. The genes *sph1* and *sph2* were heterologously expressed in *E. coli*, together with the chaperone system GroEL/GroES. Both enzymes were obtained in soluble form in high concentrations. For the substrate screening, the previously established reaction conditions, with a cofactor regeneration by flavin reductase (PrnF), and *Rhodococcus ruber* alcohol dehydrogenase (*RR*-ADH), were used [3]. Indole and L-tryptophan were the first tested substrates, because SpH1 and SpH2 are similar to BrvH in brominating indole [33]. Similar to the activity observed for BrvH, SpH1 and SpH2 catalyse the bromination of indole, but not L-tryptophan (**Figure 3**), thus revealing that both novel identified enzymes represent active halogenases.

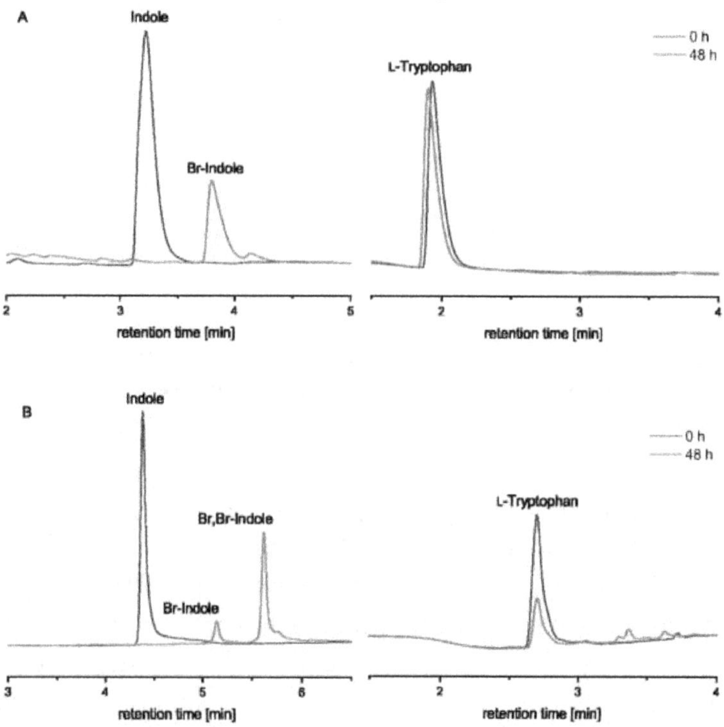

Figure 3. Reverse-phase high-performance liquid chromatography (RP-HPLC) profiles (detection at 280 nm) of the conversions of 1 mM indole or 1 mM L-tryptophan by SpH1 (A) and SpH2 (B) in presence of 100 mM NaBr as a halide source. *Rhodococcus ruber* alcohol dehydrogenase (*RR*-ADH) and flavin reductase PrnF were used for cofactor regeneration. Retention times differ due to different HPLC methods. (A) SpH1 (HPLC method 1); (B) SpH2 (HPLC method 2). SpH1 had

completely halogenated indole (retention time t_R: 3.2 min) after 48 h of incubation to give bromoindole (t_R: 3.8 min), but L-tryptophan (t_R: 2 min) was not converted. SpH2 catalyses the halogenation of indole (t_R: 4.4 min) to give bromoindole (t_R: 5.1 min) and dibromoindole (t_R: 5.6 min), without converting L-tryptophan (t_R: 2.7 min).

The newly identified brominating enzymes belong to the class of flavin-dependent halogenases. Both enzymes have the characteristic flavin binding module GxGxxG, the conserved lysine residue (K85 and K81, respectively), and the WxWxIP motif suppressing monooxygenase activity [33]. The reduced flavin ($FADH_2$) reacts with molecular oxygen, forming the FAD(C4a)-peroxide, which in turn reacts with halide ions under HOX formation [2,5,28,29]. In contrast, heme- or vanadium-dependent haloperoxidases require hydrogen peroxide and release the halogenating species hypohalous acid HOX into the medium, which leads to unselective halogenation [2,65,66]. Flavin-dependent halogenases can be distinguished from haloperoxidases by the monochlorodimedon assay [67]. In this assay, monochlorodimedon reacts with free HOX released by the haloperoxidase to give rise to a dihalogenated derivative, which can be monitored photometrically. This test was negative for both enzymes. The addition of catalase, which decomposes hydrogen peroxide and would hence stop the halogenation reaction by haloperoxidases, did not negatively affect the halogenation reaction, leading to the conclusion that SpH1 and SpH2 are not haloperoxidases. FAD reconstitution is another possibility for the classification towards flavin-dependent halogenases [33,68]. SpH1 and SpH2 were incubated with FAD overnight, and then the buffer was changed to a buffer without FAD, leading to no free FAD in the solution. UV/Vis spectroscopy can be employed for the detection of bound FAD in the enzymes. For both SpH1 and SpH2, the absorption band at 350 nm shifted to 348 nm, and the absorption band at 373 nm shifted to 360 nm, showing that both enzymes bind FAD and can therefore be classified as flavin-dependent halogenases. The SpH1 mutant where the conserved lysine residue K85 was replaced by an alanine (SpH1_K85A) was completely inactive. Hence, halogenase SpH1 depends on the lysine, which is known to be essential for the flavin-dependent halogenases.

A pharmacophore model was adapted from a virtual screening methodology for BrvH, and revealed 19 compounds halogenated by BrvH [52]. SpH1 is highly similar to BrvH and clusters within one clade with BrvH. Therefore, its substrate scope might be similar as well. In addition to indole and L-tryptophan, further substrates, such as tryptophan and indole derivatives, as well as other aromatic compounds, were tested for halogenation by SpH1 and SpH2 (Figure 4). Twenty of the 26 accepted substrates represent indole derivatives, but phenol derivatives (i.e., phenol (**24**), anthranilic acid (**25**), and 4-*n*-hexylresorcinol (**18**)), azulene (**17**), and quinoxaline (**22**) were also accepted. Enzymatic conversion was verified by reverse-phase high-performance liquid chromatography (RP-HPLC), as well as mass analysis.

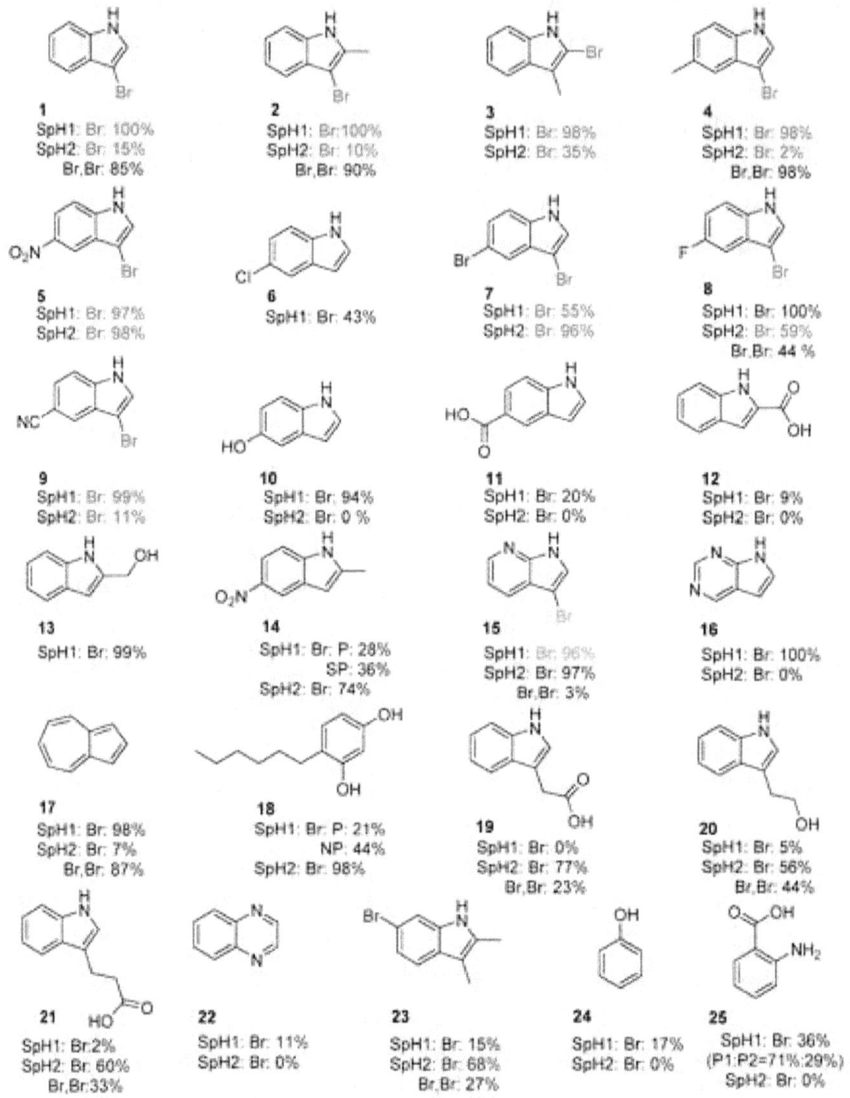

Figure 4. Halogenated substrates by SpH1 and SpH2. The conversions were calculated by determining the ratio of the peak areas of substrates and products. The position of halogenation was determined for 10 substrates by NMR analysis (red: SpH1 and SpH2, green: SpH1, and blue: SpH2). SP: different unidentified side products; P1, P2: two monobrominated products at a ratio of P1/P2.; NP: side product.

SpH1 and SpH2 were immobilised as CLEAs to convert higher amounts of substrate, in order to identify their regioselectivity. The halogenases and cofactor-regenerating enzymes were immobilised together using glutaraldehyde, and the

CLEAs were incubated for 10 to 14 days with substrate. The obtained product was purified and analysed by NMR spectroscopy. Mass spectrometry (MS) analyses revealed the correct isotopic pattern of the brominated products. All indole derivatives analysed so far are halogenated at C3, if this position is unsubstituted. Only when a non-hydrogen substituent is present at C3, halogenation occurs at C2 (**Figure 4**). The C3 and C2 positions are electronically the most favored positions for electrophilic substitution within the indole ring [**69**]. Likewise, PrnA and RebH halogenate indole and some derivatives in the most activated position. The regioselectivity strongly depends on the substrate and its position in the active site [**3,70**]. SpH2 is also able to halogenate 2,3-methylindole. In this case, it halogenates the substrate at C6, but also a minor side product halogenated at C5 was observed, leading to the conclusion that SpH2 is not strictly regiospecific for 2,3-methylindole.

Furthermore, SpH1 and SpH2 show higher bromination activity compared to chlorination activity. They brominated 1 mM indole to 100% within 48 h, while SpH1 under similar conditions chlorinated only to 10%, and SpH2 to 17%. Interestingly, dichlorinated product was not found for SpH2. The brominated, as well as the chlorinated products, were verified via ESI-MS or liquid chromatography LC-MS analyses (**SpH1:** positive mode: $[C_8H_6BrN]^+$, calcd. 195.97 (^{79}Br), 197.97 (^{81}Br), obs. $[M+H]^+$ 195.55/197.53; negative mode: $[C_8H_6ClN]^-$, calcd. 150.02 (^{35}Cl), 152.02 (^{37}Cl), obs. $[M-H]^-$ 150.01/152.01; **SpH2:** negative mode: $[C_8H_6BrN]^-$, calcd. 193.97 (^{79}Br), 195.97 (^{81}Br), obs. $[M-H]^-$ 193.96/195.95; $[C_8H_6ClN]^-$, calcd. 150.02 (^{35}Cl), 152.02 (^{37}Cl), obs. $[M-H]^-$ 150.01/152.01).

SpH1 and SpH2 are thus similar to BrvH and the Xcc halogenases Xcc4156, Xcc1333, and Xcc4345, which all prefer bromination over chlorination [**33,51**]. Noteworthy, the Xcc halogenases, SpH1, and SpH2 originate from terrestrial habitats, but favour bromination over chlorination. The Xcc halogenases were identified in *Xanthomonas campestris*, which is a plant pathogen [**51**], and SpH1 and SpH2 originate from *B. braunii* consortia, which mainly inhabit fresh water [**57**]. Usually, it is presumed that because of the higher bromide concentration in sea water, bromination is preferred in marine habitats, while halogenases from terrestrial habitat would favour chlorination [**71**].

In 2019, the Lewis group identified many flavin-dependent halogenases via family-wide activity profiling, and most of them preferred bromination over chlorination [**72**]. The group postulated that bromination activity is more widespread than chlorination activity of flavin-dependent halogenases, which matches our data. However, even RebH, which chlorinates its natural substrate tryptophan, was shown in competition experiments to prefer bromination [**72**]. Therefore, it is also possible that SpH1 and SpH2 chlorinate their natural substrate.

However, the natural substrates of SpH1 and SpH2 are yet unknown. SpH2 accepts larger substrates, such as (3-indolyl)acetic acid (**19**), 3-(3-indolyl)propionic acid (**21**), and tryptophol (**20**). This possible substrate group could be represented by auxins, pivotal plant hormones (also called phytohormones), with (3-indolyl) acetic acid (IAA) (**19**) as one of the best characterised class of regulators [73,74]. The chlorinated form of auxin, (4-chloro-3-indolyl)acetic acid (4-Cl-IAA), is a highly active hormone that is thought to play a key role in early pericarp growth [75] and fruit development in peas [76]. Auxins also represent a key modulator of plant–bacteria interactions with the bacterial ability to synthesise, e.g., IAA being an attribute for both promoting plant growth and phytopathogenic effects [77]. So far, however, there is no evidence in the literature about brominated auxins in nature, although the chemical synthetic bromination of (3-indolyl)acetic acid has been reported [78]. Another interesting hint towards possible natural substrates is represented by the gene located downstream next to *sph1* (Hit 21; J), which encodes for a catecholate siderophore receptor (**Figure 2**). Since the MAG similar to *Sphingomonas* also contains at least one metabolic gene cluster potentially responsible for siderophore synthesis, siderophores may represent another potential substrate for the halogenases SpH1 and SpH2. Siderophores possess phenol moieties, which might be halogenated by SpH1 and SpH2, since it was shown that these are able to halogenate 4-*n*-hexylresorcinol (**18**), phenol (**24**), and anthranilic acid (**25**) (**Figure 4**). In addition to iron scavenging, siderophores can have many alternative functions, including non-iron-metal transport, signaling, and antibiotic activity functions [79]. The halogenation of siderophores is common in nature; for instance, two *Streptomyces* siderophores, chlorocatechelins A and B, have been reported to contain chlorinated catecholate groups [**80**].

2.2.2. Exploration of the activity related to mono- and dibromination during the investigation

Interestingly, SpH1 only catalyses monobromination, while SpH2 is able to catalyse both mono- and dibromination of indole and its derivatives (**Figure 4**). Kinetic measurements showed that SpH2 has a negligibly higher specific activity towards indole (4.2 ± 0.03 mU/mg) than towards 3-bromoindole (3.8 ± 0.45 mU/mg). In comparison, the specific activity of SpH1 towards indole is 3.0 ± 0.7 mU/mg. The synergistic cooperation of two enzymes in the halogenation process is known in the literature [**17,19,46**]. PrnA and PrnC have been identified as catalysing the chlorination of tryptophan and monodechloroaminopyrrolnitrin in the biosynthesis cluster of pyrrolnitrin. PrnC catalyses the chlorination of monodechloroaminopyrrolnitrin, halogenated by PrnA in the first step [**17,28**]. Another example is the biosynthesis of kutzneride, which includes KtzQ and KtzR [**19**].

SpH1 and SpH2 were evaluated in silico for unveiling the dibrominating ability exhibited by SpH2. Since no crystal structure was available for either of the two enzymes, the homology model was built using YASARA structure (**www.yasara.org**; 2020) for in silico study of the novel halogenases. SpH2 exhibited a wider active site pocket with an area of 136.8 Å2, compared to SpH1 with 77.39 Å2 (**Figure 5**).

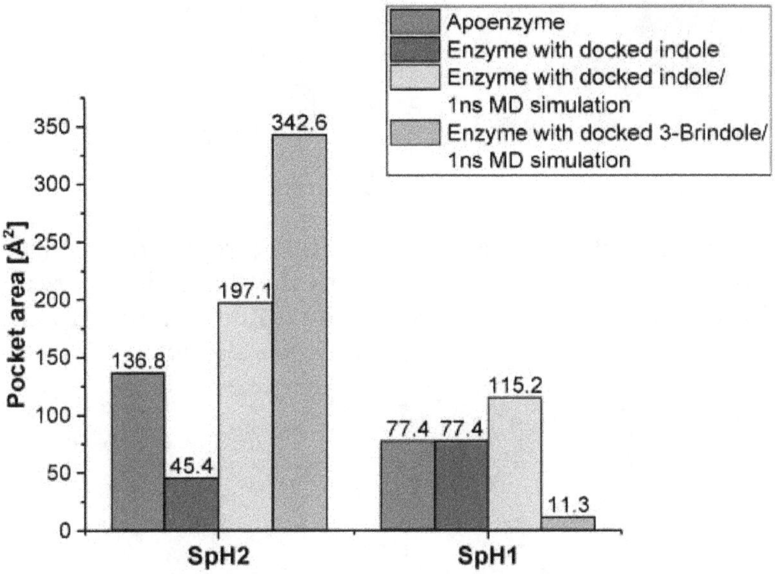

Figure 5. Active site pocket area for the apo-, indole-, and 3-bromoindole-docked SpH2 and SpH1 halogenases, showing the flexibility of SpH2 in accommodating 3-bromoindole, unlike SpH1.

The active site pockets of both halogenases differ in their geometry, as does their depth within the pocket for reaching the conserved amino acid residues (Lys and Glu). The SpH1 active site pocket is deeper compared to SpH2, with the loop containing the conserved glutamic acid residue (Glu348) being displaced backwards. On the other hand, the SpH2 active site pocket is wider and not as deep as that of SpH1; still, the pocket is extended to allow the substrate to reach the conserved Lys residue (**Figure 6**, **Figure 7**).

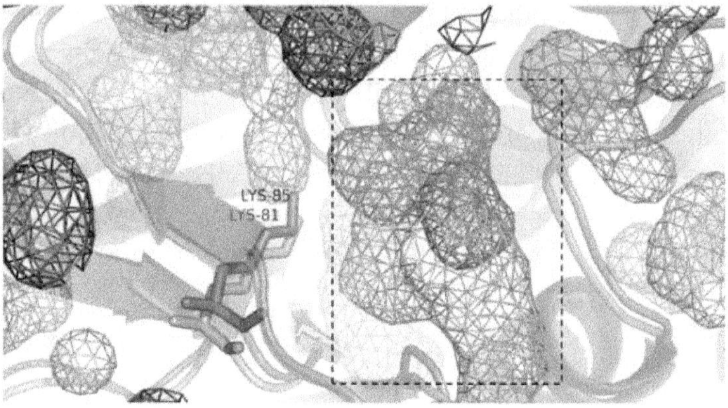

Figure 6. Superimposed homology model of SpH2 (grey) and SpH1 (red) after 1ns molecular dynamics (MD) simulation, showing the active site cavity in wireframe (in dashed-line box). The conserved active site lysine residues 81 and 85 in SpH2 and SpH1, respectively, are shown.

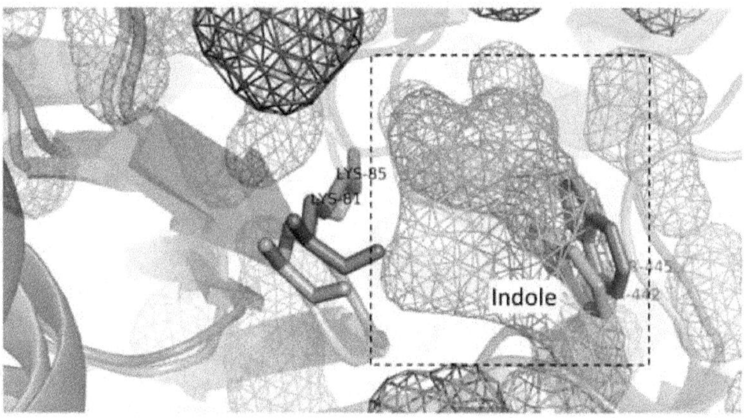

Figure 7. Superimposed homology model of SpH2 (grey) and SpH1 (red) with the docked indole after 1 ns MD simulation. The active site cavity is shown in wireframe (surrounded by dashed-line box). In both cases, an H-bond formed between indole NH and Ser442 and Ser445 in SpH2 and SpH1, respectively.

Indole was docked into both models to indicate its flexibility and expansion within the active site pocket of SpH2. The docked substrates were then subjected to MD simulation to further understand the changes that might occur as a result of indole binding to the active site. Obvious differences were detected in the size of the active site pockets containing indole after MD simulation, where the SpH2 active site was expanded to allow the binding of bulkier substrates, thus showing that the active site is flexible. However, SpH1 was not largely expanded, indicating less flexibility to accommodate larger substrates. Both cases showed H-bond formation with the indole NH group (with Ser445 in SpH1 and Ser442 in SpH2;

Figure 5 and **Figure 7**). Furthermore, the binding affinity of the docked indole towards both enzymes revealed a slight preference for SpH2 over SpH1 (**Table 2**), which corresponds to the specific activity calculated for SpH2 towards indole compared to SpH1.

Table 2. Binding affinity of indole and 3-bromoindole to the active site of SpH2 and SpH1.

Enzyme/Substrate	Mode	Binding Affinity [kJ/mol]
SpH2/indole	3	22.9
SpH1/indole	9	20.1
SpH2/3-bromoindole	5	22.5
SpH1/3-bromoindole	11	19.4

In addition, 3-bromoindole was docked into the homology models. Higher affinity towards 3-bromoindole was detected for SpH2 over SpH1, which provides further evidence for the dibromination ability of SpH2 (**Table 2**). In addition, the size of the active site pocket in the case of SpH2 was significantly expanded compared to the active site pocket of SpH1 (**Figure 5**), which is more rigid, and hence would prohibit further bromination of the 3-bromoindole (**Figure 8**).

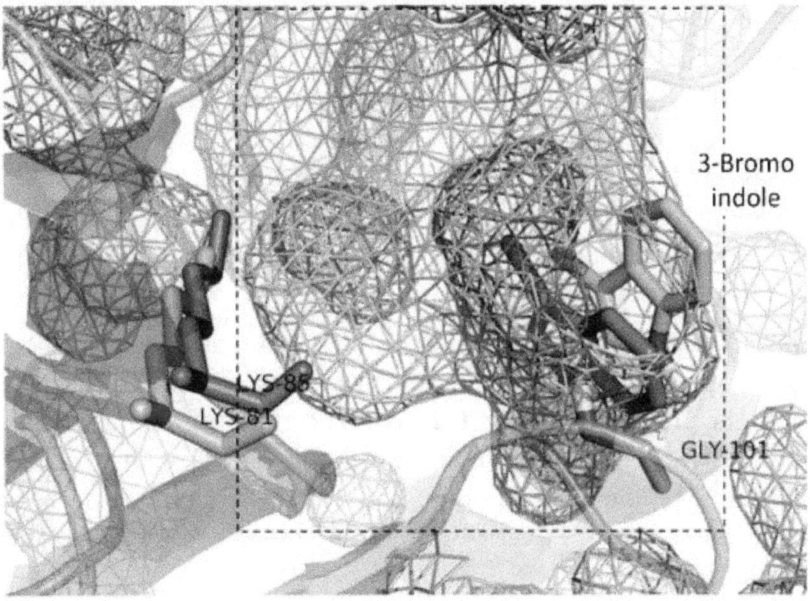

Figure 8. Superimposed homology model of SpH2 (grey) and SpH1 (red) with the docked 3-bromoindole after 1 ns MD simulation. The active site cavity is shown in wireframe (surrounded by a dashed-line box). In both cases, an H-bond is formed between indole NH and Gly442 in SpH2 and Ser445 in SpH1.

The in silico simulation of the SpH1 and SpH2 homology models underlines our findings that SpH2 is able to catalyse dihalogenation. Both SpH1 and SpH2 originate from one gene cluster, and therefore could act in concert and interact with each other. SpH1 has a deeper but less flexible active site pocket, which is likely to be suitable for the binding of smaller substrates like indole, while SpH2 features a wider and more flexible active site pocket, likely accommodating bulkier substrates, as well as the dibromination observed with some substrates.

3. MATERIALS AND METHODS

Solvents and chemicals were, unless otherwise specified, of analytical grade (p.a.) and purchased from commercial suppliers. Chemicals such as *iso*-propanol, acetonitrile, sodium chloride/bromide, ethanol, and methanol were purchased from VWR Chemicals (Darmstadt, Germany) and L-arabinose, NADH/NAD, FAD and glutaraldehyde from Carl Roth (Karlsruhe, Germany). Antibiotics were obtained from Fisher Scientific (Waltham, MA, USA; ampicillin sodium salt), AppliChem (Darmstadt, Germany; chloramphenicol) and Carl Roth (Karlsruhe, Germany; kanamycin sulfate).

Chemicals for substrate assays were acquired at Acros Organics (Fair Lawn, NJ, USA; indole, indazole, 5-cyanoindole, 3-(3-indolyl)propionic acid, indole-3-carbaldehyde, trypthophan, tryptamine, tryptophol), Sigma Aldrich (Darmstadt, Germany; 2-methylindole, 4-hydroxybenzoic acid, 5-aminoindole, 5-bromoindole, indole-2-carboxylic acid, phenylalanine, phenol, 7*H*-pyrolo[2,3-d]pyrimidine, quinoxaline), Alfa Aesar (Kandel, Germany; 2,3-dimethylindole, 3-methylindole, 4-*n*-hexylresorcinol, 5-hydroxyindole, 5-fluoroindole, 5-methylindole, azulene, indole-3-acetonitrile, (3-indolyl)acetic acid, indole-5-carbaldehyde, indole-5-carboxylic acid, 5-hydroxytryptophan hydrate, naphthylacetic acid), Maybridge (Altrincham, UK; 5-nitroindole), Carl Roth (Karlsruhe, Germany; 7-azaindole) and Fluorochem (Derbyshire, UK; benzoxazole).

The plasmid pClBhis–prnF encoding the flavin reductase PrnF from *Pseudomonas fluorescens*, was donated by Prof. Dr. Karl-Heinz van Pée, Technical University Dresden, and the plasmid vector pET-21–adh encoding the alcohol dehydrogenase from *Rhodococcus ruber* (*RR*-ADH) was kindly provided by Prof. Dr. Werner Hummel, Bielefeld University. The plasmid pGro7 coding for the chaperone system GroEL–GroES was purchased from TaKaRa Bio Inc. *E. coli* DH5 and *E. coli* BL21 (DE3) were purchased from Novagen.

3.1. Analytics

3.1.1. Analytical, Reverse-Phase High-Performance Liquid Chromatography (RP-HPLC)

For analytical, reverse-phase high performance liquid chromatography, two different methods were employed.

With method 1, reactions were monitored using a Thermo Scientific (Waltham, MA, USA) Accela 600 with NUCLEOSHELL RP 18 column 18.5 µm from Macherey-Nagel (Düren, Germany; 150 × 2.1 mm, eluent A: $H_2O/CH_3CN/TFA$ = 95.0:5.0:0.1; eluent B: $H_2O/CH_3CN/TFA$ = 5.0:95.0:0.1, flow rate 900 µl/min,; 0–1 min 100% eluent A, 1–6 min linear gradient to 100% eluent B, 6–7 min 0% eluent A, 7–7.5 min back to 100% eluent A, 7.5–9 min 100% eluent A).

With method 2 for analytical RP-HPLC analysis a Shimadzu (Duisburg, Germany) Nexera XR chromatography system was employed (Luna C18(2) column (3 µM, 100 Å, LC column, 100 × 2 mm) from Phenomenex (Aschaffenburg, Germany; eluent A: $H_2O/CH_3CN/TFA$= 95.0:5.0:0.1; eluent B: $H_2O/CH_3CN/TFA$ = 5.0:95.0:0.1; flow rate 650 µL/min; 0–5.5 min linear gradient to 95% eluent B, 5.5–6.0 min 95% eluent B, 6.0–6.1 min back to 95% eluent A, 6.1–9.0 min 95% eluent A).

3.1.2 Preparative, Reverse-Phase High-Performance Liquid Chromatography (RP-HPLC)

For the isolation and purification of products, a Merck-Hitachi (Darmstadt, Germany) LaChrom RP-HPLC system was employed. Separation of the products was conducted using a Hypersil Gold C18 column (8 µM, 250 × 21.2 mm) from Thermo Scientific (Waltham, MA, USA). Absorbance was measured simultaneously at 220 nm, 254 nm, and 280 nm. The setup was as follows: eluent A, $H_2O/CH_3CN/TFA$ = 95.0:5.0:0.1; eluent B, $H_2O/CH_3CN/TFA$ = 5.0:95.0:0.1; flow rate = 10 mL/min; 0–45 min linear gradient to 100% eluent B; 45–50 min 100% eluent B; 50–55 min 100% eluent A.

3.1.3 High-Performance Liquid Chromatography–Mass Spectrometry (HPLC-MS)

For the identification of products by electrospray ionisation, a high-performance liquid chromatography–mass spectrometry (HPLC-MS) system consisting of an Agilent 1200 HPLC system and a 6220 TOF mass spectrometer (Agilent Technologies, Santa Clara, CA, USA) was used, with the following setup: solvent A, $H_2O/CH_3CN/FA$ = 95.0:5.0:0.1; solvent B, 5.0:95.0:0.1; flow rate = 0.3 mL/min, with a linear gradient from 0% to 98% B over 10 min, 1 min at 98% B, and back to 0% B for 5 min.

3.1.4 Gas Chromatography–Mass Spectrometry (GC-MS)

For gas chromatography–mass spectrometry (GC-MS) analysis, a system consisting of a Trace GC Ultra gas chromatograph (Thermo Scientific, Waltham, MA, USA) and an ITG900 mass spectrometer from Thermo Finnigan (20 measurements per minute, 50 ± 750 *m/z*) was applied. Separation of the products was conducted using a VF-5 column (0.25 μM, 30 m × 0.25 mm, 5% diphenylsiloxan, 95% dimethylsiloxan) from Thermo Scientific (Waltham, MA, USA). Helium was used as the mobile phase, with a temperature gradient of 5 °C/min from 80 °C to 325 °C.

3.1.5 Nuclear Magnetic Resonance (NMR) Spectroscopy

NMR spectra were recorded on an Avance 500 (^{1}H: 500 MHz, ^{13}C: 126 MHz) or a DRX-500 spectrometer (^{1}H: 500 MHz, ^{13}C: 126 MHz) (Bruker, Billerica, MA, USA). Chemical shifts are reported relative to residual solvent peaks (DMSO-d_6: ^{1}H: 2.5 ppm; ^{13}C: 39.5 ppm).

3.2 BIOINFORMATIC ANALYSIS

3.2.1 Metagenomic Analysis for the Detection of Flavin-Dependent Halogenases

Metagenomic datasets of Botryococcus braunii consortia, deposited under the BioProjectID PRJEB26344 [59], were used for the search of novel flavin-dependent halogenases. The analyses were performed according to the previously described pHMM strategy [33], which is based on the PFAM (http://pfam.xfam.org, 2016) tryptophan–halogenase model (Trp_halogenase, PF04820). The dataset of the MAG21 (similar to the genus Sphingomonas) can be found under the deposited BioProject ID PRJEB26345 [59], and was used for the antiSMASH (https://antismash.secondarymetabolites.org, 2020) analysis (using default settings) [63]. Amino acid sequence alignments were performed using NCBI (www.ncbi.nlm.nih.gov, Bethesda, MD, USA, 2020) NR [81] and MultAlin [82]. The assignment of the active sites of the enzymes was accomplished on the basis of the published information for RebH and PrnA [29,32,36].

3.2.2 Phylogenetic Analysis

Phylogenetic analyses were performed with MEGA7 (**www.megasoftware.net**, accessed on 8 April 2021) [**60**], based on protein alignment (MUSCLE) [**83**], neighbour joining (NJ) method, and a bootstrap of 1000 [**61**,**62**].

4. *IN VITRO* ENZYME ASSAYS

4.1 Vector Construction and Molecular Cloning

Gene synthesis of the putative halogenase genes *spH1* and *spH2* was ordered from ThermoFisher GeneArt (Waltham, MA, USA). The genes pMA-T-spH1 and pMA-TspH2 were codon-optimised for *E. coli* and provided with restriction sites (BamHI and NdeI (New England BioLabs, Frankfurt am Main, Germany)). The synthetic genes were cloned into pET28a expression vectors and transformed in *E. coli* BL21 (DE3) pGro7.

4.2 -Heterologous Protein Expression and Purification

A total of 1.5 L of *Luria-Bertani* medium, containing 60 µg/mL kanamycin and 50 µg/mL chloramphenicol, was inoculated with 2% of an overnight culture of *E. coli* BL21 (DE3) pGro7 pET28a-spH1 or pET28a-spH2, and incubated at 37 °C until reaching an optical density (OD_{600}) of 0.4. The temperature was decreased to 25 °C for 30 min, and over expression was induced by the addition of isopropyl-β-D-thiogalactopyranoside (0.1 mM) and L-arabinose (2 g/L). After cultivation for a further 22 h at 150 rpm, cells were harvested by centrifugation (4200× *g*, 1 h, 4 °C), washed with 100 mM Na_2HPO_4 buffer (pH 7.4), and stored at −20 °C.

Expression of alcohol dehydrogenase *RR*-ADH and flavin reductase PrnF was performed as previously described [3].

The cells from 1.5 L cultivation were resuspended in 30 mL of 100 mM Na_2HPO_4 buffer (pH 7.4) and lysed by French Press (three times, 1000 psig). Cell debris was removed by centrifugation (10,000× *g*, 30 min, 4 °C), and soluble protein was filtered through a 0.2 µM Whatman filter. The purification of the hexahistidin-tagged proteins was performed via a HisTALON matrix. The lysate was loaded on HisTALON agarose affinity resin (TaKaRa Bio Inc., Saint-Germain-en-Laye, France) and washed with 10× column volume (CV) with 100 mM Na_2HPO_4 buffer (pH 7.4), and 10 × CV 50 mM Na_2HPO_4 buffer (pH 7.4) with 300 mM NaBr and 10 mM imidazole. The proteins were eluted with 50 mM Na_2HPO_4 buffer (pH 7.4), 300 mM NaBr, and 300 mM imidazole and collected in 0.75 mL fractions. After determination of the protein concentration by NanoDrop UV spectroscopy (Thermo Scientific, Waltham, MA, USA), fractions with the purified proteins were pooled and desalted via a HiTrap Desalting column, with 50 mM Na_2HPO_4 buffer (pH 7.4) and 50 mM NaBr.

4.3 Enzymatic Halogenation/Enzyme Assay with Purified Protein on an Analytical Scale

The activities of PrnF and ADH were determined as published by Frese et al. [3]. The halogenation activities of SpH1 and SpH2 towards different substrates were carried out for 48 h at 25 °C and 500 rpm in a total volume of 0.5 mL, containing 1.25 mg/mL enzyme, 1 mM substrate, 1 µM FAD, 100 µM NAD, 100 mM NaBr/NaCl, *RR*-ADH (2 U/mL) PrnF (2.5 U/mL), 5% (*v*/*v*) *iso*-propanol, and 100 mM Na_2HPO_4 buffer (pH 7.4) [33,52]. Samples were quenched by adding an equal volume of methanol. Reactions were analysed by analytical RP-HPLC and HPLC-MS.

4.5 Determination of Specific Activity

Halogenation reactions, as described in **Section 3.3.3.**, were carried out with 0.05 mM substrate and 10 μM enzyme. Samples were taken at 5 min intervals over a 20 min time period, and then the enzyme reaction was stopped with 1:1 methanol/ water (*v*/*v*). The determination was carried out in triplicate, reactions were analysed by analytical RP-HPLC, and conversion rates (*v*) were calculated. With this in hand, the specific activity can be calculated using the following:

$$specific\ activity \left[\frac{mU}{mg}\right] = \frac{v}{c_E \cdot M_E} \times 10^9 \quad (1)$$

where c_E is the enzyme concentration (10 μM) and M_E is the enzyme mass (g/ mol).

4.6 Catalase Activity Assay

Catalase from bovine liver lyophilised powder (2000–5000 units/mg protein; Sigma Aldrich, Darmstadt, Germany) at a final concentration of 100 U/mL was additionally added to the reaction mixture, and the procedure continued as described above to determine the FAD dependency of SpH1 and SpH2 [33].

4.7 Enzymatic Halogenation with Immobilised Protein on a Preparative Scale

For the isolation and analysis of halogenated products on a preparative scale, cross- linked enzyme aggregates (CLEAs), as described by Frese et al. [1], were used.

The *E. coli* BL21 (DE3) pGro7 cells from a 1.5 L cultivation containing SpH1 and SpH2 were resuspended in 30 mL 100 mM Na_2HPO_4 buffer (pH 7.4) and lysed by French Press (three times, 1000 psig). After centrifugation (30 min, 10,000× *g*, 4 °C), the cell lysate was divided in two equal parts, and PrnF (2.5 U/mL) and *RR*-ADH (1 U/mL) were added. Precipitation of the proteins was carried out by adding ammonium sulfate (8.1 g, 95% saturation), followed by incubation for 1 h at 4 °C in a tube rotator. Glutaraldehyde (1.5 mL, 0.5% (*w*/*v*)) was added for cross-linking, and the mixture was further kept for 2 h at 4 °C. The CLEAs were centrifuged and washed three times with 30 mL of 100 mM Na_2HPO_4 buffer (pH 7.4). The pH of the reaction buffer for the biocatalysis with CLEAs containing, 1.5 mM substrate, 1 μM FAD, 100 μM NAD, 30 mM NaBr/NaCl, and 5% (*v*/*v*) and 15 mM Na_2HPO_4, was adjusted with phosphoric acid to pH 7.4. Finally, CLEAs were added to reaction buffer in a final volume of 500 mL, and incubated for up to 10 days (until no further conversion was visible at the analytical RP-HPLC) at 25 °C and 150 rpm. The reaction mixture was filtered and extracted three times with 250 mL dichloromethane. The organic phase was dried over magnesium sulfate and evaporated. Before the characterisation of the halogenated products with HPLC-MS, analytical HPLC, and NMR, the remaining substance was purified over preparative RP-HPLC and lyophilised.

4.8 FAD Reconstitution

For FAD reconstitution, the buffer of the purified proteins was exchanged to FAD reconstitution buffer (50 mM Na_2HPO_4, 150 mM NaBr, 1 mM FAD, pH 7.4) using a HiTrap- Desalting column and incubated overnight at 4 °C on a rocking shaker. The proteins were washed three times with 50 mM Na_2HPO_4 (pH 7.4) and 150 mM NaBr to remove free FAD, and concentrated to a final volume of 400 µL in an Amicon Ultra-4 centrifugal filter device with a 50 kDa cutoff. To determine the degree of occupation, UV/Vis spectra of the proteins were measured in a quartz cuvette (Suprasil, Hellma, Müllheim, Germany) with a path length of 1 cm, using a Shimadzu (Duisburg, Germany) UV-2450 spectrometer [33].

4.9 In Silico Study Using YASARA

The homology models of SpH1 and SpH2 were built using YASARA structure (www.yasara.org, 2020) [84]. the built models were further energy minimised before MD simulation and the docking study [85]. The structures of indole and 3-bromoindole were built with YASARA structure and energy-minimised using YASARA 2 force field before using in the docking experiments.

MD simulations were run for the homology models of both enzymes for 1 ns using YASARA structure [86]. The simulation cell was extended by 10 Å around the whole model and filled with randomly-oriented water molecules with a density of 0.99 g/mL. The force field AMBER14, with 10 Å for cut off and particle mesh Ewald (PME) for long-range electrostatic, was used for running the simulation. The stability of the homology models was monitored by the root mean square deviation (RMSD) of the alpha carbon of the models throughout the simulation time.

Docking of either indole or 3-bromoindole was performed by YASARA structure, using Autodock for the MD simulated models of SpH1 and SpH2 [87]. The simulation cell was defined around the whole homology model of both enzymes. Docking results were analysed based on the binding energy (B-factor) calculated by YASARA structure and the interaction of the substrate with the amino acid residues of the active site pocket.

The docked models with indole or 3-bromoindole were further subjected to MD simulation using the previously mentioned parameters.

The size of the active site pocket cavity was determined using the Computed Atlas of Surface Topography of proteins (CASTp) (http://sts.bioe.uic.edu/castp/, 2020) for the free and docked homology models before and after MD simulation [88].

5. CONCLUSIONS

Flavin-dependent halogenases are the focus of research because of their enormous potential for the regioselective halogenation of organic compounds, leading to products with profound biological activity. In the present study, using our hidden Markov model (pHMM), we screened bacterial associates of the *B. braunii* consortia (PRJEB21978) and identified 25 complete and partial putative, flavin-dependent halogenase genes. Interestingly, all newly identified genes were found to form a distinct clade, together with previously characterised BrvH, potentially indicating a similar substrate preference. Two selected flavin-dependent halogenases (SpH1 and SpH2), derived from one gene cluster of *Sphingomonas* sp., were subjected to a screening with different compounds, and 26 substrates were found to become halogenated (**Figure 4**). In accordance with recent findings, as well as with the activity performance of BrvH, both enzymes were observed to prefer bromination over chlorination. Notably, SpH2 is capable of dibrominating substrates, while SpH1 can only monobrominate. Moreover, the simulation in silico using YASARA homology models revealed that SpH1 possesses a deeper but less flexible active site pocket, while SpH2 features a wider and more flexible active site pocket. This explains the ability of SpH2 to halogenate bulkier substrates, as well as catalyse dibromination. Both were identified in one gene, which suggests that the newly identified flavin-dependent halogenases could act in concert and interact with one another. Likewise, other known flavin-dependent halogenases originating from one gene cluster, such as the correlates PrnA/PrnC, Bmp2/Bmp5, or KtzR/KtzQ, act in concert for the biosynthesis of multiple halogenated compounds. SpH1 and SpH2 accepted, in total, 26 different indole and phenol derivatives. Interestingly, both accepted 4-*n*-hexylresorcinol (**18**), tryptophol (**20**), and 3-(3-indolyl)propionic acid (**21**). However, the natural substrates of SpH1 and SpH2 remain unknown so far. In the gene cluster of both enzymes, a catecholate siderophore receptor was detected, which might give a hint towards their substrate. This study shows that screening with the HMM may discover many novel flavin-dependent halogenases in natural metagenomes. The algorithm detects mostly flavin-dependent halogenases from a clade similar to BrvH. The phylogenetic tree demonstrates that the detected novel, putative, flavin-dependent halogenases cluster within the clade of BrvH.

6. CONCISE SUMMARY

Halogen substituents often lead to a profound effect on the biological activity of organic compounds. Flavin-dependent halogenases offer the possibility of regioselective halogenation at non-activated carbon atoms, while employing only halide salts and molecular oxygen. However, low enzyme activity, instability, and narrow substrate scope compromise the use of enzymatic halogenation as an economical and environmentally friendly process. To overcome these drawbacks, it is of tremendous interest to identify novel halogenases with high enzymatic activity and novel substrate scopes. Previously, Neubauer et al. developed a new hidden Markov model (pHMM) based on the PFAM tryptophan halogenase model, and identified 254 complete and partial putative flavin-dependent halogenase genes in eleven metagenomic data sets. In the present study, the pHMM was used to screen the bacterial associates of the *Botryococcus braunii* consortia (PRJEB21978), leading to the identification of several putative, flavin-dependent halogenase genes. Two of these new halogenase genes were found in one gene cluster of the *Botryococcus braunii* symbiont *Sphingomonas* sp. In vitro activity tests revealed that both heterologously expressed enzymes are active flavin-dependent halogenases able to halogenate indole and indole derivatives, as well as phenol derivatives, while preferring bromination over chlorination. Interestingly, SpH1 catalyses only monohalogenation, while SpH2 can catalyse both mono- and dihalogenation for some substrates.

REFERENCES

1. Frese, M.; Sewald, N. Enzymatic halogenation of tryptophan on a gram scale. *Angew. Chem. Int. Ed.* **2015**, *54*, 298–301. [Google Scholar] [CrossRef]
2. Schnepel, C.; Sewald, N. Enzymatic halogenation: A timely strategy for regioselective C-H activation. *Chem. Eur. J.* **2017**, *23*, 12064–12086. [Google Scholar] [CrossRef]
3. Frese, M.; Guzowska, P.H.; Voss, H.; Sewald, N. Regioselective Enzymatic Halogenation of Substituted Tryptophan Derivatives using the FAD-Dependent Halogenase RebH. *ChemCatChem* **2014**, *6*, 1270–1276. [Google Scholar]
4. Frese, M.; Schnepel, C.; Minges, H.; Voß, H.; Feiner, R.; Sewald, N. Modular Combination of Enzymatic Halogenation of Tryptophan with Suzuki-Miyaura Cross-Coupling Reactions. *ChemCatChem* **2016**, *8*, 1799–1803. [Google Scholar] [CrossRef]
5. Dachwitz, S.; Widmann, C.; Frese, M.; Niemann, H.H.; Sewald, N. Enzymatic halogenation: Enzyme mining, mechanisms, and implementation in reaction cascades. *Amino Acids Pept. Proteins* **2021**, *44*, 1. [Google Scholar]
6. Minges, H.; Sewald, N. Recent Advances in Synthetic Application and Engineering of Halogenases. *ChemCatChem* **2020**, *12*, 4450–4470. [Google Scholar] [CrossRef]
7. Mascotti,M.L.;Ayub,M.J.;Furnham,N.;Thornton,J.M.;Laskowski,R.A.Chopping and changing: The evolution of the flavin-dependent monooxygenases. *J. Mol. Biol.* **2016**, *428*, 3131–3146. [Google Scholar] [CrossRef][Green Version]

8. Badieyan, S.; Bach, R.D.; Sobrado, P. Mechanism of N-hydroxylation catalyzed by flavin-dependent monooxygenases. *J. Org. Chem.* **2015**, *80*, 2139–2147. [Google Scholar] [CrossRef] [PubMed]

9. Kemker, I.; Schnepel, C.; Schröder, D.C.; Marion, A.; Sewald, N. Cyclization of RGD Peptides by Suzuki-Miyaura Cross-Coupling. *J. Med. Chem.* **2019**, *62*, 7417–7430. [Google Scholar] [CrossRef]

10. Durak, L.J.; Payne, J.T.; Lewis, J.C. Late-stage diversification of biologically active molecules via chemoenzymatic C–H functionalization. *ACS Catal.* **2016**, 6, 1451–1454. [Google Scholar] [CrossRef][Green Version]

11. Latham, J.; Henry, J.-M.; Sharif, H.H.; Menon, B.R.K.; Shepherd, S.A.; Greaney, M.F.; Micklefield, J. Integrated catalysis opens new arylation pathways via regiodivergent enzymatic C–H activation. *Nat. Commun.* **2016**, 7, 1–8. [Google Scholar] [CrossRef] [PubMed]

12. Kemker, I.; Schröder, D.C.; Feiner, R.C.; Müller, K.M.; Marion, A.; Sewald, N. Tuning the Biological Activity of RGD Peptides with Halotryptophans. *J. Med. Chem.* **2021**, *64*, 586–601. [Google Scholar] [CrossRef] [PubMed]

13. Corr, M.J.; Sharma, S.V.; Pubill-Ulldemolins, C.; Bown, R.T.; Poirot, P.; Smith, D.R.M.; Cartmell, C.; Abou-Fayad, A.; Goss, R.J.M. Sonogashira diversification of unprotected halotryptophans, halotryptophan containing tripeptides; and generation of a new to nature bromo-natural product and its diversification in water. *Chem. Sci.* **2017**, *8*, 2039–2046. [Google Scholar] [CrossRef] [PubMed] [Green Version]

14. Gruß, H.; Sewald, N. Late-Stage Diversification of Tryptophan-Derived Biomolecules. *Chem. Eur. J.* **2020**, *26*, 5328–5340. [Google Scholar] [CrossRef] [PubMed]

15. Pubill-Ulldemolins, C.; Sharma, S.V.; Cartmell, C.; Zhao, J.; Cárdenas, P.; Goss, R.J.M. Heck Diversification of Indole-Based Substrates under Aqueous Conditions: From Indoles to Unprotected Halo-tryptophans and Halo-tryptophans in Natural Product Derivatives. *Chem. Eur. J.* **2019**, *25*, 10866–10875. [Google Scholar] [CrossRef] [PubMed]

16. Keller, S.; Wage, T.; Hohaus, K.; Hölzer, M.; Eichhorn, E.; van Pée, K.-H. Purification and partial characterization of tryptophan 7-halogenase (PrnA) from Pseudomonas fluorescens. *Angew. Chem. Int. Ed.* **2000**, *39*, 2300–2302. [Google Scholar] [CrossRef]

17. Kirner, S.; Hammer, P.E.; Hill, D.S.; Altmann, A.; Fischer, I.; Weislo, L.J.; Lanahan, M.; van Pée, K.-H.; Ligon, J.M. Functions encoded by pyrrolnitrin biosynthetic genes from Pseudomonas fluorescens. *J. Bacteriol. Res.* **1998**, *180*, 1939–1943. [Google Scholar] [CrossRef][Green Version]

18. Sánchez, C.; Butovich, I.A.; Braña, A.F.; Rohr, J.; Méndez, C.; Salas, J.A. The biosynthetic gene cluster for the antitumor rebeccamycin: Characterization and generation of indolocarbazole derivatives. *Chem. Biol.* **2002**, *9*, 519–531. [Google Scholar] [CrossRef][Green Version]

19. Heemstra, J.R., Jr.; Walsh, C.T. Tandem action of the O2-and FADH2-dependent halogenases KtzQ and KtzR produce 6, 7-dichlorotryptophan for kutzneride assembly. *J. Am. Chem. Soc.* **2008**, *130*, 14024–14025. [Google Scholar] [CrossRef] [Green Version]

20. Seibold, C.; Schnerr, H.; Rumpf, J.; Kunzendorf, A.; Hatscher, C.; Wage, T.; Ernyei, A.J.; Dong, C.; Naismith, J.H.; van Pée, K.-H. A flavin-dependent tryptophan 6-halogenase and its use in modification of pyrrolnitrin biosynthesis. *Biocatal. Biotransform.* **2006**, *24*, 401–408. [Google Scholar] [CrossRef]

21. Zeng, J.; Zhan, J. Characterization of a tryptophan 6-halogenase from Streptomyces toxytricini. *Biotechnol. Lett.* **2011**, *33*, 1607–1613. [Google Scholar] [CrossRef]

22. Xu, L.; Han, T.; Ge, M.; Zhu, L.; Qian, X. Discovery of the new plant growth-regulating compound LYXLF2 based on manipulating the halogenase in Amycolatopsis orientalis. *Curr. Microbiol.* **2016**, *73*, 335–340. [Google Scholar] [CrossRef]

23. Zehner, S.; Kotzsch, A.; Bister, B.; Süssmuth, R.D.; Méndez, C.; Salas, J.A.; van Pée, K.-H. A regioselective tryptophan 5-halogenase is involved in pyrroindomycin biosynthesis in Streptomyces rugosporus LL-42D005. *Chem. Biol.* **2005**, *12*, 445–452. [Google Scholar] [CrossRef][Green Version]

24. Ryan, K.S. Biosynthetic gene cluster for the cladoniamides, bis-indoles with a rearranged scaffold. *PLoS ONE* **2011**, *6*, e23694. [Google Scholar] [CrossRef]

25. Moritzer, A.-C.; Minges, H.; Prior, T.; Frese, M.; Sewald, N.; Niemann, H.H. Structure-based switch of regioselectivity in the flavin-dependent tryptophan 6-halogenase Thal. *J. Biol Chem.* **2019**, *294*, 2529–2542. [Google Scholar] [CrossRef] [PubMed][Green Version]

26. Minges, H.; Schnepel, C.; Böttcher, D.; Weiß, M.S.; Sproß, J.; Bornscheuer, U.T.; Sewald, N. Targeted Enzyme Engineering Unveiled Unexpected Patterns of Halogenase Stabilization. *ChemCatChem* **2020**, *12*, 818–831. [Google Scholar] [CrossRef][Green Version]

27. Van Pée, K.-H. Enzymatic chlorination and bromination. *Meth. Enzymol.* **2012**, *516*, 237–257. [Google Scholar]

28. Dong, C.; Flecks, S.; Unversucht, S.; Haupt, C.; van Pee, K.-H.; Naismith, J.H. Tryptophan 7-halogenase (PrnA) structure suggests a mechanism for regioselective chlorination. *Science* **2005**, *309*, 2216–2219. [Google Scholar] [CrossRef]

29. Zhu, X.; de Laurentis, W.; Leang, K.; Herrmann, J.; Ihlefeld, K.; Van Pée, K.-H.; Naismith, J.H. Structural insights into regioselectivity in the enzymatic chlorination of tryptophan. *J. Mol. Biol.* **2009**, *391*, 74–85. [Google Scholar] [CrossRef] [PubMed][Green Version]

30. Dong, C.; Kotzsch, A.; Dorward, M.; van Pée, K.-H.; Naismith, J.H. Crystallization and X-ray diffraction of a halogenating enzyme, tryptophan 7-halogenase, from *Pseudomonas fluorescens*. *Acta Cryst.* **2004**, *60*, 1438–1440. [Google Scholar]

31. Ainsley, J.; Mulholland, A.J.; Black, G.W.; Sparagano, O.; Christov, C.Z.; Karabencheva-Christova, T.G. Structural insights from molecular dynamics simulations of tryptophan 7-halogenase and tryptophan 5-halogenase. *ACS Omega* **2018**, *3*, 4847–4859. [Google Scholar] [CrossRef]

32. Yeh, E.; Blasiak, L.C.; Koglin, A.; Drennan, C.L.; Walsh, C.T. Chlorination by a long-lived intermediate in the mechanism of flavin-dependent halogenases. *Biochemistry* **2007**, *46*, 1284–1292. [Google Scholar] [CrossRef]

33. Neubauer, P.R.; Widmann, C.; Wibberg, D.; Schröder, L.; Frese, M.; Kottke, T.; Kalinowski, J.; Niemann, H.H.; Sewald, N. A flavin-dependent halogenase from metagenomic analysis prefers bromination over chlorination. *PLoS ONE* **2018**, *13*, e0196797. [Google Scholar] [CrossRef] [PubMed][Green Version]

34. Menon, B.R.K.; Brandenburger, E.; Sharif, H.H.; Klemstein, U.; Shepherd, S.A.; Greaney, M.F.; Micklefield, J. RadH: A Versatile Halogenase for Integration into Synthetic Pathways. *Angew. Chem. Int. Ed.* **2017**, *56*, 11841–11845. [Google Scholar] [CrossRef] [PubMed]

35. Flecks, S.; Patallo, E.P.; Zhu, X.; Ernyei, A.J.; Seifert, G.; Schneider, A.; Dong, C.; Naismith, J.H.; van Pée, K.-H. New insights into the mechanism of enzymatic chlorination of tryptophan. *Angew. Chem. Int. Ed.* **2008**, *47*, 9533–9536. [Google Scholar] [CrossRef][Green Version]

36. Bitto, E.; Huang, Y.; Bingman, C.A.; Singh, S.; Thorson, J.S.; Phillips, G.N. The structure of flavin-dependent tryptophan 7-halogenase RebH. *Proteins* **2008**, *70*, 289–293. [Google Scholar] [CrossRef]

37. Podzelinska, K.; Latimer, R.; Bhattacharya, A.; Vining, L.C.; Zechel, D.L.; Jia, Z. Chloramphenicol biosynthesis: The structure of CmlS, a flavin-dependent halogenase showing a covalent flavin-aspartate bond. *J. Mol. Biol.* **2010**, *397*, 316–331. [Google Scholar] [CrossRef] [PubMed]

38. Roy, A.D.; Grüschow, S.; Cairns, N.; Goss, R.J.M. Gene expression enabling synthetic diversification of natural products: Chemogenetic generation of pacidamycin analogs. *J. Am. Chem. Soc.* **2010**, *132*, 12243–12245. [Google Scholar] [CrossRef] [PubMed]

39. Glenn, W.S.; Nims, E.; O'Connor, S.E. Reengineering a tryptophan halogenase to preferentially chlorinate a direct alkaloid precursor. *J. Am. Chem. Soc.* **2011**, *133*, 19346–19349. [Google Scholar] [CrossRef]

40. Runguphan, W.; Qu, X.; O'connor, S.E. Integrating carbon–halogen bond formation into medicinal plant metabolism. *Nature* **2010**, *468*, 461–464. [Google Scholar] [CrossRef]

41. Veldmann, K.H.; Dachwitz, S.; Risse, J.M.; Lee, J.-H.; Sewald, N.; Wendisch, V.F. Bromination of L-tryptophan in a Fermentative Process with Corynebacterium glutamicum. *Front. Bioeng. Biotechnol.* **2019**, *7*, 219. [Google Scholar] [CrossRef] [PubMed]

42. Hammer, P.E.; Hill, D.S.; Lam, S.T.; Van Pée, K.-H.; Ligon, J.M. Four genes from Pseudomonas fluorescens that encode the biosynthesis of pyrrolnitrin. *Appl. Environ. Microbiol.* **1997**, *63*, 2147–2154. [Google Scholar] [CrossRef] [PubMed] [Green Version]

43. Smith, D.R.M.; Uria, A.R.; Helfrich, E.J.N.; Milbredt, D.; van Pée, K.-H.; Piel, J.; Goss, R.J.M. An unusual flavin-dependent halogenase from the metagenome of the marine sponge *Theonella swinhoei* WA. *ACS Chem. Biol.* **2017**, *12*, 1281–1287. [Google Scholar] [CrossRef] [PubMed][Green Version]

44. Menon, B.R.K.; Latham, J.; Dunstan, M.S.; Brandenburger, E.; Klemstein, U.; Leys, D.; Karthikeyan, C.; Greaney, M.F.; Shepherd, S.A.; Micklefield, J. Structure and biocatalytic scope of thermophilic flavin-dependent halogenase and flavin reductase enzymes. *Org. Biomol. Chem.* **2016**, *14*, 9354–9361. [Google Scholar] [CrossRef][Green Version]

45. Ortega, M.A.; Cogan, D.P.; Mukherjee, S.; Garg, N.; Li, B.; Thibodeaux, G.N.; Maffioli, S.I.; Donadio, S.; Sosio, M.; Escano, J. Two flavoenzymes catalyze the post-translational generation of 5-chlorotryptophan and 2-aminovinyl-cysteine during NAI-107 biosynthesis. *ACS Chem. Biol.* **2017**, *12*, 548–557. [Google Scholar] [CrossRef] [PubMed][Green Version]

46. Agarwal, V.; el Gamal, A.A.; Yamanaka, K.; Poth, D.; Kersten, R.D.; Schorn, M.; Allen, E.E.; Moore, B.S. Biosynthesis of polybrominated aromatic organic compounds by marine bacteria. *Nat. Chem. Biol.* **2014**, *10*, 640. [Google Scholar] [CrossRef][Green Version]

47. Zeng, J.; Zhan, J. A Novel Fungal Flavin-Dependent Halogenase for Natural Product Biosynthesis. *ChemBioChem* **2010**, *11*, 2119–2123. [Google Scholar] [CrossRef]

48. Chooi, Y.-H.; Cacho, R.; Tang, Y. Identification of the viridicatumtoxin and griseofulvin gene clusters from *Penicillium aethiopicum*. *Chem. Biol.* **2010**, *17*, 483–494. [Google Scholar] [CrossRef][Green Version]

49. Wang, S.; Xu, Y.; Maine, E.A.; Wijeratne, E.K.; Espinosa-Artiles, P.; Gunatilaka, A.L.; Molnár, I. Functional characterization of the biosynthesis of radicicol, an Hsp90 inhibitor resorcylic acid lactone from Chaetomium chiversii. *Chem. Biol.* **2008**, *15*, 1328–1338. [Google Scholar] [CrossRef] [PubMed][Green Version]

50. Bayer, K.; Scheuermayer, M.; Fieseler, L.; Hentschel, U. Genomic mining for novel FADH 2-dependent halogenases in marine sponge-associated microbial consortia. *Mar. Biotechnol.* **2013**, *15*, 63–72. [Google Scholar] [CrossRef]

51. Ismail, M.; Frese, M.; Patschkowski, T.; Ortseifen, V.; Niehaus, K.; Sewald, N. Flavin-Dependent Halogenases from Xanthomonas campestris pv. campestris B100 Prefer Bromination over Chlorination. *Adv. Synth. Catal.* **2019**, *361*, 2475–2486. [Google Scholar] [CrossRef][Green Version]

52. Neubauer, P.R.; Pienkny, S.; Wessjohann, L.A.; Brandt, W.; Sewald, N. Predicting the substrate scope of the flavin-dependent halogenase BrvH. *ChemBioChem* **2020**, *21*, 3282–3288. [Google Scholar] [CrossRef]

53. Widmann, C.; Ismail, M.; Sewald, N.; Niemann, H.H. Structure of apo flavin-dependent halogenase Xcc4156 hints at a reason for cofactor-soaking difficulties. *Acta Cryst.* **2020**, *76*, 687–697. [Google Scholar] [CrossRef] [PubMed]

54. Rojo, F. Degradation of alkanes by bacteria. *Environ. Microbiol.* **2009**, *11*, 2477–2490. [Google Scholar] [CrossRef] [PubMed]

55. Löwe, J.; Blifernez-Klassen, O.; Baier, T.; Wobbe, L.; Kruse, O.; Gröger, H. Type II flavoprotein monooxygenase PsFMO_A from the bacterium Pimelobacter sp. Bb-B catalyzes enantioselective Baeyer-Villiger oxidations with a relaxed cofactor specificity. *J. Biotechnol.* **2019**, *294*, 81–87. [Google Scholar] [CrossRef]

56. Metzger, P.; Largeau, C. *Botryococcus braunii*: A rich source for hydrocarbons and related ether lipids. *Appl. Microbiol. Biotechnol.* **2005**, *66*, 486–496. [Google Scholar] [CrossRef] [PubMed]

57. Blifernez-Klassen, O.; Chaudhari, S.; Klassen, V.; Wördenweber, R.; Steffens, T.; Cholewa, D.; Niehaus, K.; Kalinowski, J.; Kruse, O. Metabolic survey of *Botryococcus braunii*: Impact of the physiological state on product formation. *PLoS ONE* **2018**, *13*, e0198976. [Google Scholar] [CrossRef] [PubMed][Green Version]

58. Chirac, C.; Casadevall, E.; Largeau, C.; Metzger, P. Bacterial influence upon growth and hydrocarbon production of the green alga *Botryococcus braunii* 1. *J. Phycol.* **1985**, *21*, 380–387. [Google Scholar] [CrossRef]

59. Blifernez-Klassen, O.; Klassen, V.; Wibberg, D.; Cebeci, E.; Henke, C.; Rückert, C.; Chaudhari, S.; Rupp, O.; Blom, J.; Winkler, A. *Phytoplankton consortia* as a blueprint for mutually beneficial eukaryote-bacteria ecosystems based on the

biocoenosis of *Botryococcus consortia*. *Sci. Rep.* **2021**, *11*, 1–13. [Google Scholar] [CrossRef] [PubMed]

60. Kumar, S.; Stecher, G.; Li, M.; Knyaz, C.; Tamura, K. MEGA X: Molecular evolutionary genetics analysis across computing platforms. *Mol. Biol. Evol.* **2018**, *35*, 1547–1549. [Google Scholar] [CrossRef]
61. Saitou, N.; Nei, M. The neighbour-joining method: A new method for reconstructing phylogenetic trees. *Mol. Biol. Evol.* **1987**, *4*, 406–425. [Google Scholar]
62. Felsenstein, J. Confidence limits on phylogenies: An approach using the bootstrap. *Evolution* **1985**, *39*, 783–791. [Google Scholar] [CrossRef] [PubMed]
63. Blin, K.; Wolf, T.; Chevrette, M.G.; Lu, X.; Schwalen, C.J.; Kautsar, S.A.; Suarez Duran, H.G.; de los Santos, E.L.C.; Kim, H.U.; Nave, M. antiSMASH 4.0—Improvements in chemistry prediction and gene cluster boundary identification. *Nucleic Acids Res.* **2017**, *45*, W36–W41. [Google Scholar] [CrossRef] [PubMed]
64. Cotter, P.D.; Ross, R.P.; Hill, C. Bacteriocins—A viable alternative to antibiotics? *Nat. Rev. Microbiol.* **2013**, *11*, 95–105. [Google Scholar] [CrossRef] [PubMed]
65. Ligtenbarg, A.G.J.; Hage, R.; Feringa, B.L. Catalytic oxidations by vanadium complexes. *Coord. Chem. Rev.* **2003**, *237*, 89–101. [Google Scholar] [CrossRef] [Green Version]
66. Crichton, R.R. *Biological Inorganic Chemistry: A New Introduction to Molecular Structure and Function*, 2nd ed.; Elsevier: Amsterdam, The Netherlands, 2012; ISBN 9780444537829. reprinted. [Google Scholar]
67. Hager, L.P.; Morris, D.R.; Brown, F.S.; Eberwein, H. Chloroperoxidase: II. Utilization of halogen anions. *J. Biol. Chem.* **1966**, *241*, 1769–1777. [Google Scholar] [CrossRef]
68. Schroeder, L.; Frese, M.; Müller, C.; Sewald, N.; Kottke, T. Photochemically Driven Biocatalysis of Halogenases for the Green Production of Chlorinated Compounds. *ChemCatChem* **2018**, *10*, 3336–3341. [Google Scholar] [CrossRef]
69. Sundberg, R.J. Electrophilic Substitution Reactions of Indoles. In *Heterocyclic Scaffolds II: Topics in Heterocyclic Chemistry*; Gribble, G.W.W., Ed.; Springer: Berlin/Heidelberg, Germany, 2010; pp. 47–115. [Google Scholar] [CrossRef]
70. Hölzer, M.; Burd, W.; Reißig, H.-U.; van Pée, K.-H. Substrate specificity and regioselectivity of tryptophan 7-halogenase from *Pseudomonas fluorescens* BL915. *Adv. Synth. Catal.* **2001**, *343*, 591–595. [Google Scholar] [CrossRef]
71. Van Pée, K.H. Microbial biosynthesis of halometabolites. *Arch. Microbiol.* **2001**, *175*, 250–258. [Google Scholar] [CrossRef]

72. Fisher, B.F.; Snodgrass, H.M.; Jones, K.A.; Andorfer, M.C.; Lewis, J.C. Site-Selective C–H Halogenation Using Flavin-Dependent Halogenases Identified via Family-Wide Activity Profiling. *ACS Cent. Sci.* **2019**, *5*, 1844–1856. [Google Scholar] [CrossRef][Green Version]

73. Kunkel, B.N.; Harper, C.P. The roles of auxin during interactions between bacterial plant pathogens and their hosts. *J. Exp. Bot.* **2018**, *69*, 245–254. [Google Scholar] [CrossRef] [PubMed]

74. Woodward, A.W.; Bartel, B. Auxin: Regulation, action, and interaction. *Ann. Bot.* **2005**, *95*, 707–735. [Google Scholar] [CrossRef] [PubMed][Green Version]

75. Ozga, J.A.; Reinecke, D.M.; Ayele, B.T.; Ngo, P.; Nadeau, C.; Wickramarathna, A.D. Developmental and hormonal regulation of gibberellin biosynthesis and catabolism in pea fruit. *Plant Physiol.* **2009**, *150*, 448–462. [Google Scholar] [CrossRef][Green Version]

76. Tivendale, N.D.; Davidson, S.E.; Davies, N.W.; Smith, J.A.; Dalmais, M.; Bendahmane, A.I.; Quittenden, L.J.; Sutton, L.; Bala, R.K.; Le Signor, C. Biosynthesis of the halogenated auxin, 4-chloroindole-3-acetic acid. *Plant Physiol.* **2012**, *159*, 1055–1063. [Google Scholar] [CrossRef][Green Version]

77. Duca, D.; Lorv, J.; Patten, C.L.; Rose, D.; Glick, B.R. Indole-3-acetic acid in plant–microbe interactions. *Antonie Leeuwenhoek* **2014**, *106*, 85–125. [Google Scholar] [CrossRef]

78. Murphy, K. Applying Green Chemistry Principles in the Electrophilic Bromination of Indole-3-Acetic Acid. *Undergrad. Rev.* **2014**, *10*, 111–115. [Google Scholar]

79. Kramer, J.; Özkaya, Ö.; Kümmerli, R. Bacterial siderophores in community and host interactions. *Nat. Rev. Microbiol.* **2020**, *18*, 152–163. [Google Scholar] [CrossRef]

80. Kishimoto, S.; Nishimura, S.; Hattori, A.; Tsujimoto, M.; Hatano, M.; Igarashi, M.; Kakeya, H. Chlorocatechelins A and B from Streptomyces sp.: New siderophores containing chlorinated catecholate groups and an acylguanidine structure. *Org. Lett.* **2014**, *16*, 6108–6111. [Google Scholar] [CrossRef] [PubMed]

81. O'Leary, N.A.; Wright, M.W.; Brister, J.R.; Ciufo, S.; Haddad, D.; McVeigh, R.; Rajput, B.; Robbertse, B.; Smith-White, B.; Ako-Adjei, D. Reference sequence (RefSeq) database at NCBI: Current status, taxonomic expansion, and functional annotation. *Nucleic Acids Res.* **2016**, *44*, D733–D745. [Google Scholar] [CrossRef][Green Version]

82. Corpet, F. Multiple sequence alignment with hierarchical clustering. *Nucleic Acids Res.* **1988**, *16*, 10881–10890. [Google Scholar] [CrossRef] [PubMed]

83. Edgar, R.C. MUSCLE: Multiple sequence alignment with high accuracy and high throughput. *Nucleic Acids Res.* **2004**, *32*, 1792–1797. [Google Scholar] [CrossRef][Green Version]

84. Krieger, E.; Vriend, G. YASARA View—Molecular graphics for all devices—From smartphones to workstations. *Bioinformatics* **2014**, *30*, 2981–2982. [Google Scholar] [CrossRef] [PubMed][Green Version]

85. Krieger, E.; Joo, K.; Lee, J.; Lee, J.; Raman, S.; Thompson, J.; Tyka, M.; Baker, D.; Karplus, K. Improving physical realism, stereochemistry, and side-chain accuracy in homology modeling: Four approaches that performed well in CASP8. *Proteins* **2009**, *77*, 114–122. [Google Scholar] [CrossRef] [PubMed] [Green Version]

86. Krieger, E.; Vriend, G. New ways to boost molecular dynamics simulations. *J. Comput. Chem.* **2015**, *36*, 996–1007. [Google Scholar] [CrossRef] [PubMed]

87. Trott, O.; Olson, A.J. AutoDock Vina: Improving the speed and accuracy of docking with a new scoring function, efficient optimization, and multithreading. *J. Comput. Chem.* **2010**, *31*, 455–461. [Google Scholar] [CrossRef][Green Version]

88. Tian, W.; Chen, C.; Lei, X.; Zhao, J.; Liang, J. CASTp 3.0: Computed atlas of surface topography of proteins. *Nucleic Acids Res.* **2018**, *46*, W363–W367. [Google Scholar] [CrossRef][Green Version]

CHAPTER-3

UTILIZING EXOGENOUS ENZYMES AS ADDITIVES WITH ZOOTECHNICAL BENEFITS IN ANIMAL FEED: A REVIEW

1. INTRODUCTION: AN OVERVIEW

In the ever-evolving landscape of animal nutrition, the incorporation of exogenous enzymes as additives in feed has emerged as a pivotal area of research and application. This comprehensive review explores the multifaceted realm of utilizing exogenous enzymes to enhance the zootechnical aspects of animal feeding. As advancements in nutritional science continue to unravel the complexities of optimal animal diets, the strategic integration of enzymes stands out as a promising avenue for improving feed efficiency, nutrient utilization, and overall animal performance. This review delves into the current state of knowledge, shedding light on the mechanisms, benefits, and potential challenges associated with incorporating exogenous enzymes into animal feed formulations.

Food security is a current challenge in most parts of the world, where the need to increase the intensive production of farm animals for the generation of meat, milk and eggs is a priority; however, to achieve this, it is necessary for the animals to consume nutritious and highly digestible feed, and their diet should not compete with humans' [1]. Traditionally, depending on the geographic location, animal feed has been based on grains, forages and silage, among others, but the need to improve the costs of feeding and animal production has led to the search for new ingredients for this purpose. The use of agro-industrial waste and agro-forestry is a current trend; however, its use has disadvantages because the nutritional components are unbalanced or unavailable; therefore, they must be supplemented with grains of cereals, legumes or additives to meet the nutritional needs [2].

The use of exogenous enzymes has been shown to exert positive effects on the agro-industrial and agro-forestry wastes used as animal feed by increasing the bioavailability of nutrients and digestibility as well as helping to eliminate some anti-nutritional factors. Although animals have endogenous enzymes involved in digestion, they do not have the ability to degrade them and to take advantage of all their nutritional components; therefore, their treatment with exogenous enzymes is a trend with beneficial results on production and animal yield [3,4]. An example of this is the use of fibrolytic enzymes that, when added to fibrous substrates, produce small amounts of oligomers, and, therefore, will degrade both soluble and insoluble fiber. This causes breaks in the insoluble fiber, increasing the amorphous nature of the fiber and reducing the time for the attachment of fibrolytic bacteria, thus improving fiber digestibility and the ability of the microbiome to degrade fiber [5]. The addition of enzymes improves the availability of nutrients (starch, proteins, amino acids and minerals, etc.) [6]. Additionally, they offer a beneficial performance response most of the time, but their catalytic efficiency is associated with factors such as the types of food and environmental factors [7]. The enzymes with carbohydrase, protease, phytase and lipase activities are the most used in the improvement of animal feed. For its application, it must be ensured that the biocatalyst is capable of resisting feed processing extrusion and granulation as well as changes in the gastrointestinal tract [8].

Although there are several studies on the application of enzyme additives in the animal feed industry, it is necessary to search for new enzymes, as well as to make use of bioinformatics tools for the design of specific enzymes that work in certain environmental conditions and substrates, allowing for the improvement of the productive parameters in the animals, reducing the costs and making the processes more efficient. This review compiles the progress in obtaining enzymes and their use in the feeding of ruminants and monogastric animals to evaluate their effect on the productive performance.

2. ENZYMES EMPLOYED AS ADDITIVES IN ANIMAL HUSBANDRY FOR ZOOTECHNICAL PURPOSES

Enzymes, considered biological catalysts, are proteins capable of accelerating the speed of chemical reactions, which are essential for the proper cellular functioning of all living beings. Due to their diversity, specificity and catalytic capacity, they have been widely accepted by the scientific and industrial community. Their use has shown benefits in various production processes, traditionally, in the food industry in the production of beer, bread, cheese, juice, etc. However, its use in the livestock feed industry was limited until a few years ago [2]; in the 1980s, the poultry industry was the first one to show interest in their use and over the years their use in animal feed grew remarkably, with an estimated commercial value of 1280 million dollars in 2019 [9].

The enzymes used in animal feed are considered zootechnical additives, which improve the consistency and nutritional value of the feed, increase digestibility, animal performance and reduce the effect of antinutrients. They also maintain intestinal health; in addition, the digestion process overcomes the growth of pathogenic microorganisms. They are added separately or as multienzyme preparations at all stages of ruminant and non-ruminant growth [1,4,9,10].

The effects of the action of enzymes used in animal feed processing have not been fully clarified; however, their success can be attributed to any of the following mechanisms [3,4,11].

- Action on the bonds or components that cannot be hydrolyzed by endogenous enzymes.
- Degradation of anti-nutritional factors that reduce digestibility and increase the viscosity of feed.
- Cell wall rupture and the release of nutrients attached to the cell wall.
- Digestion of nutrients.
- Reduction in secretions and the loss of endogenous proteins in the intestine, reducing maintenance needs.
- Increase in digestive enzymes, which are insufficient or non-existent in the animal, resulting in better digestion, especially in young animals with immature digestive systems.

The mode of action of each enzyme is different and interdependent, its use in combination with feed formulations must be carried out rationally and carefully to achieve maximum positive effects. Enzymes act directly or indirectly on nutrients, having main effects on the substrate to which it is directed as well as having side effects. For example, in the lignocellulosic complex, lignin degrading enzymes will attack their substrate as the main effect and, consequently, they will access the nutrients linked to lignin (carbohydrates or proteins) as a side effect [12].

In addition, the catalytic activity of enzymes is influenced by temperature, pH, substrate specificity, among others; therefore, the enzymes used as additives in animal feed processing must be heat-resistant, stable and capable of preserving their activity through the digestive system of the animal, all of which are important factors in the catalytic response to pH, retention time, resistance to endogenous digestive proteases, microbial enzymes, water content and ionic strength [4]. One way to ensure that the exogenous enzymes are conserved until the place and moment where they will act is through their encapsulation, in order to stabilize them for the processing of the feed or their passage through the gastrointestinal tract, protecting them from adverse conditions and triggering their release at the action site [13,14]. The purpose of adding enzymes in animal feed is to improve food efficiency, production requirements and, consequently, reduce the cost of feeding [15].

Enzymes are generally classified according to the substrate on which they act; commercially, in animal nutrition, they are divided into three categories according to their purpose (Table 1), those directed to carbohydrates (fiber and starch), proteins and phytates [4,9]. Phytases act on phytate and release phosphorus from phytate, while beta-glucanase acts on non-starch polysaccharides and breaks down the fiber. Proteases act on protein and improve its digestibility. Cellulases act on cellulose polysaccharides and break down the fiber and alpha-amylase act on starch and improve its digestibility [10].

Table 1. Enzymes used in animal feed processing.

Enzymes	Substrates	Effect	Example	Ref.
Carbohydrases	Carbohydrates (fiber and/or starch)	Improves digestibility of plant biomass and increases energy. Beneficial effect on poultry and pig diets.	Xylanases and β-glucanases (degrade cell walls, used in poultry) β-mannanases Pectinases β-galactosidases	[16,17]
			β-amylase (improves digestibility of starch, body weight gain has been observed in poultry)	
Proteases	Proteins	Some proteases increased apparent ileal nitrogen, digestibility and apparent nitrogen retention across the whole digestive tract in broiler chicks and broiler cockerels. Exogenous proteases can further improve protein digestibility of ingredients through solubilization and hydrolysis of dietary proteins. Antinutritional factor levels decrease. They can be of animal, vegetable or microbial origin.	Proteases isolated from microorganisms such as *Aspergillus niger* and *Bacillus spp*.Chymosin, pepsin A Bromelain, papain, ficine, aminopeptidase, bacillolysin 1, dipeptidyl peptidase III, chymotrypsin, subtilisin, trypsin.	**[18, 19, 20]**

Enzymes	Substrates	Effect	Example	Ref.
Phytase	Phytates	Degrade phytate bonds releasing trapped nutrients. Improves cattle efficiency It increases the absorption of phosphorus, reducing the possibility of contamination of soil and water through excreta. Increase amino acid availability.	Acid phytases of histidine (pH 5.0) mainly applied to feed for poultry or pigs.	[4]

3. APPLICATION OF ENZYMES IN THE DIETARY PLANS FOR ANIMALS

3.1. Enzymes in Poultry Feed

Poultry is a domestic species used as a source of high-quality meat and eggs. In its production, the feed represents between 70 and 75% of the total cost, and is mainly constituted by grains of cereals (corn, wheat, sorghum and proteins of vegetable flours) that provide energy to the animal; however, due to their high costs, producers have replaced them with cheaper ingredients such as barley, oats, rye, sunflower flour, etc., but these usually contain anti-nutritional factors (ANF) such as antigenic components, raffinose oligosaccharides, saponins, protease inhibitors, tannins, lectins and phytic acid, which are unable to be digested by monogastric animals. The presence of ANF can increase the digesta viscosity, decrease the absorption of nutrients and has even been associated with the incidence of pathogenic infections such as necrotic enteritis affecting the health of poultry and increasing production costs [21,22]. Poultry, when fed with cereals, cannot hydrolyze the starch-free polysaccharides present in the cell wall due to the lack of enzymes, which causes low feed efficiency; these effects can be counteracted by modifying the diet or adding exogenous enzymes. These have been a useful option to improve nutritional and economic aspects in poultry production, in addition to the decrease in the effects caused by ANF [23].

The poultry industry is the most experienced in the application of exogenous enzymes with more than 30 years of research and application [1]. In poultry supplementation, the most commonly used enzymes are xylanases, glucanases, pectinases, cellulases, proteases, amylases, phytases and galactosidases. Their use not only represents a nutritional improvement; it also allows the use of raw materials to be expanded [24]. In the feed industry, in order to neutralize the effects of the viscous, non-starch polysaccharides in cereals such as barley, wheat, rye and triticale have been mostly used. These antinutritive carbohydrates are undesirable, as they reduce digestion and the absorption of all nutrients in the diet, especially fat and protein [23].

The use of enzymes has been extended until the majority of intensively produced poultry diets contain carbohydrases to increase the bioavailability and assimilation of nutrients, as well as the reduction in digestive problems due to the decrease in viscosity [4,24]. In organoleptic matters, egg yolk colour, specifically, has been better observed when xylanases and β-glucanases are added to the diet, and a greater energy generation and a better production of meat and eggs with diets supplemented with β-amylases [4], as well as a greater performance of poultry fed with a diet consisting of corn and soybeans added with β-amylase [25]. The phytases increase the utilization of phytate phosphorus. The ability of phytase to improve the digestion of phytate phosphorus and, subsequently, to reduce the output of organic phosphorus to the environment has attracted a great deal of scientific and commercial interest [23].

Café et al. [26] incorporated a multienzyme complex (Avizyme) composed of xylanases, proteases and amylases into the poultry diet, observing an increase in body weights, a decrease in mortality and a greater amount of net energy compared to the group without Avizyme. On the other hand, Babalola et al. [27] observed a better apparent absorption of nitrogen and fiber in poultry fed diets containing xylanases.

Furthermore, the addition of glucanases to whole barley (52.5 U/kg of barley) improves the nutritional value by decreasing the digesta viscosity in diets based on this grain, due to the depolymerization of glucans, in vivo results suggest that 1,4-glucanases act preferentially on cellulosic substrates and not on glucans mixed together [28]. The beneficial effects of the enzymes are reflected in the performance of broilers up to a 75% inclusion of treated barley [29]. Treatment with multi-glucanases in wheat and barley (180 U/g) was shown to improve the growth rate and physicochemical properties of the broiler carcass; however, the feed conversion was decreased [30]. However, when combined with xylanases, they help reduce between 30 and 50% of viscosity in diets made from wheat and barley, respectively. In addition, they favour the increase in body weight and nutritional conversion [31]. Generally, xylanases are obtained from microorganisms such as *Bacillus spp.* and *Trichoderma reesei* [32]. Xylanases, β-glucanases, pentosanases and phytases not only improve the nutritional quality and viscosity of the diets and productive parameters of the poultry, but also improve the intestinal health of the animal, since they restrict the proliferation of fermentative microorganisms in the small intestine [33].

Another exogenous enzyme of importance in the feeding of poultry are the phytases that, until the 1990s, were little used for their high cost. Currently, its application in Europe and countries such as the United States, Mexico and Brazil has been increasingly frequent in the main farms of broilers, not only for the economic benefits but also for the additional effects on the nutritional use of the diet, especially in the absorption of Ca, Zn, Mg and amino acids [34]. The importance of using these enzymes is due to the fact that phosphorus is a mineral associated with important metabolic functions and its excess or deficiency can cause problems in animal productivity. Poultry diets are made up of ingredients in which the phosphorus is mainly in the form of phytate, which, being poorly assimilated, is almost completely discarded through feces, which is why it is considered a source of environmental pollution [34]. To improve the use of phosphorus and reduce the environmental impact, the addition of microbial phytases is an option because it reduces waste and allows smaller amounts of inorganic phosphorus to be used in the diet [35].

Some studies have shown that the effects of phytases are not only limited to the improvement of the digestibility of phosphorus in monogastric animals, but that there has also been a greater retention of P, Ca and N in chicks fed with corn-soybeans and 600 U of phytase/kg compared to those with a diet to which the enzyme was not added [36]. Namkung and Leeson [37] showed that phytase supplementation (1149 IU/kg) created a positive effect of approximately 2% in the digestibility of protein and total amino acids in broilers.

3.2. Enzymes in Swine Feeding

The increase in the consumption of pork has led to a need to find alternatives that reduce the maturation and fattening time of piglets. Unfortunately, the increase in the number of piglets per litter results in animals with low birth weights, immune systems and immature digestive systems, in addition to the limited enzyme secretion reflected in the slow maturation of the animal [1,38]. Although the enzymes supplemented in the feeding of poultry and swine is similar, the results differ due to the digestive physiology of each species [39].

The addition of exogenous enzymes in pig feeding can improve the digestibility of feed, degrading their complex matrix. Carbohydrase supplementation (xylanase, β-glucanase, β-mannanase, β-galactosidase) increases substrate digestibility [38]. Furthermore, phytase is added to improve the digestibility of phytate with the consequent reduction to inorganic phosphorus, as well as improvements in the growth of piglets that are supplemented with this enzyme [38,40]. Although pigs have endogenous phytases present in the intestinal mucosa, they have almost zero activity, and although these enzymes are excreted by microorganisms in the large intestine, the released phosphorus is not absorbed and is almost completely excreted, which is why it is very important to supplement the diet with these type enzymes that can be obtained from fungi of the *Aspergillus* and *Peniophora* genera [41].

Tiwari et al. [42] showed that xylanase and mannanase can be used together or separately in diets rich in arabinoxylans and mannans (depending on the composition of the feed and the amount of the substrates) to improve their digestibility and, subsequently, the animal's intestinal health. Although the supplementation of the diets of weaned piglets with individual or combined mannanases, phytases, proteases and carbohydrases have shown positive effects on digestibility, growth and an improvement of the intestinal structure, the ability of enzymes to improve intestinal maturity and the health of weaned piglets is inconsistent [38,43].

3.3. Enzymes in Ruminant Feeds

The application of enzymes in the feeding of ruminants has developed slowly, the complexity of the digestive system (four compartments) of these animals and the existence of ruminal microorganisms that excrete enzymes and perform fermentation processes make it difficult to interpret the data obtained. Most of the research conducted in ruminants has been based on the use of fibrolytic enzymes, amylases and proteases; mainly multienzyme complexes composed of cellulases, xylanases, amylases and pectinases [1]. They are generally used to improve the digestibility of forage cell walls, increase the availability of the starch present in cereals and improve the performance of dairy cattle [44]. In the search to increase milk production and decrease its cost, the use of enzymes in ruminant feeding has been shown to have a positive effect on milk yield [45].

Enzymes such as xylanases have been used in the previous treatment of forages to improve their digestibility in ruminants [32] and to facilitate composting with glucanases, pectinases, cellulases, proteases, amylases, phytases, galactosidases and lipases, to break down feed components, reducing the viscosity of the raw material [46].

It is common to use agricultural waste in the feeding of ruminants; however, these wastes are not nutritious, have little protein, high amounts of fiber, low digestibility and contain anti-nutritional factors; therefore, the use of exogenous enzymes to improve the quality of these materials is a current trend. The use of cellulases in diets of farm animals has been shown to improve feed utilization and animal performance in vitro, in situ and in vivo through fiber degradation, as well as improved milk production in cows and small ruminants [47]. In dairy cows, the digestibility of dry matter and milk production has increased in diets with 34% silage of barley treated with fibrolytic enzymes (mixture of cellulase and xylanase) from *Trichoderma reesei* [45]. Recently, Golder et al. [48] characterized the response in the field of the application of fibrolytic enzymes in supplemented dairy cows before delivery and for 200 days from the start of lactation in three dairy farms in the United States; eight randomly assigned pens were controlled without enzyme administration and eight more pens received a dose of 750 mL/t of feed for five months. The results showed that milk production increased with the enzymatic treatment to 0.70–0.80 kg/day, this could be due to the higher digestibility of the feed. The body weight of the cows that were supplemented with enzymes did not show increases; however, a higher dry matter intake was observed (0.20 kg/head per day). Furthermore, in goats, it was possible to show an effect on daily weight gain, milk production and feed consumption with the inclusion of enzyme extract obtained from the spent substrate of *Pleurotus ostreatus* in the diet [49]. However, despite the benefits offered by fibrolytic enzymes, the use of these in the diet of livestock, specialized in meat production, does not show significant results [50].

In regard to amylolytic enzymes, they are potential additives to improve starch digestion in ruminant diets. Some amylases from *B. licheniformis* and *Aspergillus niger* may increase the digestibility of cereal starch such as sorghum and corn [44,51]; these enzymes have been able to act on the final non-reducing group of amylose and amylopectin, specifically on the α-1,4 or α-1,6 glycosidic bonds releasing glucose, maltotriose and maltose, which can be used as a substrate by ruminals microorganisms such as *Megasphaera elsdenii*, *Prevottella ruminicola* and *Selenomonas ruminantium*, which implies a greater degradation of dietary starch [52]. An isolated amylolytic enzyme of *B. licheniformis* recorded an activity of 4.19 mM/min, which was 69 times more active than the enzymes found in the rumen; their positive effects were demonstrated in studies 'in vivo' in sheep fed with a diet based on sorghum (70%) treated with these enzymes, achieving a decrease in the consumption of dry matter, organic matter and starch. However, the degree of use of starch is determined by the type or source, chemical and nutritional composition of the diet, the amount of food consumed per unit of time, mechanical alterations (degree of chewing) and physicochemical properties (degree of hydration and gelatinization) and the adaptation of ruminal microorganisms to the substrate consumed in order to degrade it [44,52].

On the other hand, the use of phytase in ruminants improves the use of phosphorus and reduces the need to supplement the feed with inorganic phosphate, and it also contributes to the decrease in phosphorus in feces, which represents an environmental benefit [13].

3.4. Enzymes in Fish Feeds

Another sector where exogenous enzymes have been applied is aquaculture. This industry uses fish meal with ingredients of vegetable origin in its processes; the addition of biocatalysts allows for a reduction in anti-nutritional factors, such as phytin, non-starch polysaccharides and inhibitors of proteases that affect nutrient utilization and interfere with fish performance and health [53]. In the larval stage, nutrition is a crucial factor that affects survival and performance, to achieve this it is necessary to have feed that meets nutritional requirements. An alternative can be the manipulation of the diet that allows for the optimization of the performance of the fish larvae. The addition of exogenous enzymes to the formulated diet can increase the assimilation by the larvae. On the other hand, it has been reported that the greater growth of fish larvae fed with live food is attributed to the activity of digestive enzymes present in live food [54,55].

The use of zootechnical additives in this industry is low; increasing research on this topic could be a useful tool to improve and sustain commercial aquaculture [56]. Phytases and carbohydrases are the most used in aquaculture; however, the latter have not been as common in aquatic species. Despite their promising effects to improve nutrient digestibility by hydrolyzing the non-starch polysaccharides present in plants food, their effects are not yet clear due to the difficulty of comparisons between studies [57,58]. The fish species that have been used as a study model to determine the effect of exogenous enzymes on larval growth are Dorada *Sparus auratus* [54], sea bass *Dicentrarchus labrax* [59], rainbow trout *Oncorhynchus mykiss* [60,61], Japanese sea bass *Lateolabrax japonicus* [62] and African catfish *Clarias gariepinus* [63].

Yigit and Olmes [64] mention that the cellulase supplementation obtained from *Aspergillus niger* in the diets for tilapia fingerlings (*Oreochromis niloticus*) did not show effects on growth, concluding that the addition of enzymes to the fish diet will depend on the enzymes, the species used and the source of the feed ingredients. In addition, an enzyme complex (hemicellulase, pectinase, cellulase) that allows the components of the cell wall, such as hemicellulose and pectin bound to cellulose, to degrade must be used.

In this sense, in another study, they evaluated the addition of multiple enzymes (Natuzyme® and Hemicell®) on the diet of Caspian salmon (*Salmo trutta caspius*), finding improvements in body weight gain and feeding efficiency. In addition to this, the authors mention that it is necessary to consider the effects of the enzymatic supplement on the intestinal microbial flora and the improvement of growth through the release of a growth-enhancing factor that cannot be ignored [65].

Adeoye et al. [66] found greater growth when using phytase, protease and carbohydrases in the diet of Tilapia (*Oreochromis niloticus*), in contrast to the fish fed with the control diet. Furthermore, they do not report changes in the hematological, intestinal morphological or intestinal microbiological parameters. However, carbohydrases showed a significant difference in the gut microbiota of the fish that were fed the tilapia diet compared to those fed the control diet. Although the species diversity parameters of the microbiota were not affected by dietary treatment, the analysis revealed differences in community profiles. Therefore, research is needed to confirm how exogenous enzymes (especially carbohydrase) modulate the gut microbiota and whether these modulations contribute to enhancing the host growth performance.

The aquaculture industry must face several challenges before applying exogenous enzymes in the fish diet, although phytase is the most used and has the best results. Research is needed on the effect of enzymes on amino acid availability, specific enzymatic modes of action, including interactions with endogenous enzymes during digestion, the effects of enzymes on target substrates, the consistency and predictability of the effects of enzyme doses and the effects of the quality of the ingredients on response predictability [67].

3.5. Enzymes in Dog's Feeds

The dog feed industry has also considered the use of exogenous enzymes in the diet, due to the exocrine pancreatic insufficiency (EPI) that these can present. The most widely used are amylase, protease and lipase from the porcine pancreas. Moreover, some vets recommend the use of plant and animal enzyme supplements for all pets, including those without EPI The possible benefits that are expected range from an increase in the digestibility of nutrients to the support of the immune system; however, they have not been proven thus far [68].

The type of processing that dog feed has inactivates the enzymatic activity; therefore, it is necessary to superficially apply enzymes, finding freeze-dried enzymes on the market, which must be added to the feed. One of the parameters evaluated has been digestibility, where the results do not show significant changes when amylase has been added, which does not increase starch digestion with respect to the control diets [69,70,71]. The use of proteases has also not shown effects on protein digestibility, which may be due to the type of protease used, the low concentrations of enzymes and their low specificity [72].

The use of amylases during the extrusion process in the production of feed for dogs has been evaluated, showing that they do not interfere with the final texture of the feed, the gelatinization of the starch, the digestibility of the nutrients nor the palatability of the feed, but it does increase production in extrusion and a possible reduction in electrical energy use, but its use must be evaluated in various feed extrusion and formulation systems to determine the optimal balance between enzyme cost and increased feed productivity and energy costs savings during extrusion [71].

In the preparation of feed for dogs, it is common to use barley and wheat, but due to their composition, these have negative effects on digestibility and fecal consistency due to the high content of dietary fiber. To counteract this, the effect of glycanases in the formulations has been evaluated, finding that they digest the soluble fraction of the fiber, which increases the fermentation of these substrates when they reach the large intestine. Moreover, the addition of the enzyme improves the antinutritive effects, but it did not improve the problems associated with increased bacterial fermentation and an accumulation of lactic acid in the large intestine, yet the use of these enzymes is recommended due to the acceptable quality of the feces and good digestion in general [73].

On the other hand, Sá et al. [70] mention that adding wheat bran in the formulation increases the dietary fiber, promotes a reduction in nutrient and energy digestibility and there is an increase in fecal production when using a mixture of enzymes (b-glucanase, xylanase, cellulase, glucoamylase, phytase) before and after extrusion, concluding that enzyme supplementation, during feed processing or as exogenous enzyme supplementation for the animal, does not reduce the negative effects of wheat bran on digestibility. However, other combinations or doses of enzymes need to be tested to promote more extensive use of wheat bran in dog diets.

4. PRODUCTION OF ENZYMES FOR ANIMAL FEEDING

Enzymes can be obtained from animals, plants or microorganisms. Initially their use was limited by the difficulties of their production and recovery as well as poor stability; however, current biotechnological advances allow for the production of biologically active enzymes in large volumes [74].

The development of recombinant DNA technology (DNAr) has allowed for the isolation and expression of genes of some microorganisms, such as *Bacillus subtilis*, *Bacillus amyloliquefaciens*, *Aspergillus oryzae*, *Aspergillus niger*, *Kluyveromyces lactis*, *Trichoderma reesei*, among others and the production of enzymes for industrial use, including animal feed processing [13,74,75]. Protein engineering is used to optimize enzyme performance characteristics for specific industrial applications such as changing the pH optimum, thermal stability and stability against chemical oxidation and altering the requirement for cofactors such as metal ions [13].

To ensure the safety, along with the chemical and microbiological purity, of the enzymes used in animal feed, these are produced under established standards by FAO/WHO and Codex Alimentarius, using safe documented microbial strains and GRAS (Generally Recognized as Safe) raw materials. **Table 2** shows the enzymes commonly used in animal feed and the source from which they were obtained [13].

Table 2. Enzyme producing organisms used in animal feed.

Common Name	Classification	Function	Producing Organism	Ref.
α-**Amylase**	Carbohydrase	Starch Hydrolysis	*Bacillus licheniformis, Bacillus stearothermophilus, Bacillus amyloliquefaciens*	[76]
			Aspergillus niger *Aspergillus oryzae*	[77]
			Thermomyces lanuginosus	[76]
			Pseudomonas flourescens	[78]
Maltogenic α-amylase	Carbohydrase	Starch hydrolysis with maltose production.	*Bacillus subtilis*	[78]
Cellulase	Carbohydrase	Breaks down cellulose	*Aspergillus niger*	[79,80]
			Humicola insolens	[81]
			Trichoderma	[80]
			Pleurotus ostreatus	[49]
α-**Galactosidasa**	Carbohydrase	Hydrolyzes oligosaccharides	*Morteirella vinaceae* var *raffinoseutilizer* *Saccharomyces cerevisiae* *Saccharomyces carlsbergensis*	[82]
β-**Glucanase**	Carbohydrase	Hydrolyzes β-glucans	*Trichoderma reesei*	[83]
Glucoamylase (amyloglucosidas)	Carbohydrase	Hydrolyzes starch with glucose production.	*Aspergillus niger*	[79]
Hemicellulase	Carbohydrase	Breaks down the hemicellulose	*Humicola insolens* *Aspergillus niger*	[81] [79]
Pectinase	Carbohydrase	Breaks downs the pectin	*Aspergillus niger*	[79]
Pullulanase	Carbohydrase	Hydrolyzes starch	*Bacillus licheniformis*	[78]

Common Name	Classification	Function	Producing Organism	Ref.
Xylanase	Carbohydrase	Hydrolyse xylan	*Pleurotus ostreatus*	[49]
			Bacillus circulans *Bacillus Stearothermophilus* *Bacillus polymyxa* *Bacillus subtilis* *Bacillus amyloliquifaciens* *Bacillus acidocaldarius* *Bacillus thermoalkalophilus*	[84]
			Fusarium venenatum	[78]
Laccases	Oxidase	Oxidation of an organic or inorganic substrate and the reduction of molecular oxygen to water.	*Aspergillus oryzae*	[78]
			Pleurotus ostreatus	[49]
Lipase	Lipase	Hydrolyzes triglycerides, diglycerides and monoglycerides.	*Aspergillus niger*	[79]
			Rhizopus oryzae	[85]
			Candida rugosa	[86]
Papain	Protease	Hydrolyzes proteins	*Carica papaya*	[87]
Pepsin	Protease	Hydrolyzes proteins	Animal stomach	[13]
Trypsin	Protease	Hydrolyzes proteins	Animal pancreas	[13]
Chymosin	Protease	Hydrolyzes proteins	*Aspergillus niger*	[78]
			Escherichia coli K-12	[78]
			Kluyveromyces marxianus var. *Lactis*	[78]
Catalase	Oxidoreductase	Hydrogen peroxide is needed for oxidation of compounds.	*Aspergillus niger*	[79]
Glucose oxidase	Oxidoreductase	It degrades glucose to hydrogen peroxide and gluconic acid.	*Aspergillus niger*	[79]
			Aspergillus oryzae	[78]
Phytase	Phosphatase	Hydrolyse phytate.	*Penicillium funiculosum*	[88]

In response to environmental problems and considering the need to improve the nutritional quality of feed for animals, current research is committed to the use of ruminal fluid from slaughterhouses to obtain enzymes. Ruminal fluid is rich in ammonia and phosphorus, which represents an environmental problem; however, it is also rich in cellulase and xylanase enzymes with a potential use in the industry [89].

2Sarteshnizi et al. [90] evaluated the use of ruminal fluid as a potential source of exogenous enzymes to improve the nutritional quality of ground corn, barley grain, soy flour and alfalfa hay to be used in ruminant feed. Cherdthong et al. [91] have proposed the use of dry ruminal fluid in animal feed. On the other hand, Sarteshnizi et al. [89] proposed spray drying of the ruminal fluid for subsequent encapsulation, using various hydrocolloids.

On the other hand, the limited stability and functional capacity of enzymes in difficult conditions, such as industrial processes, has long been recognized as a major problem. A current trend to obtain enzymes of industrial interest is made from organisms' extremities. Enzymes obtained from this type of organism have a potential use in the feed industry, since their ability to work under unfavourable conditions is well known, offering better enzymatic activities [92].

The fungal mannanases are attractive enzymes for the animal feed industry, which operate at a pH that ranges from 2.4 to 6. The conditions of their production from the fungi of the genera *Aspergillus* and *Trichoderma* are currently thoroughly investigated; however, they have successfully cloned and expressed mannanases in *Pichia pastoris*, which were evaluated in animal feed, managing to establish a significant source of β-mannanases [92].

The bacteria *Citrobacter braakii* and *Escherichia coli* as well as the fungi *Aspergillus niger*, *Aspergillus* sp. and *Peniophora* sp. have been reported to be producers of the phytase enzyme, which has been used in pig diets to facilitate the release of the P bound to phytate. [93,94,95]. Phytate-degrading enzymes, through stepwise dephosphorylation, can release P-phytate, thus improving the absorption of P and reducing its excretion; therefore, the use of these enzymes has nutritional and ecological benefits. In addition, it has been observed that minerals such as Ca, Zn, Fe, Cu and Mg are bound to phytate in mineral-phytate complexes, which reduces the degradation of phytate, thus reducing the digestibility of the minerals [93]. However, it has been reported that supplementation with phytase from *Escherichia coli* can increase the digestibility of Ca, Mg, Mn, Zn, Cu and Fe in pigs with doses of 500 to 1500 FTU of phytase/kg of feed [94]. It is worth mentioning that bacterial or fungal phytases have been shown to be effective when added to the feed of poultry and pigs, increasing the availability of P, the use of energy and reducing the excretion of P in feces [96].

Although many microorganisms have been molecularly characterized and the function and application of their enzymes have been described, work is being conducted to improve their biochemical properties that entail expanding their application. The development of molecular genetics, cell biology, genetic engineering, sequencing techniques and high-throughput omics have allowed modifications at the amino acid sequence level through rational design or molecular evolution [78,97]. With the above, the exploitation of numerous microbial enzymes with a greater pH and temperature stability is sought, as well as an increase in their yield with the use of promoters, which have been introduced as multiple copies in the gene that codes for the enzyme [98,99,100].

The catalytic activity of enzymes at different temperatures and pH values can be improved using directed evolution [101], which is a tool that, together with mutagenesis and screening, helps to characterize new proteins from a parent protein under a particular evolutionary pressure, obtaining more efficient biocatalysts [102,103,104]. On the other hand, the use of protein engineering has contributed to generating biocatalysts with greater activity and stability at high temperatures and extreme pH values through directed mutagenesis. This has been achieved due to the advancement of bioinformatics, which allows the protein structure to be visualized in a three-dimensional way and, thus, achieve a rational design that allows the introduction of disulfide bridges, replacing the N terminal and increasing the number of hydrogen bonds [105]. Table 3 shows some enzymes designed from these techniques, taken from Victorino da Silva Amatto et al. [106].

Table 3. Engineered enzyme.

Enzymes	Source	Purpose of Modification	Method	Ref.
Alfa-amylase	*Bacillus sp.* TS-23	Improve thermostability, change glutamic acid 219, crucial for the thermostability	Site-directed mutagenesis	[107]
Xylanase (glycoside hydrolase -GH11)	*Neocalli mastix patriciarum*	Improve thermostability. Single mutants Gln87Arg, Asn88Gly, Ser89His and Ser90Thr	Site-directed mutagenesis	[108]
Phytase	*Aspergillus niger*	Improve thermostability and catalytic efficiency. Change in Thr195Leu/Gln368Glu/Phe376Tyr; Gln172Arg/Lys432Arg/Gln368Glu; Gln172Arg/Lys432Arg/Gln368Glu/Phe376Tyr and Gln172Arg/Lys432Arg/Gln368Glu/Phr376 Tyr/Thr195Leu; phyA: Gln172Arg; Gln172Arg/Lys432Arg; Gln368Glu/Lys432Arg	Error-prone PCR/Directed evolution	[109]
Endo-1,4 betaxylanase II	*Trichoderma reessei*	Improve thermostability, substituting Thr2 and Thr28 by cysteine	Error-prone PCR	[110]

The design of enzymes has also led to the choice of the expression system for the production of recombinant proteins, using mainly bacteria (*Escherichia coli*, *Bacillus* spp., *Lactobacillus lactis*), filamentous fungi (*Aspergillus* spp.) and yeasts (*Pichia pastoris*). Each of these microorganisms have characteristics that allow them to be used as hosts. In particular, *Escherichia coli* quickly and easily overexpress recombinant enzymes; however, they cannot express very large proteins that require post-translational modifications, in addition to producing toxins that reduce their use [111]. Despite this, bacterial expression systems remain attractive due to their rapid growth, use of inexpensive culture media, genetic characterization, number of cloning vectors and mutant host strains [112]. Recombinant enzyme production is still a promising field that will help meet the demand in the animal feed industry; but, to achieve this, it is necessary to use high-throughput expression technologies and to know the proteins at the proteome level to understand them at the systems level. Current advances in post-genomic technology make it possible to design improved cost-effective expression systems to meet the growing demand for enzymes [111].

Therefore, current processes in animal nutrition are using genetically modified microorganisms with the aim of increasing the productive capacity of the fermentation unit for the production of enzymes and avoiding undesirable activities [96]. In ruminant feeding, one approach to improve fiber digestion is by modifying the fiber inoculants, using genetically modified bacteria to produce enzymes that give new properties to the silage and/or pre-digest the plant material. Recombinant *Lactobacillus plantarum* is a strain used as a silage starter, which was constructed to express the alpha-amylase, cellulase or xylanase genes [113]. The possibility of genetically modifying rumen microorganisms to increase the degree of degradation of fiber components has also been proposed, and a possible strategy is through the establishment of recombinant organisms. One of the options has been the manipulation of the predominant bacterial species in the rumen such as *Prevotella ruminicola*, *Butyrivibrio fibrisolvens* or *Streptococuccus bovis*. In *Prevotella ruminicola*, one of the objectives has been to increase its fibrolytic activities, while in *Ruminococcus flavefaciens*, it has been chosen to induce the expression of endoglucanase/xylanase genes, but it should be noted that these studies are still in progress [113].

5. PROSPECTIVE ADVANCEMENTS IN THE UTILIZATION OF ENZYMES FOR ANIMAL NUTRITION

The industry dedicated to the production of food for animals should focus its efforts on investing in research and development and in the design of new enzymes that can play a key role in the improvement and nutritional quality of feed. The development of thermostable enzymes will simplify the application of the pre-granulation of dry product and will promote the use of the enzyme in granulated diets [114]. To achieve this, it is necessary to continue work on the rational redesign of existing biocatalysts and standardize combinatorial methods that seek the desired functionality in randomly generated libraries as well as employing robust computational methods combined with screening technologies and directed evolution to improve the properties of enzymes, the application of multistep reactions using multifunctional catalysts, the de novo design and the selection of specific catalytic proteins for a desired chemical reaction and, thus, meet process perspectives [115].

Suplatov et al. [116] mentioned that future computational advances will be the key to achieving success in the design of new enzymes. Currently, bioinformatics is used to predict structural changes that can be applied to wild proteins and produce more stable variants. The techniques used can be classified into stochastic approaches, empirical or systematic rational design strategies and chimeric protein design. Bioinformatic analysis can be used efficiently to study large protein super-families in a systematic way, as well as to predict particular structural changes that increase the stability of the enzyme. However, further development of systematic bioinformatics procedures is needed to organize and analyze protein sequences and structures within large super-families and link them to their function, as well as to provide knowledge-based predictions for experimental evaluation. Therefore, bioinformatics can become the cornerstone for the design of more stable and functionally diverse enzymes [116].

The next generation sequencing technologies and bioinformatics has facilitated the collection and analysis of a large amount of genomic, transcriptomic, proteomic and metabolomic data from different organisms that have allowed predictions to be made on the regulation of expression, transcription, translation, structure and the mechanisms of action of proteins as well as homology, mutations and evolutionary processes that generate structural and functional changes over time. Although the amount of information in the databases is greater every day, all the bioinformatics tools continue to be constantly modified to improve performance that leads to more accurate predictions regarding protein functionality [117].

The main databases used in computational biology are NCBI, GenBank, Protein Data Bank, Swiss-Prot, PIR, Flybase, TrEMBL, Enzyme, Prosite, InterPro, UniProt and PDB [117,118]. The high number of sequences that are stored in the different databases, have allowed the evolutionary relationships of different proteins to be inferred, which retain their function during long evolutionary times when presenting homology; however, homologous proteins can perform the same activity, but the substrates they use can come from different routes [117,119].

The study of ancestral enzymes has suggested that these presented a high thermostability, due to the Precambrian era that was thermophilic, in addition to the fact that most microorganisms and other organisms adapted to these environments with high temperatures. The ancestral protein alignments with the current ones show evidence of a slow evolution in structure, but not in amino acids [117,120]. Álvarez-Cervantes et al. [121] performed the phylogenetic analysis of β-xylanase SRXL1 from *Sporisorium reilianum* and its relationship with families (GH10 and GH11) from Ascomycetes and Basidiomycetes, demonstrating that groupings analysis of a higher-level in the Pfam database allowed the proteins under study to be classified into families GH10 and GH11, based on the regions of highly conserved amino acids, 233–318 and 180–193, respectively, where glutamate residues are responsible for the catalysis. The phylogenetic relationship of xylanase SRXL1 of *S. reilianum* with the xylanases analyzed shows a monophyly and a relationship is observed with respect to their status as plant pathogens or saprophytic fungi, in this case the functionality of these enzymes is related to its adaptation to their ecological niche.

To analyze these changes in the sequences, bioinformatics programs use algorithms and mathematical models, based on empirical matrices of amino acid substitution, as well as those that incorporate structural properties of the native state, such as secondary structure and accessibility [117,122]. Protein phylogeny studies are currently necessary to know protein-protein interactions in biological systems. Molecular or structural analyzes on proteins will require more information to respond if a protein is present in one or several species, as well as to predict the common ancestor and evolution times [123].

The bioinformatics tools are TOPAL, Hennig86 and PAML; the computational packages that are allowed to occupy any of these are PHYLIP and PAUP, as well as MOLPHY, PASSML, PUZZLE and TAAR [117,124].

On the other hand, one of the challenges of protein engineering and biology is to improve industrial processes; to achieve this it is necessary to determine the tertiary structure of proteins from the amino acid sequence in order to design new proteins. Many of the protein structures that we know today have been obtained using X-ray crystallography, Nuclear magnetic resonance spectroscopy (NMR) or cryo-EM [117,125].

The use of algorithms and computer programs have achieved the design of new catalysts with the ability to use substrates, this considering their structural characteristics, structural dynamics and structural remodelling [126]. Computational design has obviously improved the possibility of finding active enzymes. However, it must be combined with optimized experimental protocols to obtain efficient biocatalysts [127].

The powerful and revolutionary techniques developed for protein engineering thus far provide excellent opportunities for the design of industrial enzymes with specific properties and the production of high-value products at lower production costs [127]. **Table 4** shows the biochemical properties of some enzymes that the animal feed industry reported in the UniProt and PDB databases.

6. CONCLUSIONS

The use of enzymes in animal feed is a dynamic research and development field; current research and its future application aim to assess its effect on the health and intestinal flora of animals, and to check whether or not there is a synergistic effect between the composition of the diet and the enzymatic action to improve the digestibility, to contribute to intestinal development and their effect on the productive performance, or if its application can diminish the use of antibiotics, by means of a decrease in the incidence of diseases or mortality. Furthermore, the search for new sources of obtaining enzymes for their use in animal feed is a current trend with a view to evolve in future work. The modification of commercial enzyme-producing strains that resist the conditions of the gastrointestinal tract or the search for new ways to encapsulate and protect enzymes are important aspects to be investigated and applied in the agro-industry for the production of animal feed.

7. CONCISE SUMMARY

Enzymes are widely used in the food industry. Their use as a supplement to the raw material for animal feed is a current research topic. Although there are several studies on the application of enzyme additives in the animal feed industry, it is necessary to search for new enzymes, as well as to utilize bioinformatics tools for the design of specific enzymes that work in certain environmental conditions and substrates. This will allow the improvement of the productive parameters in animals, reducing costs and making the processes more efficient. Technological needs have considered these catalysts as essential in many industrial sectors and research is constantly being carried out to optimize their use in those processes. This review describes the enzymes used in animal nutrition, their mode of action, their production and new sources of production as well as studies on different animal models to evaluate their effect on the productive performance intended for the production of animal feed.

Table 4. Biochemical properties of some enzymes reported in the UniProt and PDB databases.

Protein	Gen	Organism	Catalytic Activity	pH	Temperature °C	Active Site	Glycosylation	Access Code UniProtKB	Access Code PDBe	Ref.
Endo-beta-1,4-glucanase B	eglb	*Aspergillus niger*	Endohydrolysis of (1,4)-beta-D-glucosidic linkages in cellulose, lichenin and cereal beta-D-glucans. EC:3.2.1.4	6.0	70	160, 266	38, 100, 211, 288	O74706 (EGLB_ASPNG)	5i77, 5i78, 5i79	**[128,129]**
Xyloglucan-specific endo-beta-1,4-glucanase A	**xgeA**	*Aspergillus aculeatus*	Xyloglucan + H_2O = xyloglucan oligosaccharides. EC:3.2.1.151	3.4	30	-	-	O94218 (XGEA_ASPAC)	3VL8, 3VL9, 3VLB	**[130]**
Pancreatic alpha-amylase	AMY2	*Sus scrofa (Pig)*	Endohydrolysis of (1,4)-alpha-D-glucosidic linkages in polysaccharides containing three or more (1,4)-alpha-linked D-glucose units. EC:3.2.1.1	-	-	212, 248, 315	427	P00690 (AMYP_PIG)	1BVN	**[131]**
Alpha-amylase	amyS	*Bacillus licheniformis*		11	100	260, 290	-	P06278 (AMY_BACLI)	1ob0	**[132,133]**
Beta-amylase	spoII	*Bacillus cereus*	Hydrolysis of (1,4)-alpha-D-glucosidic linkages in polysaccharides to remove successive maltose units from the non-reducing ends of the chains. EC:3.2.1.2	-	-	202, 397	-	P36924 (AMYB_BACCE)	1J0Z	**[134]**

Endo-1,4-beta-xylanase	xylC	*Talaromyces cellulolyticus* CF-2612	Endohydrolysis of (1,4)-beta-D-xylosidic linkages in xylans	-	-	119, 210	-	W8VR85 (W8VR85_9EURO)	5HXV	**[135]**
Endo-1,4-beta-xylanase 2	Xyn2	*Trichoderma reesei*		4.5–5.5	40	119, 210	71, 94	P36217 (XYN2_HYPJR)	4HKW	**[136]**
Endopolygalacturonase I	pgaI	*Aspergillus niger*	(1,4-alpha-D-galacturonosyl)(n + m) + H_2O = (1,4-alpha-D-galacturonosyl)(n) + (1,4-alpha-D-galacturonosyl)(m). EC:3.2.1.15	-	-	207, 229	44, 46, 246	P26213 (PGLR1_ASPNG)	5ONK	**[137]**
3-phytase A	phyA	*Aspergillus niger*	Catalyzes the hydrolysis of inorganic orthophosphate from phytate.	-	-	82, 362	27, 59, 105, 120, 207, 230, 339, 352, 376, 388	P34752 (PHYA_ASPNG)	3K4P	**[138]**
3- phytase	phyC	*Bacillus subtillis*		7	55	-	-	O31097 (PHYC_BACIU)	3AMS	**[139]**
3-phytase	phy	*Bacillus* sp.		-	-	-	-	O66037 (PHYT_BACSD)	2POO	**[140]**

-: data not reported, EC: Enzyme commission.

REFERENCES

1. Ugwuanyi, J.O. Enzymes for nutritional enrichment of agro-residues as livestock feed. In *Agro-Industrial Wastes as Feedstock for Enzyme, Production*; Gurpreet, S.D., Surinder, K., Eds.; Academic Press: Cambridge, MA, USA, 2016; pp. 233–260. [**Google Scholar**]
2. Walsh, G.A.; Ronan, F.P.; Denis, R.H. Enzymes in the animal-feed industry. *Trends Biotechnol.* **1993**, *11*, 424–430. [**Google Scholar**] [**CrossRef**]
3. Ravindran, V. Feed enzymes: The science, practice, and metabolic realities. *J. Appl. Poult. Res.* **2013**, *22*, 628–636. [**Google Scholar**] [**CrossRef**]
4. Ojha, B.K.; Singh, P.K.; Shrivastava, N. Enzymes in the Animal Feed Industry. In *Enzymes in Food Biotechnology*; Mohammed, K., Ed.; Academic Press: Cambridge, MA, USA, 2019; pp. 93–109. [**Google Scholar**]
5. Wang, Y.; McAllister, T.A.; Rode, L.M.; Beauchemin, K.A.; Morgavi, D.P.; Nsereko, V.L.; Iwaasa, D.A.; Yang, W. Effects of an exogenous enzyme preparation on microbial protein synthesis, enzyme activity and attachment to feed in the Rumen Simulation Technique (Rusitec). *Br. J. Nutr.* **2001**, *85*, 325–332. [**Google Scholar**] [**CrossRef**][**Green Version**]
6. Greiner, R.; Konietzny, U. Phytase for food applications. *Food Technol. Biotechnol.* **2006**, *44*, 125–140. [**Google Scholar**]
7. Rosen, G.D. Effects of genetic, managemental and dietary factors on the efficacy of exogenous microbial phytase in broiler nutrition. *Br. Poult. Sci.* **2003**, *44*, 25–26. [**Google Scholar**] [**CrossRef**]
8. Debyser, W.; Peumans, W.J.; Van Damme, E.J.M.; Delcour, J.A. *Triticum aestivum* xylanase inhibitor (TAXI), a new class of enzyme inhibitor affecting breadmaking performance. *J. Cereal Sci.* **1999**, *30*, 39–43. [**Google Scholar**] [**CrossRef**]
9. Brufau, J. Introducción al uso de enzimas en la alimentación animal un proceso de innovación. *nutriNews* **2014**, *1*, 17–21. [**Google Scholar**]
10. Imran, M.; Nazar, M.; Saif, M.; Khan, M.A.; Sanaullah, V.M.; Javed, O. Role of Enzymes in Animal Nutrition: A Review. *PSM Vet Res.* **2016**, *1*, 38–45. [**Google Scholar**]
11. Bedford, M.R.; Cowieson, A.J. Exogenous enzymes and their effects on intestinal microbiology. *Anim. Feed Sci. Tech.* **2012**, *173*, 76–85. [**Google Scholar**] [**CrossRef**]
12. Beauchemin, K.A.; Colombatto, D.; Morgavi, D.P.; Yang, W.Z.; Rode, L.M. Mode of action of exogenous cell wall degrading enzymes for ruminants. *Can. J. Anim. Sci.* **2004**, *84*, 13–22. [**Google Scholar**] [**CrossRef**]

13. Pariza, M.W.; Cook, M. Determining the safety of enzymes used in animal feed. *Regul. Toxicol. Pharmacol.* **2010**, *56*, 332–342. [Google Scholar] [CrossRef] [PubMed]

14. De Vos, P.; Faas, M.M.; Spasojevic, M.; Sikkema, J. Encapsulation for preservation of functionality and targeted delivery of bioactive food components. *Int. Dairy J.* **2010**, *20*, 292–302. [Google Scholar] [CrossRef]

15. Bedford, M.R. The evolution and application of enzymes in the animal feed industry: The role of data interpretation. *Br. Poult. Sci.* **2018**, *59*, 486–493. [Google Scholar] [CrossRef] [PubMed][Green Version]

16. Nortey, T.N.; Patience, J.F.; Sands, J.S.; Zijlstra, R.T. Xylanase supplementation improves energy digestibility of wheat by-products in grower pigs. *Livest. Sci.* **2007**, *109*, 96–99. [Google Scholar] [CrossRef]

17. Yin, Y.L.; Baidoo, S.K.; Jin, L.Z.; Liu, Y.G.; Schulze, H.; Simmins, P.H. The effect of different carbohydrase and protease supplementation on apparent (ileal and overall) digestibility of nutrients of five hulless barley varieties in young pigs. *Livest. Prod. Sci.* **2001**, *71*, 109–120. [Google Scholar] [CrossRef]

18. Ghazi, S.; Rooke, J.A.; Galbraith, H.; Bedford, M.R. The potential for the improvement of the nutritive value of soya-bean meal by different proteases in broiler chicks and broiler cockerels. *Br. Poult. Sci.* **2002**, *43*, 70–77. [Google Scholar] [CrossRef]

19. Marsman, G.J.; Gruppen, H.; Van der Poel, A.F.; Kwakkel, R.P.; Verstegen, M.W.; Voragen, A.G. The effect of thermal processing and enzyme treatments of soybean meal on growth performance, ileal nutrient digestibilities, and chyme characteristics in broiler chicks. *Poult. Sci.* **1997**, *76*, 864–872. [Google Scholar] [CrossRef]

20. Philipps, W.P. Proteases–animal feed. In *Enzymes in Human and Animal Nutrition*; Simões, N.C., Kumar, V., Eds.; Academic Press: Cambridge, MA, USA, 2018; pp. 279–297. [Google Scholar]

21. Kaldhusdal, M.I. Necrotic enteritis as affected by dietary ingredients. *World Poult.* **2000**, *16*, 42–43. [Google Scholar]

22. Raza, A.; Bashir, S.; Tabassum, R. An update on carbohydrases: Growth performance and intestinal health of poultry. *Heliyon* **2019**, *5*, e01437. [Google Scholar] [CrossRef][Green Version]

23. Khattak, F.M.; Pasha, T.N.; Hayat, Z.; Mahmud, A. Enzymes in poultry nutrition. *J. Anim. Plant Sci.* **2006**, *16*, 1–7. [Google Scholar]

24. Wenk, C. Recent advances in animal feed additives such as metabolic modifiers, antimicrobial agents, probiotics, enzymes and highly available minerals-review. Asian-Australas. *J. Anim. Sci.* **2000**, *13*, 86–95. [Google Scholar]

25. Gracia, M.I.; Aranibar, M.; Lazaro, R.; Medel, P.; Mateos, G.G. Alpha-amylase supplementation of broiler diets based on corn. *Poult. Sci.* **2003**, *82*, 436–442. [Google Scholar] [CrossRef] [PubMed]

26. Café, M.B.; Borges, C.A.; Fritts, C.A.; Waldroup, P.W. Avizyme improves performance of broilers fed corn-soybean meal-based diets. *J. Appl. Poult. Res.* **2002**, *11*, 29–33. [Google Scholar] [CrossRef]

27. Babalola, T.O.; Apata, D.F.; Atteh, J.O. Effect of β-xylanase supplementation of boiled castor seed meal-based diets on the performance, nutrient absorbability and some blood constituents of pullet chicks. *Trop. Sci.* **2006**, *46*, 216–223. [Google Scholar] [CrossRef]

28. Fernandes, V.O.; Costa, M.; Ribeiro, T.; Serrano, L.; Cardoso, V.; Santos, H.; Lordelo, M.; Ferreira, L.M.; Fontes, C.M. 1,3-1,4-β-Glucanases and not 1,4-β-glucanases improve the nutritive value of barley-based diets for broilers. *Anim. Feed Sci. Technol.* **2016**, *211*, 153–163. [Google Scholar] [CrossRef]

29. Teymouri, H.; Zarghi, H.; Golian, A. Evaluation of hull-less barley with or without enzyme cocktail in the finisher diets of broiler chickens. *J. Agric. Sci. Technol.* **2018**, *20*, 469–483. [Google Scholar]

30. Kalantar, M.; Khajali, F.; Yaghobfar, A. Different dietary source of non-starch polysaccharides supplemented with enzymes affected growth and carcass traits, blood parameters and gut physicochemical properties of broilers. *Glob. J. Anim. Sci. Res.* **2015**, *3*, 412–418. [Google Scholar]

31. Juanpere, J.; Perez, V.A.M.; Angulo, E.; Brufau, J. Assessment of potential interactions between phytase and glycosidase enzyme supplementation on nutrient digestibility in broilers. *Poult. Sci.* **2005**, *84*, 571–580. [Google Scholar] [CrossRef] [PubMed]

32. Álvarez, C.J.; Domínguez, H.E.M.; Mercado, F.Y.; O'Donovan, A.; Diaz, G.G. Mycosphere Essay 10: Properties and characteristics of microbial xylanases. *Mycosphere* **2016**, *7*, 1600–1619. [Google Scholar] [CrossRef]

33. Ohimain, E.I.; Ofongo, R. Enzyme supplemented poultry diets: Benefits so far—A review. *Int. J. Adv. Biotechnol. Res.* **2014**, *3*, 31–39. [Google Scholar]

34. Acosta, A.; Cárdenas, M. Enzimas en la alimentación de las aves. Fitasas. *Rev. Cuba. Cienc. Agríc.* **2006**, *40*, 377–387. [Google Scholar]

35. Waldroup, P.W.; Kersey, J.H.; Saleh, E.A.; Fritts, C.A.; Yan, F.; Stilborn, H.L.; Crum, R.C., Jr.; Raboy, V. Nonphytate phosphorus requirement and phosphorus excretion of broiler chicks fed diets composed of normal or high available phosphate corn with and without microbial phytase. *Poult. Sci.* **2000**, *79*, 1451–1459. [Google Scholar] [CrossRef]

36. Sebastian, S.; Touchburn, S.P.; Chavez, E.R.; Lague, P.C. Efficacy of supplemental microbial phytase at different dietary calcium levels on growth performance and mineral utilization of broiler chickens. *Poult. Sci.* **1996**, *75*, 1516–1523. [Google Scholar] [CrossRef]

37. Namkung, H.; Leeson, S. Effect of phytase enzyme on dietary nitrogen-corrected apparent metabolizable energy and the ileal digestibility of nitrogen and amino acids in broiler chicks. *Poult. Sci.* **1999**, *78*, 1317–1319. [Google Scholar] [CrossRef]

38. Torres, P.A.; Hermans, D.; Manzanilla, E.G.; Bindelle, J.; Everaert, N.; Beckers, Y.; Torrallardona, D.; Bruggeman, G.; Gardiner, G.E.; Lawlor, P.G. Effect of feed enzymes on digestibility and growth in weaned pigs: A systematic review and meta-analysis. *Anim. Feed Sci. Technol.* **2017**, *233*, 145–159. [Google Scholar] [CrossRef]

39. Hernández, J.F.P. Utilización de enzimas en la alimentación del ganado porcino. *Anaporc Rev. Asoc. Porc. Cient.* **2006**, *3*, 32–36. [Google Scholar]

40. Dersjant, L.Y.; Awati, A.; Schulze, H.; Partridge, G. Phytase in non-ruminant animal nutrition: A critical review on phytase activities in the gastrointestinal tract and influencing factors. *J. Sci. Food Agric.* **2015**, *95*, 878–896. [Google Scholar] [CrossRef] [PubMed][Green Version]

41. Rodehutscord, M.; Rosenfelder, P. Update on phytate degradation pattern in the gastrointestinal tract of pigs and broiler chickens. In *Phytate Destruction-Consequences for Precision Animal Nutrition*; Walk, C.L., Kühn, I., Stein, H.H., Kidd, M.T., Rodehutscord, M., Eds.; Wageningen Academic Publishers: Wagingen, The Netherlands, 2016; pp. 237–246. [Google Scholar]

42. Tiwari, U.P.; Chen, H.; Kim, S.W.; Jha, R. Supplemental effect of xylanase and mannanase on nutrient digestibility and gut health of nursery pigs studied using both in vivo and in vitro models. *Anim. Feed Sci. Technol.* **2018**, *245*, 77–90. [Google Scholar] [CrossRef]

43. Nortey, T.N.; Owusu, A.A.; Zijlstra, R.T. Effects of xylanase and phytase on digestion site of low-density diets fed to weaned pigs. *Livest. Res. Rural. Dev.* **2015**, *27*, 133. [Google Scholar]

44. Rojo, R.; Mendoza, G.D.; González, S.S.; Landois, L.; Bárcena, R.; Crosby, M.M. Effects of exogenous amylases from *Bacillus licheniformis* and *Aspergillus niger* on ruminal starch digestion and lamb performance. *Anim. Feed Sci. Technol.* **2005**, *123*, 655–665. [Google Scholar] [CrossRef]

45. Refat, B.; Christensen, D.A.; McKinnon, J.J.; Yang, W.; Beattie, A.D.; McAllister, T.A.; Eun, J.S.; Abdel, R.G.A.; Yu, P. Effect of fibrolytic enzymes on lactational performance, feeding behavior, and digestibility in high-producing dairy cows

fed a barley silage-based diet. *Int. J. Dairy Sci.* **2018,** *101*, 7971–7979. [Google Scholar] [CrossRef] [PubMed]

46. Gilbert, H.J.; Hazlewood, G.P. Bacterial cellulases and xylanases. *J. Gen. Microbiol.* **1993,** *139*, 187–194. [Google Scholar] [CrossRef][Green Version]
47. Murad,H.A.;Azzaz,H.H.Cellulaseanddairyanimalfeeding. *Biotechnology* **2010,** *9*, 238–256. [Google Scholar] [CrossRef][Green Version]
48. Golder, H.M.; Rossow, H.A.; Lean, I.J. Effects of in-feed enzymes on milk production and components, reproduction, and health in dairy cows. *J. Dairy Sci.* **2019,** *102*, 8011–8026. [Google Scholar] [CrossRef]
49. Trejo, L.T.; Zepeda, B.A.; Franco, F.J.; Soto, S.S.; Ojeda, R.D.; Ayala, M.M. Uso de extracto enzimático de *Pleurotus ostreatus* sobre los parámetros productivos de cabras. *Abanico Vet.* **2017**, *7*, 14–21. [Google Scholar]
50. Zinn, R.A.; Salinas, J. Influence of Fibrozyme on digestive function and growth performance of feedlot steers fed a 78 % concentrate growing diet. In *Biotechnology in the Feed Industry Proceedings of the Fifteenth Annual Symposium Nottingham*; Lyons, T.P., Jacques, K.A., Eds.; Nottingham University Press: Nottingham, UK, 1999; pp. 313–319. [Google Scholar]
51. Rojo, R.R.; Martínez, G.D.M.; Galván, M.M.C. Uso de la amilasa termoestable de *Bacillus licheniformis* en la digestibilidad in vitro del almidón de sorgo y maíz. *Agrociencia* **2001,** *35*, 423–427. [Google Scholar]
52. Rojo, R.R.; Mendoza, M.G.D.; Montañez, V.O.D.; Rebollar, R.S.; Cardoso, J.D.; Hernández, M.J.; González, R.F.J. Enzimas amilolíticas exógenas en la alimentación de rumiantes. *Univ. Cienc.* **2007,** *23*, 173–182. [Google Scholar]
53. Francis, G.; Makkar, H.; Becker, K. Antinutritional factors present in plant- derived alternate fish feed ingredients and their effects in fish. *Aquaculture* **2001,** *199*, 197–227. [Google Scholar] [CrossRef]
54. Kolkovski, S.; Tandler, A.; Kissil, G.; Gertler, A. The effect of dietary exogenous digestive enzymes on ingestion, assimilation, growth and survival of gilthead seabream (*Sparus auratus, Sparidae, Linnaeus*) larvae. *Fish Physiol. Biochem.* **1993,** *12*, 203–209. [Google Scholar] [CrossRef]
55. Ghomi, M.R.; Shahriari, R.; Langroudi, H.F.; Nikoo, M.; von Elert, E. Effects of exogenous dietary enzyme on growth, body composition, and fatty acid profiles of cultured great sturgeon Huso huso fingerlings. *Aquac. Int.* **2012,** *20*, 249–254. [Google Scholar] [CrossRef]
56. Castillo, S.; Gatlin, D.M., III. Dietary supplementation of exogenous carbohydrase enzymes in fish nutrition: A review. *Aquaculture* **2015,** *435*, 286–292. [Google Scholar] [CrossRef]

57. Adeola, O.; Cowieson, A.J. Board-Invited Review: Opportunities and challenges in using exogenous enzymes to improve nonruminant animal production. *Anim. Sci. J.* **2011**, *89*, 3189–3218. [Google Scholar] [CrossRef] [PubMed]

58. Kumar, V.; Sinha, A.K.; Makkar, H.P.S.; De Boeck, G.; Becker, K. Phytate and phytase in fish nutrition. *J. Anim. Physiol. Anim. Nutr.* **2012**, *96*, 335–364. [Google Scholar] [CrossRef] [PubMed]

59. Kolkovski, S.; Tandler, A.; Izquierdo, M.S. Effects of live food and dietary digestive enzymes on the efficiency of microdiets for seabass (*Dicentrarchus labrax*) larva. *Aquaculture* **1997**, *148*, 313–322. [Google Scholar] [CrossRef]

60. Farhangi, M.; Carter, C. Effect of enzyme supplementation to dehulled lupin-based diets on growth, feed efficiency, nutrient digestibility and carcass composition of rainbow trout, *OnWcorhynchus mykiss* (Walbaum). *Aquac. Res.* **2007**, *38*, 1274–1282. [Google Scholar] [CrossRef]

61. Ogunkoya, A.E.; Page, G.I.; Adewolu, M.A.; Bureau, D.P. Dietary incorporation of soybean meal and exogenous enzyme cocktail can affect physical characteristics of faecal material egested by rainbow trout (*Oncorhynchus mykiss*). *Aquaculture* **2006**, *254*, 466–475. [Google Scholar] [CrossRef]

62. Ai, Q.; Mai, K.; Zhang, W.; Xu, W.; Tan, B.; Zhang, C.; Li, H. Effects of exogenous enzymes (phytase, non- starch polysaccharide enzyme) in diets on growth, feed utilization, nitrogen and phosphorus excretion of Japanese seabass, *Lateolabrax japonicas*. *Comp. Biochem. Physiol. Part A Mol. Integr. Physiol.* **2007**, *147*, 502–508. [Google Scholar] [CrossRef]

63. Yildirim, Y.B.; Turan, F. Effects of exogenous multienzyme supplementation in diets on growth and feed utilization of African catfish, *Clarias gariepinus*. *J. Anim. Vet. Adv.* **2010**, *9*, 327–331. [Google Scholar]

64. Yigit, N.O.; Olmez, M. Effects of cellulase addition to canola meal in tilapia (*Oreochromis niloticus* L.) diets. *Aquac. Nutr.* **2011**, *17*, e494–e500. [Google Scholar] [CrossRef]

65. Ali, Z.A.; Kanani, H.G.; Azam, E.A.; Ramezani, S.; Zoriezahra, S.J. Effects of two dietary exogenous multi-enzyme supplementation, Natuzyme® and beta-mannanase (Hemicell®), on growth and blood parameters of Caspian salmon (Salmo trutta caspius). *Comp. Clin. Path.* **2014**, *23*, 187–192. [Google Scholar]

66. Adeoye, A.A.; Jaramillo, T.A.; Fox, S.W.; Merrifield, D.L.; Davies, S.J. Supplementation of formulated diets for tilapia (*Oreochromis niloticus*) with selected exogenous enzymes: Overall performance and effects on intestinal histology and microbiota. *Anim. Feed. Sci. Technol.* **2016**, *215*, 133–143. [Google Scholar] [CrossRef]

67. Dalsgaard, J.; Verlhac, V.; Hjermitslev, N.H.; Ekmann, K.S.; Fischer, M.; Klausen, M.; Pedersen, P.B. Effects of exogenous enzymes on apparent nutrient digestibility in rainbow trout (*Oncorhynchus mykiss*) fed diets with high inclusion of plant-based protein. *Anim. Feed Sci. Technol.* **2012**, *171*, 181–191. [Google Scholar] [CrossRef]

68. Villaverde, C.; Manzanilla, E.G.; Molina, J.; Larsen, J.A. Effect of enzyme supplements on macronutrient digestibility by healthy adult dogs. *J. Nutr. Sci.* **2017**, *6*, e12. [Google Scholar] [CrossRef][Green Version]

69. Twomey, L.N.; Pluske, J.R.; Rowe, J.B.; Choct, M.; Brown, W.; Pethick, D.W. The replacement value of sorghum and maize with or without supplemental enzymes for rice in extruded dog foods. *Anim. Feed Sci. Technol.* **2003**, *108*, 61–69. [Google Scholar] [CrossRef]

70. Sá, F.C.; Vasconcellos, R.S.; Brunetto, M.A.; Filho, F.O.R.; Gomes, M.O.S.; Carciofi, C.A. Enzyme use in kibble diets formulated with wheat bran for dogs: Effects on processing and digestibility. *J. Anim. Physiol. Anim. Nutr.* **2013**, *97*, 51–59. [Google Scholar] [CrossRef] [PubMed][Green Version]

71. Carciofi, A.C.; Palagiano, C.; Sá, F.C.; Martins, M.S.; Gonçalves, K.N.V.; Bazolli, R.S.; Souza, D.F.; Vasconcellos, R.S. Amylase utilization for the extrusion of dog diets. *Anim. Feed Sci. Technol.* **2012**, *177*, 211–217. [Google Scholar] [CrossRef]

72. Pacheco, G.F.E.; Marcolla, C.S.; Machado, G.S.; Kessler, A.M.; Trevizan, L. Effect of full-fat rice bran on palatability and digestibility of diets supplemented with enzymes in adult dogs. *Anim. Sci. J.* **2014**, *92*, 4598–4606. [Google Scholar] [CrossRef]

73. Twomey, L.N.; Pluske, J.R.; Rowe, J.B.; Choct, M.; Brown, W.; McConnell, M.F.; Pethick, D.W. The effects of increasing levels of soluble non-starch polysaccharides and inclusion of feed enzymes in dog diets on faecal quality and digestibility. *Anim. Feed Sci. Technol.* **2003**, *108*, 71–82. [Google Scholar] [CrossRef]

74. McAllister, T.A.; Hristov, A.N.; Beauchemin, K.A.; Rode, L.M.; Cheng, K.J. Enzymes in ruminant diets. *Enzym. Farm Anim. Nutr.* **2001**, 273–298. [Google Scholar] [CrossRef][Green Version]

75. Sarder, N.U.; Hasan, A.M.; Anower, M.R.; Salam, M.A.; Alam, M.J.; Islam, S. Commercial Enzymes Production by Recombinant DNA Technology: A Conceptual Works. *Pak. J. Biol. Sci.* **2005**, *8*, 345–355. [Google Scholar]

76. Souza, P.M. Application of microbial α-amylase in industry—A review. *Braz. J. Microbiol.* **2010**, *41*, 850–861. [Google Scholar] [CrossRef] [PubMed]

77. Gupta, A.; Gupta, V.K.; Modi, D.R.; Yadava, L.P. Production and characterization of α-amylase from *Aspergillus niger*. *Biotechnol.* **2008**, *7*, 551–556. [Google Scholar] [CrossRef]

78. Olempska, B.Z.S.; Merker, R.I.; Ditto, M.D.; DiNovi, M.J. Food-processing enzymes from recombinant microorganisms–a review. *Regul. Toxicol. Pharmacol.* **2006**, *45*, 144–158. [Google Scholar] [CrossRef] [PubMed][Green Version]

79. Schuster, E.; Dunn, C.N.; Frisvad, J.C.; Van Dijck, P.W. On the safety of *Aspergillus niger*–a review. *Appl. Microbiol. Biotechnol.* **2002**, *59*, 426–435. [Google Scholar] [PubMed]

80. Zhang, X.Z.; Zhang, Y.H. Cellulases: Characteristics, sources, production, and applications. In *Bioprocessing Technologies in Biorefinery for Sustainable Production of Fuels, Chemicals, and Polymers*; Shang-Tian, Y., Hesham, A., El-Enshasy, N.T., Eds.; John Wiley & Sons: Hoboken, NJ, USA, 2013; pp. 131–146. [Google Scholar]

81. Schülein, M. Enzymatic properties of cellulases from *Humicola insolens*. *J. Biotechnol.* **1997**, *57*, 71–81. [Google Scholar] [CrossRef]

82. De Andrade, T.P.C.S.; Da Costa, C.L.A.; Karp, S.G. □-Galactosidases: Characteristics, production and immobilization. *J. Food Nutr. Res.* **2016**, *55*, 195–204. [Google Scholar]

83. Coenen, T.M.; Schoenmakers, A.C.; Verhagen, H. Safety evaluation of □-glucanase derived from *Trichoderma reesei*: Summary of toxicological data. *Food Chem. Toxicol.* **1995**, *33*, 859–866. [Google Scholar] [CrossRef]

84. Shahi, N.; Hasan, A.; Akhtar, S.; Siddiqui, M.H.; Sayeed, U.; Khan, M.K. Xylanase: A promising enzyme. *J. Chem. Pharm. Res.* **2016**, *8*, 334–339. [Google Scholar]

85. Flood, M.T.; Kondo, M. Safety evaluation of lipase produced from *Rhizopus oryzae*: Summary of toxicological data. *Regul. Toxicol. Pharmacol.* **2003**, *37*, 293–304. [Google Scholar] [CrossRef]

86. Flood, M.T.; Kondo, M. Safety evaluation of lipase produced from *Candida rugosa*: Summary of toxicological data. *Regul. Toxicol. Pharmacol.* **2001**, *33*, 157–164. [Google Scholar] [CrossRef][Green Version]

87. De Melo, S.K.; Sá, F.C.; de Souza, D.F.; da Silva, F.L.; Urrego, M.I.; Vendramini, T.H.; Brunetto, M.A.; Carciofi, A.C. Effect of the addition of papain enzyme on digestive parameters and palatability of extruded diets for dogs. *Braz. J. Vet. Res. Anim. Sci.* **2017**, *54*, 357–365. [Google Scholar]

88. Awad, G.E.; Helal, M.M.; Danial, E.N.; Esawy, M.A. Optimization of phytase production by *Penicillium purpurogenum* GE1 under solid state fermentation by using Box–Behnken design. *Saudi J. Biol. Sci.* **2014**, *21*, 81–88. [Google Scholar] [CrossRef]

89. Sarteshnizi, F.R.; Benemar, H.A.; Seifdavati, J.; Greiner, R.; Salem, A.Z.; Behroozyar, H.K. Production of an environmentally friendly enzymatic feed additive for agriculture animals by spray drying abattoir's rumen fluid in the

presence of different hydrocolloids. *J. Clean. Prod.* **2018**, *197*, 870–874. [Google Scholar] [CrossRef]

90. Sarteshnizi, F.R.; Seifdavati, J.; Abdi-Benemar, H.; Salem, A.Z.; Sharifi, R.S.; Mlambo, V. The potential of rumen fluid waste from slaughterhouses as an environmentally friendly source of enzyme additives for ruminant feedstuffs. *J. Clean. Prod.* **2018**, *195*, 1026–1031. [Google Scholar] [CrossRef]
91. Cherdthong, A.; Wanapat, M.; Saenkamsorn, A.; Waraphila, N.; Khota, W.; Rakwongrit, D.; Anantasook, N.; Gunun, P. Effects of replacing soybean meal with dried rumen digesta on feed intake, digestibility of nutrients, rumen fermentation and nitrogen use efficiency in Thai cattle fed on rice straw. *Livest. Sci.* **2014**, *169*, 71–77. [Google Scholar] [CrossRef]
92. Hassan, N.; Rafiq, M.; Rehman, M.; Sajjad, W.; Hasan, F.; Abdullah, S. Fungi in acidic fire: A potential source of industrially important enzymes. *Fungal Biol. Rev.* **2019**, *33*, 58–71. [Google Scholar] [CrossRef]
93. Kristoffersen, S.; Gjefsen, T.; Svihus, B.; Kjos, N.P. The effect of reduced feed pH, phytase addition and their interaction on mineral utilization in pigs. *Livest. Sci.* **2021**, *248*, 104498. [Google Scholar] [CrossRef]
94. Zeng, Z.K.; Wang, D.; Piao, X.S.; Li, P.F.; Zhang, H.Y.; Shi, C.X.; Yu, S.K. Efectos de la adición de superdosis de fitasa a las dietas deficientes en fósforo de los cerdos jóvenes sobre el rendimiento del crecimiento, la calidad ósea, la digestibilidad de minerales y aminoácidos. *Rev. Cienc. Anim. Asia Aust.* **2014**, *27*, 237–246. [Google Scholar] [CrossRef] [PubMed][Green Version]
95. Cassie, L.H.; Dean, B.R.; Koehler, D.; Stacie, A.G.; Qingyun, L.; Patience, J.F. The impact of "super-dosing" phytase in pig diets on growth performance during the nursery and grow-out periods. *Transl. Anim. Sci.* **2019**, *1*, 419–428. [Google Scholar]
96. Valdivia, A.L.; Matos, M.M.; Rodríguez, Z.; Pérez, Y.; Rubio, Y.; Vega, J. Los aditivos enzimáticos, su aplicación en la crianza animal. *Cuba. J. Agric. Sci.* **2019**, *53*, 341–352. [Google Scholar]
97. Roodveldt, C.; Aharoni, A.; Tawfik, D.S. Directed evolution of pro- teins for heterologous expression and stability. *Curr. Opin. Struct. Biol.* **2005**, *15*, 50–56. [Google Scholar] [CrossRef]
98. Chapman, J.; Ismail, A.; Dinu, C. Industrial applications of enzymes: Recent advances, techniques, and outlooks. *Catalysts* **2018**, *8*, 238. [Google Scholar] [CrossRef][Green Version]
99. Currin, A.; Swainston, N.; Day, P.J.; Kell, D.B. Synthetic biology for the directed evolution of protein biocatalysts: Navigating sequence space

intelligently. *Chem. Soc. Rev.* **2015**, *44*, 1172–1239. [Google Scholar] [CrossRef] [Green Version]

100. Thapa, S.; Li, H.; OHair, J.; Bhatti, S.; Chen, F.C.; Al Nasr, K.; Zhou, S. Biochemical characteristics of microbial enzymes and their significance from industrial perspectives. *Mol. Biotechnol.* **2019**, *61*, 579–601. [Google Scholar] [CrossRef]

101. Soo, V.W.; Yosaatmadja, Y.; Squire, C.J.; Patrick, W.M. Mechanistic and evolutionary insights from the reciprocal promiscuity of two pyridoxal phosphate-dependent enzymes. *J. Biol. Chem.* **2016**, *291*, 19873–19887. [Google Scholar] [CrossRef] [PubMed][Green Version]

102. Packer, M.S.; Liu, D.R. Methods for the directed evolution of proteins. *Nat. Rev. Genet.* **2015**, *16*, 379. [Google Scholar] [CrossRef] [PubMed]

103. Tiwari, V. In vitro engineering of novel bioactivity in the natural enzymes. *Front. Chem.* **2016**, *4*, 39. [Google Scholar] [CrossRef] [PubMed][Green Version]

104. Yang, J.; Ruff, A.J.; Arlt, M.; Schwaneberg, U. Casting epPCR (cepPCR): A simple random mutagenesis method to generate high quality mutant libraries. *Biotechnol. Bioeng.* **2017**, *114*, 1921–1927. [Google Scholar] [CrossRef]

105. Wang, K.; Luo, H.; Tian, J.; Turunen, O.; Huang, H.; Shi, P.; Yao, B. Thermostability improvement of a *Streptomyces* xylanase by introducing proline and glutamic acid residues. *Appl. Environ. Microbiol.* **2014**, *80*, 2158–2165. [Google Scholar] [CrossRef][Green Version]

106. Victorino da Silva, A.I.; Gonsales da Rosa-Garzon, N.; Antônio de Oliveira Simões, F.; Santiago, F.; Pereira da Silva Leite, N.; Raspante Martins, J.; Cabral, H. Enzyme engineering and its industrial applications. *Appl. Biochem. Biotechnol.* **2021**. [Google Scholar] [CrossRef]

107. Lin, L.L.; Liu, J.S.; Wang, W.C.; Chen, S.H.; Huang, C.C.; Lo, H.F. Glutamic acid 219 is critical for the thermostability of a truncated α-amylase from alkaliphilic and thermophilic *Bacillus* sp. strain TS23-. *World J. Microbiol. Biotechnol.* **2008**, *24*, 619–626. [Google Scholar] [CrossRef]

108. Han, N.; Ma, Y.; Mu, Y.; Tang, X.; Li, J.; Huang, Z. Enhancing thermal tolerance of a fungal GH11 xylanase guided by B-factor analysis and multiple sequence alignment. *Enzyme Microb. Technol.* **2019**, *131*, 109422. [Google Scholar] [CrossRef]

109. Tang, Z.; Jin, W.; Sun, R.; Liao, Y.; Zhen, T.; Chen, H.; Li, C. Improved thermostability and enzyme activity of a recombinant phyA mutant phytase from *Aspergillus niger* N25 by directed evolution and site-directed mutagenesis. *Enzyme Microb. Technol.* **2018**, *108*, 74–81. [Google Scholar] [CrossRef]

110. Fenel, F.; Leisola, M.; Jänis, J.; Turunen, O. A de novo designed N-terminal disulphide bridge stabilizes the *Trichoderma reesei* endo-1,4-β-xylanase II. *J. Biotech.* **2004**, *108*, 137–143. [Google Scholar] [CrossRef] [PubMed]

111. Liu, L.; Yang, H.; Shin, H.D.; Chen, R.R.; Li, J.; Du, G.; Chen, J. How to achieve high-level expression of microbial enzymes: Strategies and perspectives. *Bioengineered* **2013**, *4*, 212–223. [Google Scholar] [CrossRef] [PubMed][Green Version]

112. Terpe, K. Overview of bacterial expression systems 18. for heterologous protein production: From molecular and biochemical fundamentals to commercial systems. *Appl. Microbiol. Biotechnol.* **2006**, *72*, 211–222. [Google Scholar] [CrossRef]

113. Forano, E.; Flint, H. Genetically modified organisms: Consequences for ruminant health and nutrition. *Ann. Zootech.* **2000**, *49*, 255–271. [Google Scholar] [CrossRef][Green Version]

114. Ravindran, V.; Son, J.H. Feed enzyme technology: Present status and future developments. *Recent Pat. Food Nutr. Agric.* **2011**, *3*, 102–109. [Google Scholar]

115. Adrio, J.L.; Demain, A.L. Microbial enzymes: Tools for biotechnological processes. *Biomolecules* **2014**, *4*, 117–139. [Google Scholar] [CrossRef] [PubMed][Green Version]

116. Suplatov, D.; Voevodin, V.; Švedas, V. Robust enzyme design: Bioinformatic tools for improved protein stability. *Biotechnol. J.* **2015**, *10*, 344–355. [Google Scholar] [CrossRef]

117. Hernández-Domínguez, E.M.; Castillo-Ortega, L.S.; García-Esquivel, Y.; Mandujano-González, V.; Díaz-Godínez, G.; Álvarez-Cervantes, J. Bioinformatics as a Tool for the Structural and Evolutionary Analysis of Proteins. In *Computational Biology and Chemistry*; IntechOpen: London, UK, 2020; pp. 37–64. [Google Scholar]

118. Escobar, C.A.M.; Murillo, L.V.R.; Soto, J.F. Tecnologías bioinformáticas para el análisis de secuencias de ADN. *Sci. Tech.* **2011**, *3*, 116–121. [Google Scholar]

119. Robertson, G.; Schein, J.; Chiu, R.; Corbett, R.; Field, M.; Jackman, S.D.; Mungall, K.; Lee, S.; Okada, H.M.; Qian, J.Q.; et al. De novo assembly and analysis of RNA-seq data. *Nat. Methods.* **2010**, *7*, 909–912. [Google Scholar] [CrossRef] [PubMed]

120. Merkl, R.; Sterner, R. Ancestral protein reconstruction: Techniques and applications. *Biol. Chem.* **2016**, *397*, 1–21. [Google Scholar] [CrossRef] [PubMed]

121. Álvarez-Cervantes, J.; Díaz-Godínez, G.; Mercado-Flores, Y.; Gupta, V.K.; Anducho-Reyes, M.A. Phylogenetic analysis of β-xylanase SRXL1 of Sporisorium reilianum and its relationship with families (GH10 and GH11) of Ascomycetes and Basidiomycetes. Sci. Rep. 2016, 6, 24010. [Google Scholar] [CrossRef][Green Version]

122. Bastolla, U.; Arenas, M. The influence of protein stability on sequence evolution: Applications to phylogenetic inference. In *Computational Methods in Protein Evolution*; Sikosek, T., Ed.; Humana Press: New York, NY, USA, 2019; pp. 215–231. [Google Scholar]

123. Szurmant, H.; Weigt, M. Interresidue, inter-protein and inter-family coevolution: Bridging the scales. *Curr. Opin. Struct. Biol.* **2018**, *50*, 26–32. [Google Scholar] [CrossRef]

124. Xu, D.; Xu, Y.; Uberbacher, C.E. Computational tools for protein modeling. *Curr. Protein Pept. Sci.* **2000**, *1*, 1–21. [Google Scholar] [CrossRef][Green Version]

125. Cheung, N.J.; Yu, W. De novo protein structure prediction using ultra-fast molecular dynamics simulation. *PLoS ONE.* **2018**, *13*, e0205819. [Google Scholar] [CrossRef][Green Version]

126. Eiben, C.B.; Siegel, J.B.; Bale, J.B.; Cooper, S.; Khatib, F.; Shen, B.W.; Players, F.; Stoddard, B.L.; Popovic, Z.; Baker, D. Increased Diels-Alderase activity through backbone remodeling guided by Foldit players. *Nat. Biotechnol.* **2012**, *30*, 190–192. [Google Scholar] [CrossRef][Green Version]

127. Jemli, S.; Ayadi-Zouari, D.; Hlima, H.B.; Bejar, S. Biocatalysts: Application and engineering for industrial purposes. *Crit. Rev. Biotechnol.* **2014**, *36*, 246–258. [Google Scholar] [CrossRef]

128. Hong, J.; Tamaki, H.; Akiba, S.; Yamamoto, K.; Kumagai, H. Cloning of a gene encoding a highly stable endo-β-1, 4-glucanase from *Aspergillus niger* and its expression in yeast. *J. Biosci. Bioeng.* **2001**, *92*, 434–441. [Google Scholar] [CrossRef]

129. Yan, J.; Liu, W.; Li, Y.; Lai, H.L.; Zheng, Y.; Huang, J.W.; Chen, C.-C.; Chen, Y.; Jin, J.; Li, H.; et al. Functional and structural analysis of Pichia pastoris-expressed *Aspergillus niger* 1,4-β-endoglucanase. Biochem. *Biophys.* Res. Commun. 2016, 475, 8–12. [Google Scholar] [CrossRef]

130. Yoshizawa, T.; Shimizu, T.; Hirano, H.; Sato, M.; Hashimoto, H. Structural basis for inhibition of xyloglucan-specific endo-β-1, 4-glucanase (XEG) by XEG-protein inhibitor. *J. Biol. Chem.* **2012**, *287*, 18710–18716. [Google Scholar] [CrossRef][Green Version]

131. Wiegand, G.; Epp, O.; Huber, R. The crystal structure of porcine pancreatic α-amylase in complex with the microbial inhibitor Tendamistat. *J. Mol. Biol.* **1995**, *247*, 99–110. [Google Scholar] [CrossRef]

132. Shahhoseini, M.; Ziaee, A.A.; Ghaemi, N. Expression and secretion of an α-amylase gene from a native strain of *Bacillus licheniformis* in *Escherichia coli* by T7 promoter and putative signal peptide of the gene. *J. Appl. Microbiol.* **2003**, *95*, 1250–1254. [Google Scholar] [CrossRef]

133. Machius, M.; Declerck, N.; Huber, R.; Wiegand, G. Kinetic stabilization of *Bacillus licheniformis* α-amylase through introduction of hydrophobic residues at the surface. *J. Biol. Chem.* **2003**, *278*, 11546–11553. [Google Scholar] [CrossRef] [Green Version]

134. Oyama, T.; Miyake, H.; Kusunoki, M.; Nitta, Y. Crystal structures of β-amylase from *Bacillus cereus* var. *mycoides* in complexes with substrate analogs and affinity-labeling reagents. *J. Biochem.* **2003**, *133*, 467–474. [Google Scholar] [CrossRef]

135. Watanabe, M.; Fukada, H.; Ishikawa, K. Construction of thermophilic xylanase and its structural analysis. *Biochemistry* **2016**, *55*, 4399–4409. [Google Scholar] [CrossRef] [PubMed]

136. Wan, Q.; Zhang, Q.; Hamilton-Brehm, S.; Weiss, K.; Mustyakimov, M.; Coates, L.; Langan, P.; Graham, D.; Kovalevsky, A. X-ray crystallographic studies of family 11 xylanase Michaelis and product complexes: Implications for the catalytic mechanism. *Acta Crystallogr. Sect. D Biol. Crystallogr.* **2014**, *70*, 11–23. [Google Scholar] [CrossRef][Green Version]

137. Ramaswamy, S.; Rasheed, M.; Morelli, C.F.; Calvio, C.; Sutton, B.J.; Pastore, A. The structure of PghL hydrolase bound to its substrate poly-γ-glutamate. *FEBS J.* **2018**, *285*, 4575–4589. [Google Scholar] [CrossRef] [PubMed]

138. Oakley, A.J. The structure of *Aspergillus niger* phytase PhyA in complex with a phytate mimetic. *Biochem. Biophys. Res. Commun.* **2010**, *397*, 745–749. [Google Scholar] [CrossRef] [PubMed]

139. Zeng, Y.F.; Ko, T.P.; Lai, H.L.; Cheng, Y.S.; Wu, T.H.; Ma, Y.; Chen, C.-C.; Yang, C.-S.; Cheng, K.-J.; Huang, C.-H.; et al. Crystal structures of Bacillus alkaline phytase in complex with divalent metal ions and inositol hexasulfate. *J. Mol. Biol.* **2011**, *409*, 214–224. [Google Scholar] [CrossRef] [PubMed]

140. Ha, N.C.; Oh, B.-C.; Shin, S.; Kim, H.J.; Oh, T.K.; Kim, Y.O.; Choi, K.Y.; Oh, B.H. Crystal structures of a novel, thermostable phytase in partially and fully calcium-loaded states. *Nat. Struct. Biol.* **2000**, *7*, 147–153. [Google Scholar] [PubMed]

CHAPTER-4

EMPLOYING A SEQUENTIAL FERMENTATION APPROACH TO GENERATE PROTEASE FROM *ASPERGILLUS TAMARRI* URM4634, ACCOMPANIED BY AN INVESTIGATION INTO THE ENZYME'S KINETIC AND THERMODYNAMIC PROPERTIES

1. INTRODUCTION: AN OVERVIEW

In the realm of enzyme production, the utilization of innovative fermentation methodologies has become paramount for optimizing yields and enhancing the properties of the generated enzymes. This study focuses on the application of a sequential fermentation approach for the purposeful synthesis of protease derived from *Aspergillus tamarri* URM4634. Concurrently, the investigation extends its scope to a thorough exploration of the kinetic and thermodynamic characteristics of the resulting enzyme. By delving into these aspects, this research aims to contribute valuable insights into the efficient generation and understanding of the enzymatic properties, paving the way for advancements in both bio-processing and enzyme utilization domains.

Proteases are enzymes that catalyze the hydrolysis of proteins resulting in their breakdown into peptides and amino acids. They represent an important group of industrial enzymes, accounting for almost 60% of the entire enzyme market and 40% of the global enzyme sales [1]. Proteases are in fact used in various industrial sectors, including those of detergents, leather goods, food, fine chemicals, pharmaceuticals, and cosmetics, and can be obtained from different sources such as plants, animal, and microorganisms, with the latter source contributing to nearly two-third of industrially used proteases. This is mainly due to the possibility of producing them quickly in large amounts by established fermentation processes that also allow for their regular supply [2,3]. Filamentous fungi have an extraordinary capability to secrete different types of proteases, with the main fungal strains for their industrial production belonging to the genera *Trichoderma* and *Aspergillus* [2].

Commonly, microbial proteases are produced by submerged fermentation (SmF), a process in which microorganisms grow in a liquid medium with high free water content, or by solid state fermentation (SSF), where they grow in an environment without free water or with a very low content of free water [4]. Both processes have their own advantages and disadvantages depending on the operating conditions. SmF has advantages related to the presence of instrumentation and process control, and consequently it is the preferred process for industrial enzyme production. On the other hand, SSF can be particularly advantageous to cultivate filamentous fungi, because it simulates their natural habitat leading to higher enzyme productivity compared to SmF [5,6].

A combination of these two cultivation methods, called sequential fermentation (SF or SeqF), was proposed initially by Cunha et al. [7] and applied effectively to produce cellulolytic enzymes. It is characterized by a preculture preparation with an initial step of fungal growth under solid state, followed by a submerged step [8]. This fermentation technique, so far, has been poorly documented and used exclusively to produce lignocellulolytic enzymes using filamentous fungi such as *Aspergillus niger* and *Trichoderma reesei* [9,10] or the white rot fungus *Pleurotus ostreatus* [11]. Different configurations were validated for this approach, using shake flasks, stirred tanks and bubble column bioreactors, which allowed an increase in enzyme productivity compared to the conventional SmF [8]. Most studies have reported the use of sugarcane bagasse as a substrate, nonetheless other substrates such as wheat bran and soybean bran have also been used successfully [12].

Although SF is used for a specific category of enzymes [8], it could be adapted to produce different enzymes of industrial interest. Based on that background, sequential fermentation is used for the first time in this study to produce *Aspergillus tamarii* protease. Particularly, the aims of this study were: (a) to test the sequential fermentation method for protease production by an *Aspergillus tamarii* strain using wheat bran as a substrate, (b) to biochemically characterize the crude extract, and (c) to determine and investigate the kinetic and thermodynamic parameters related to both substrate hydrolysis and enzyme thermal denaturation.

2. RESULTS AND DISCUSSION

2.1. Protease Production by Sequential Fermentation

Protease production by *Aspergillus tamarii* URM4634 in submerged fermentation (SmF) and sequential fermentation (SF) was evaluated after 48 h of cultivation for comparison purposes. Even though other fermentation times were evaluated (data not shown), 48 h was shown to be the most suitable for enzyme production in both fermentation processes. SF allowed an almost double enzyme production (39.08 U mL^{-1}) compared to SmF (20.0 U mL^{-1}), confirming the positive trend observed by several authors who used this approach to produce cellulolytic enzymes [8,10,12]. A possible explanation for this increase is related to the morphology of filamentous fungi, since in SF there was a predominance of dispersed filamentous mycelium, while in the conventional SmF early formation of fungal pellets took place [9]. However, the dispersed hyphal morphology results in a more viscous medium that can make mixing and aeration difficult [8]; furthermore, other factors should be considered for a complete understanding of SF, especially if intended for the production of proteolytic enzymes.

Glucose concentration, medium volume and inoculum size were selected as the independent variables for protease production by SF, because they are among those that mostly influence the conventional fermentation processes. Their effects on such a response were investigated through cultivations carried out according to a 2^3-full factorial design whose results are listed in **Table 1**. The highest protease production (127.67 U mL^{-1}) was obtained in the run 6 performed using a glucose concentration of 50 g L^{-1}, a medium volume of 15 mL g^{-1} of substrate and an inoculum size of 10^8 spores g^{-1} of substrate. According to the statistical analysis (**Table 2**), all the individual variables and their interactions were statistically significant. In particular, glucose concentration showed a positive effect probably because, as known, in fermentation processes the carbon source not only acts as the major constituent for the construction of cell material but is also used in the synthesis of polysaccharides and as an energy source [13]. However, it should be remembered that glucose can have a repressive effect on the synthesis of other enzymes such as pectinases [14]. A similar positive effect was exerted by the inoculum size, while the medium volume had a negative influence on protease production. A larger inoculum size reduces the time that cells take to consume all the substrate and produce the enzyme of interest. However, beyond a certain limit, enzyme production can decrease due to the depletion of nutrients needed for biomass growth, which results in a decrease in metabolic activity [15,16]. To the best of our knowledge, this is the first report on the evaluation of the influence of medium volume on SF, and the trend observed at lower medium volumes may be related to the prevention of enzyme dilution in the culture medium itself.

Table 1. Experimental conditions and results of protease production by *Aspergillus tamarii* URM4634 in sequential fermentations using wheat bran as a substrate. Runs were carried out according to a 2^3-full factorial design plus three central points.

Run	Glucose Concentration(g L^{-1})	Medium Volume (mL g^{-1} of Substrate)	Inoculum Size (Spores g^{-1} of Substrate)	Protease Activity (U mL^{-1})
1	30	15	10^6	33.33 ± 0.94
2	50	15	10^6	25.33 ± 3.53
3	30	25	10^6	18.00 ± 4.95
4	50	25	10^6	12.67 ± 1.17
5	30	15	10^8	90.33 ± 4.95
6	50	15	10^8	127.67 ± 11.66
7	30	25	10^8	80.67 ± 0.47
8	50	25	10^8	78.67 ± 1.64
9	40	20	10^7	35.67 ± 1.88
10	40	20	10^7	39.67 ± 3.29
11	40	20	10^7	43.00 ± 3.06
12	40	20	10^7	38.00 ± 3.30

Table 2. Effects of the independent variables and their interactions on protease production by *Aspergillus tamarii* URM4634 in sequential fermentations carried out according to the 2^3-full factorial design shown in Table 1 using wheat bran as a substrate.

Variable or Interaction	Estimates	p-Value
(1) Glucose concentration	4.74 *	0.0047
(2) Medium volume	-18.68 *	0.0021
(3) Inoculum size	62.06 *	0.0006
1 × 2	-7.90 *	0.0245
1 × 3	10.49 *	0.0113
2 × 3	-6.61 *	0.0390
1 × 2 × 3	-9.05 *	0.0170

* Statistically significant estimates at 95% confidence level ($p < 0.05$).

Interaction effects are interpreted differently from what is usually done with the individual variables. A positive interaction between two variables means that they have a synergistic effect, i.e., the levels of both must increase or decrease together to enhance the response. A negative interaction, on the other hand, means that the variables have an antagonistic effect, i.e., the level of one variable must increase and that of the other decrease to enhance the response. In the case of interaction among three variables, the negative sign means that one of them has an antagonistic effect with the others. More specifically, as shown by the results of **Table 2**, the medium volume had a negative effect, while the other variables a positive effect. The interaction had a negative sign, but the behaviour of each variable involved in the interaction must be evaluated. The interaction among the three factors can be better seen in the cubic graph depicted in **Figure 1**.

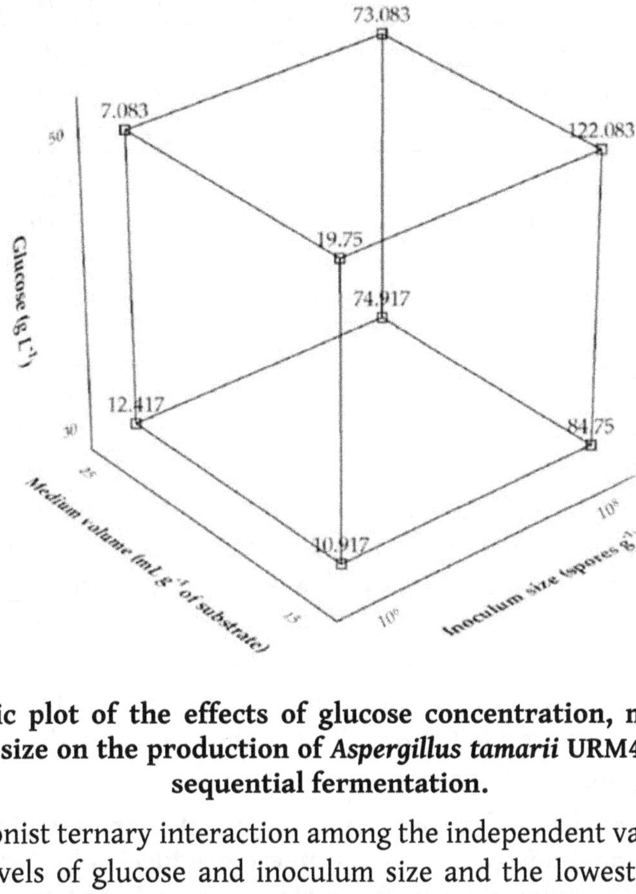

Figure 1. Cubic plot of the effects of glucose concentration, medium volume, and inoculum size on the production of *Aspergillus tamarii* URM4634 protease by sequential fermentation.

The antagonist ternary interaction among the independent variables confirms the highest levels of glucose and inoculum size and the lowest one of medium volume as the best fermentation conditions. For optimization purposes and due to the difficulty of carrying out industrial production at higher spore concentrations, a 2^2-CCRD was performed using only glucose concentration and medium volume as the independent variables, while keeping the inoculum size constant at the highest level (10^8 spores g^{-1}), whose experimental conditions and results are listed in **Table 3**. The highest protease activity (180.17 U mL^{-1}) was observed in the run 4 carried out at a glucose concentration of 55 g L^{-1} and a medium volume of 17.5 mL per g of substrate.

Table 3. Experimental conditions and results of protease production by *Aspergillus tamarii* URM4634 in sequential fermentations using wheat bran as a substrate. Optimization tests were carried out according to a central composite rotational design.

Run	Glucose Concentration(g L-1)	Medium Volume (mL g-1 of Substrate)	Protease Activity (U mL-1)
1	45	12.5	151.00 ± 14.66
2	45	17.5	89.17 ± 6.36
3	55	12.5	169.50 ± 10.13
4	55	17.5	180.17 ± 4.94
5	42.05	15.0	135.50 ± 7.77
6	57.05	15.0	94.00 ± 8.66
7	50	11.475	156.67 ± 11.66
8	50	18.525	103.83 ± 5.42
9	50	15.0	134.17 ± 16.5
10	50	15.0	128.44 ± 2.35
11	50	15.0	139.50 ± 10.78

The analysis of variance (ANOVA) (**Table 4**) indicates that the quadratic model showed a significant lack of fit and an unsatisfactory adjustment to the experimental data (R^2 = 0.44), therefore, the response surface was not shown. Nonetheless, the experiments performed were effective in increasing protease production by as much as 4.6 times compared to the initial SF runs. Only the medium volume showed a significant linear negative effect (-8.05), and its interaction with glucose concentration, contrary to the statistical analysis of the 2^3-full factorial design, was significant and synergistic (6.55) (results not shown), likely due to the different experimental regions investigated.

Table 4. Results of the analysis of variance applied to the results of the central composite rotational design used for protease production by *Aspergillus tamarii* URM4634 in sequential fermentations using wheat bran as a substrate.

Source	Sum of Squares	Degrees of Freedom	Mean Square	F-Value	p-Value
GC(L)	322.71	1	322.71	10.55	0.083
GC(Q)	65.40	2	65.40	2.14	0.281
MV(L)	1980.86	1	1980.86	64.75	0.015
MV(Q)	106.70	1	106.70	3.49	0.203
GC(L) × MV(L)	1314.06	1	1314.06	42.95	0.022

Source	Sum of Squares	Degrees of Freedom	Mean Square	F-Value	p-Value
Lack of fit	4851.15	3	1617.05	52.85	0.019
Pure error	61.19	2	30.59		
Total	8772.15	10			

GC = Glucose concentration; MV = Medium volume; L = Linear effect; Q = Quadratic effect.

2.2. Influence of Temperature and pH on the Activity of Protease

The rate of enzyme-catalyzed reactions can be influenced by various environmental factors through reversible or irreversible changes in protein structure, of which temperature and pH are the most important [17]. The effect of a temperature rise on enzyme activity is the result of two phenomena acting simultaneously but in opposite directions, i.e., an increase in the reaction rate and a progressive enzyme inactivation due to its denaturation. On the other hand, the enzyme activity is influenced by pH because the ionization state of amino acids constituting the enzyme structure depends on the concentration of hydrogen ions [17,18]. In the present study, the optimal temperature and pH of *A. tamarii* URM4634 protease were shown to be 50 °C (Figure 2A) and 7.0 (Figure 2B), respectively.

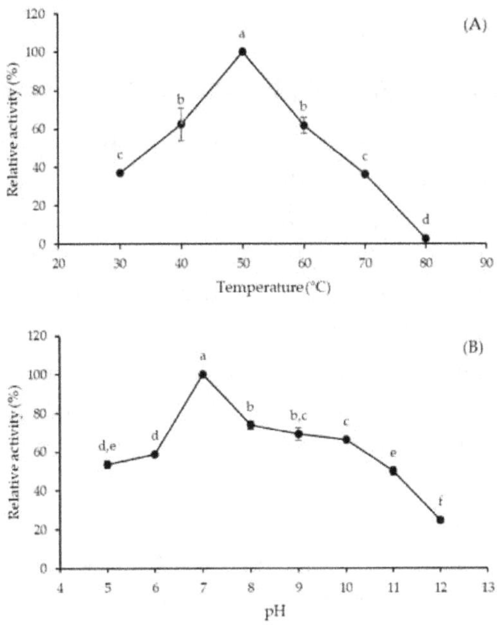

Figure 2. Effects of temperature (A) and pH (B) on the activity of *Aspergillus tamarii* URM4634 protease produced by sequential fermentation using wheat bran as a substrate. Different letters (a–f) indicate statistically significant differences (p < 0.05).

In a previous study by our research group, the protease obtained by SSF using the same fungal strain and substrate (wheat bran) showed different optimal conditions (40°C and pH 8.0) [19], suggesting that the fermentation method may directly influence the characteristics of the biocatalyst. A similar optimum temperature was observed for proteases obtained from different *Aspergillus* species, such as *A. heteromorphus* [20], *A. niger* [21] and *A. oryzae* [22], in conventional fermentation processes (SSF and SmF).

Based on these and previous results, the *A. tamarii* URM4634 protease under investigation was characterized as a neutral protease, being active under neutral, weakly acidic or weakly alkaline conditions. Proteases belonging to this family are considered especially important for the food industry, mainly for baking and brewing processes, as they lead to less bitter food protein hydrolysates thanks to an intermediate reaction rate and are poorly sensitive to plant proteinase inhibitors [23,24]. Fungal neutral proteases, of which *A. oryzae* is the most important producer [25], are constituents of various enzyme preparations, and the search for new sources of proteases has aroused considerable interest. Therefore, the neutral protease from *A. tamarii* URM4634 obtained in this study by the SF method may be an interesting alternative for further studies involving applications in the food industry. A neutral protease from another *A. tamarii* strain was already produced by our research group by extractive fermentation [26], which provides confirmation of the potential of this species as a producer of neutral proteases.

2.3. Kinetic and Thermodynamic Parameters of Azocasein Hydrolysis

The crude extract obtained by SF showed a typical Michaelis-Menten kinetic profile, and the kinetic parameters of azocasein hydrolysis were estimated from the slope of Lineweaver-Burk plot (**Figure 3**A) with satisfactory correlation (R^2 = 0.993). As shown in **Table 5**, the enzyme exhibited values of the Michaelis constant (K_m), maximum rate (V_{max}) and turnover number (k_{cat}) of 16.26 mg mL^{-1}, 147.06 mg mL^{-1} min^{-1} and 195.37 s^{-1}, respectively. Such a K_m value is lower than that obtained in SSF (18.7 mg mL^{-1}) [19] using the same fungal strain and substrate, while that of V_{max} is 5.15-fold higher (28.57 mg mL^{-1} min^{-1}) [19]. These results indicate an increase in the affinity for substrate and the catalytic potential of protease produced by SF compared to that produced by SSF. The value of k_{cat} (195.37 s^{-1}), which corresponds to the maximum number of substrate molecules converted into product per active site per unit of time, is almost 20% higher than that reported by Lee et al. [27] for purified *A. oryzae* protease obtained by SmF using casein as a substrate (163.5 s^{-1}).

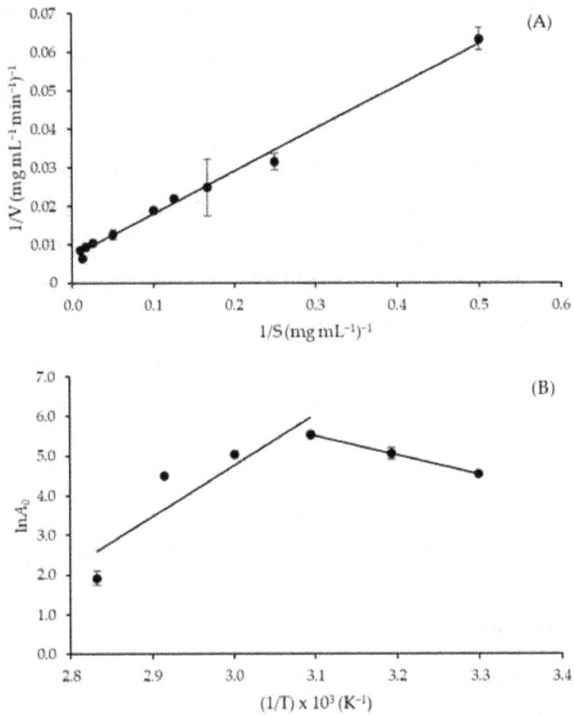

Figure 3. (A) Lineweaver-Burk plot to calculate K_m and V_{max} of *Aspergillus tamarii* URM4634 protease produced by sequential fermentation using azocasein as a substrate (R^2 = 0.993), and (B) Arrhenius-type plot (on the right) to estimate the activation energy of the azocasein hydrolysis reaction (R^2 = 0.998).

Table 5. Kinetic and thermodynamic parameters of the azocasein hydrolysis reaction catalyzed by the *Aspergillus tamarii* URM4634 protease produced by sequential fermentation using wheat bran as a substrate. 25°C was selected as a reference temperature.

Parameter	Value
[a] K_m (mg mL^{-1})	16.26
[b] V_{max} (mg mL^{-1} min^{-1})	147.06
[c] k_{cat} (s^{-1})	195.37
[d] E^*_a (kJ mol^{-1})	40.38
[e] ΔG^* (kJ mol^{-1})	59.94
[f] ΔH^* (kJ mol^{-1})	37.90
[g] ΔS^* (J K^{-1} mol^{-1})	-73.94

[a] Michaelis constant; [b] Maximum rate; [c] Turnover number; [d] Activation energy; [e] Activation Gibbs free energy; [f] Activation enthalpy; [g] Activation entropy.

From the slope of the straight line to the right of the Arrhenius-type plot (**Figure 3**B), an activation energy (E^*_a) of 40.38 kJ mol^{-1} was estimated for azocasein hydrolysis (**Table 5**). This value is lower than that reported for casein hydrolysis by *A. fumigatus* protease (62 kJ mol^{-1}) [28] and slightly higher than that of bread protein hydrolysis by *A. awamori* protease (36.8 kJ mol^{-1}) [29]. It is worth remembering, for comparative purposes, that a lower E^*_a value indicates that less energy is required to form the activated complex of azocasein hydrolysis, so the reaction is favored. Another important parameter to evaluate the temperature influence on a catalyzed reaction is the temperature quotient (Q_{10}), which indicates the rate increase resulting from a 10 °C temperature rise. At temperatures between 30 and 50 °C, the mean value of Q_{10} (1.17) was very close to those reported by de Castro et al. [30] for *A. niger* protease (1.20 to 1.28) in the temperature range of 35–55 °C. It has been stressed that deviations of Q_{10} values from the typical range of enzyme reactions (1.0 to 2.0) are indicative of the involvement of factors other than temperature in the control of the reaction rate [31]. Therefore, it can be inferred that azocasein hydrolysis by *A. tamarii* protease was kinetically controlled by temperature within the entire temperature range investigated.

The other thermodynamic parameters of azocasein hydrolysis, namely activation enthalpy (ΔH^*), entropy (ΔS^*) and Gibbs free energy (ΔG^*), were also calculated by Equations (2) to (4) and listed in **Table 5**.

The relatively low ΔH^* value (37.90 kJ mol^{-1}) indicates that the formation of the transition state or activated enzyme-substrate complex occurred effectively; however, this value is higher than that reported by Fernandes et al. [20] for azocasein hydrolysis by *A. heteromorphus* protease (21.8 kJ mol^{-1}). As known, ΔS^* is correlated to the order degree of a reaction system; therefore, in enzyme-catalyzed reactions negative values such as that estimated in this study (–73.94 J K^{-1} mol^{-1}) suggest that the structure of enzyme-substrate at transition state is more ordered than that of the enzyme-substrate complex. Similar qualitative behaviour was observed for proteases [32] and other hydrolases such as β-fructofuranosidases [31], pectinases [33] and levansucrases [34]. As known, ΔG^* is the most suitable thermodynamic parameter to measure the feasibility and extent of a reaction, as the lower the ΔG^* value the more feasible the reaction, i.e., the conversion of the reactant into the product is more spontaneous [35]. The value of ΔG^* estimated for azocasein hydrolysis catalyzed by *A. tamarii* URM4634 protease (59.94 kJ mol^{-1}) is lower than those of casein and azocasein hydrolysis catalyzed by other fungal (*A. heteromorphus*: ΔG^* = 65.8 kJ mol^{-1} [20]) and bacterial (*Bacillus stearothermophilus*: ΔG^* = 91.71 kJ mol^{-1} [32]; *Bacillus cereus*: ΔG^* = 68.68–69.42 kJ mol^{-1} [36]) proteases, which indicates that the conversion of the transition state of enzyme-substrate complex into the product is more spontaneous.

2.4. Kinetic and Thermodynamic Parameters of Protease Thermal Denaturation

The time-dependent thermal inactivation of biocatalysts (denaturation) can be a major problem in enzyme-catalyzed processes, and knowledge of the impact of this phenomenon is particularly useful before running any industrial-scale bioprocess. Then, to evaluate the protease thermostability, residual activity tests were performed in the temperature range of 50 to 80 °C, the results of which are shown in semi-log plots (**Figure 4**).

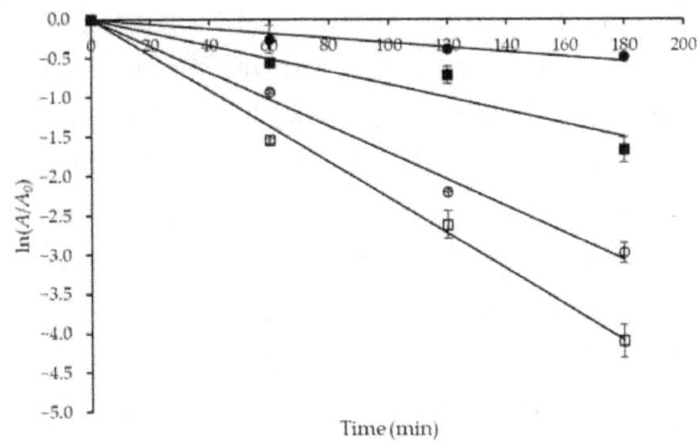

Figure 4. Semi-log plots of the coefficient of residual activity (A/A_0) of *Aspergillus tamarii* URM4634 protease versus time used to estimate the first-order inactivation rate constant (k_d) at different temperatures (°C): (•) 50; (■) 60; (○) 70; (□) 80.

The slopes of the straight lines obtained at different temperatures were used to estimate the first-order inactivation rate constant (k_d), whose values were estimated with a satisfactory correlation ($0.969 \le R^2 \le 0.998$) and are listed in **Table 6** together with those of the other kinetic and thermodynamic parameters. One can see that k_d progressively increased with temperature from 0.0030 to 0.0227 min^{-1}, which means that denaturation became progressively more significant, likely due to the breaking of an increasing number of strong electrostatic bonds.

Table 6. Kinetic and thermodynamic parameters of thermal denaturation of crude *Aspergillus tamarii* URM4634 protease produced by sequential fermentation using wheat bran as a substate.

T (°C)	[a] k_d (min^{-1})	R^2	[b] $t_{1/2}$ (min)	[c] D-Value (min)	[d] Z-Value (°C)	[e] E^*_d (kJ mol^{-1})	[f] ΔG^*_d (kJ mol^{-1})	[g] ΔH^*_d (kJ mol^{-1})	[h] ΔS^*_d (J mol^{-1} K^{-1})
50	0.0030	0.979	231.05	767.53	33.89	64.78	105.97	62.09	-135.76
60	0.0083	0.969	83.51	277.42			106.51	62.01	-133.57
70	0.0170	0.997	40.77	135.45			107.75	61.93	-133.52
80	0.0227	0.998	30.53	101.43			110.12	61.84	-136.71

[a] First-order inactivation rate constant; [b] Half-life; [c] Decimal reduction time; [d] Thermal resistance constant; [e] Activation energy of thermal denaturation; [f] Activation Gibbs free energy of thermal denaturation; [g] Activation enthalpy of thermal denaturation, and [h] Activation entropy of thermal denaturation.

At 50 °C, a temperature commonly used in several industrial processes, the crude protease obtained by SF exhibited much longer half-life ($t_{1/2}$ = 231.05 min) and decimal reduction time (D-value = 767.53 min) than that produced by the same fungal strain and substrate by SSF, either in crude ($t_{1/2}$: 96.3 min; D-value: 319.9 min) or purified ($t_{1/2}$: 70 min; D-value: 232.6 min) form [37]. These results indicate that the protease produced by SF is more thermostable than that obtained by SSF; therefore, one of the next efforts will be an enzyme structural analysis that can provide some additional information for a better understanding of the thermostabilization mechanism under SF conditions. Moreover, the protease produced in the present study appears to be more thermostable than other *Aspergillus* proteases at same temperature (50 °C), such as those produced by *A. fumigatus* ($t_{1/2}$: 96.3 min) [28], *A. niger* ($t_{1/2}$: 135.91 min; D-value: 451.49 min) [30] and *A. heteromorphus* ($t_{1/2}$: 54.6 min; D-value: 181.3 min) [20].

As proof of protease thermostability, a thermal resistance constant (Z-value) of 33.89 °C was estimated with good correlation (R^2 = 0.945) from the slope of the semi-log plot of D-value versus temperature (data not shown). In general, high Z-values, such as that obtained in the present study, indicate a greater sensitivity to the duration of the heat treatment than to the temperature increase [35]. The Z-value obtained is lower than that detected for the protease produced by SSF either in crude (44.2 °C) or purified (75.8 °C) form [37], which suggests a noticeable reduction in sensitivity to duration of thermal treatment when the enzyme is produced by SF.

As for the thermodynamic parameters of protease thermal denaturation, an activation energy (E^*_d) of 64.78 kJ mol^{-1} was estimated from the Arrhenius-type plot illustrated in **Figure 5** with satisfactory correlation (R^2 = 0.955). As known, E^*_d is related to the activation enthalpy of thermal denaturation (ΔH^*_d), i.e., the total amount of thermal energy required to denature the enzyme through disruption of non-covalent bonds, including hydrophobic interactions [31], which varied between 61.84 and 62.09 kJ mol^{-1}. Comparison of the above E^*_d and ΔH^*_d values with those of protease obtained by SSF either in crude (E^*_d = 49.7 kJ mol^{-1}; ΔH^*_d = 46.7–47.0 kJ mol^{-1}) or purified (E^*_d = 28.8 kJ mol^{-1}; ΔH^*_d = 25.9–26.1 kJ mol^{-1}) form [37] confirms the higher thermostability of the enzyme produced by SF. To the best of our knowledge, studies comparing the thermostability of enzymes produced by different fermentation processes are very scarce. Nonetheless, the work of Saqib et al. [38] can be cited, in which the authors reported, on the basis of kinetic and thermodynamic parameters, a greater thermostability of an *A. fumigatus* endoglucanase produced by SSF rather than by SmF. Therefore, further studies are needed to investigate possible structural differences in enzymes produced in different fermentation processes, including unconventional methods such as SF.

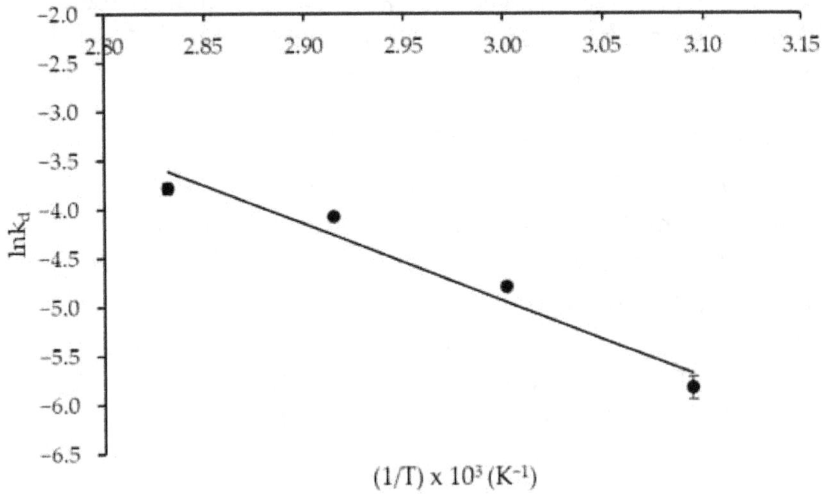

Figure 5. Arrhenius-type plot used to estimate the activation energy of thermal denaturation of *Aspergillus tamarii* URM4634 protease produced by sequential fermentation using wheat bran as a substrate.

Usually, a breakdown of the enzyme structure occurs during thermal denaturation, which is accompanied by an increase in the degree of disorder or randomness, i.e., a positive value of the activation entropy (ΔS^*_d) [28]. However, negative values of ΔS^*_d of thermal denaturation ranging from -133.52 to -136.71 J mol^{-1} K^{-1} were estimated for *A. tamarii* URM4634 protease produced by SF, suggesting an even higher order degree of the transition state compared to the separate reactants and then, high thermostability in the investigated range of temperature [39]. When ΔS is negative and ΔH positive, a process is not spontaneous at any temperature, but the reverse process is [29], which means that protease denaturation was reversible and not spontaneous at all temperatures investigated. As for the Gibbs free energy of thermal denaturation (ΔG^*_d), small or negative values of this parameter are related to a more spontaneous process; contrariwise, large or positive values like those estimated in this study ($105.97 \leq \Delta G^*_d \leq 110.12$ kJ mol^{-1}) indicate special resistance to denaturation, confirming the thermostability indicated by the other kinetic and thermodynamic parameters.

3. MATERIALS AND METHODS

3.1. Materials

Azocasein was purchased from Sigma-Aldrich (St. Louis, MO, USA). Potato dextrose agar (PDA) medium, glucose, yeast extract and all other reagents used were analytical grade. Wheat bran used as a substrate in SmF and SF was acquired in a local market in the city of Garanhuns, PE, Brazil.

3.2. Microorganism and Inoculum Preparation

The *Aspergillus tamarii* URM4634 strain was provided by the "Micoteca-URM" of Mycology Department, Centre of Biosciences of Federal University of Pernambuco (UFPE), Recife, PE, Brazil, preserved in mineral oil and maintained at 25 ± 1 °C in Czapek Dox agar medium. For sporulation, the strain was inoculated in PDA medium for 7 days at 30 °C. The spores were suspended in NaCl solution (0.9%) containing Tween 80 (0.01%), and the inoculum was adjusted to a final concentration of 10^5 spores mL^{-1} and 10^7 spores g^{-1} of substrate for the initial SmF and SF runs, respectively.

3.3. Protease Production by Sequential Fermentation

The sequential fermentation (SF) was performed at 30 °C in two steps. First the solid-state fermentation (SSF) was conducted for 24 h, then the nutrient medium was added, and the submerged fermentation (SmF) was carried out for the same time. The fermentation was performed in 250 mL-Erlenmeyer flasks containing 5 g of substrate (wheat bran standardized to a particle size between 0.5 and 2.0 mm), nutrition solution (1.0% glucose and 0.5% yeast extract) and a spore suspension (10^7 spores g^{-1}), corresponding to a 60% moisture content. After 24 h of growth, we added, in the proportion of 20 mL g^{-1} of substrate, a nutrient medium adapted from Cunha et al. [7] having the following composition: 0.14% $(NH_4)_2SO_4$, 0.20% KH_2PO_4, 0.03% $CaCl_2$, 0.02% $MgSO_4{\cdot}7H_2O$, 0.50% peptone, 0.20% yeast extract, 0.03% urea, 0.10% Tween 80, 0.10% mineral solution (5 mg L^{-1} $FeSO_4{\cdot}7H_2O$, 1.6 mg L^{-1} $MnSO_4{\cdot}H_2O$ and 1.4 mg L^{-1} $ZnSO_4{\cdot}7H_2O$) and supplemented with glucose (40.0 g L^{-1}). After 24 h of SmF carried out under stirring (150 rpm), the fermented medium was filtered with a vacuum pump, and the crude enzyme extract was stored at -20 °C for further analysis.

Protease production by SF was investigated using a 2^3-full factorial design plus four central points, in which glucose concentration (30, 40 and 50 g L^{-1}), medium volume (15, 20 and 25 mL g^{-1} of substrate) and inoculum size (10^6, 10^7 and 10^8 spores g^{-1} of substrate) were selected as the independent variables (**Table 1**). Based on the results obtained with this statistical design (**Table 2**), a central composite rotational design (CCRD) was performed to better identify the experimental conditions to maximize protease production, in which the independent variables were glucose concentration and medium volume (**Table 3**).

The statistical analysis of both experimental designs was performed using the software package Statistica 7.0 (Statsoft Inc., Tulsa, OK, USA). The best fitting quadratic model equation was determined by the Fisher's test. The fitting ability of the model was assessed by the coefficient of determination (R^2) and the analysis of variance (ANOVA).

3.4. Protease Production by Submerged Fermentation

For comparison purposes, submerged fermentations were also performed at 130 rpm and 30°C for 48 h, according to Porto et al. [40] with some modifications, using wheat flour as a substrate in 250 mL-Erlenmeyer flasks containing 50 mL of culture medium with the following composition (%): filtered wheat flour, 4.0; NH_4Cl, 0.1; $MgSO_4 \cdot 7H_2O$, 0.06; yeast extract, 0.1; K_2HPO_4, 0.435; glucose, 0.5; and 1.0% (v v^{-1}) mineral solution (100 mg $FeSO_4 \cdot 7H_2O$; 100 mg $MnCl_2 \cdot 4H_2O$; and 100 mg $ZnSO_4 \cdot H_2O$ in 100 mL of distilled water). The medium pH was adjusted to 7.0 before fermentation.

3.5. Analytical Determinations

Protease activity was determined as described by Ginther [41] with adaptations using azocasein as a substrate. For this purpose, a 1.0% azocasein solution was prepared in Tris-HCl buffer (0.2 M, pH 7.2) containing 1.0 mM $CaCl_2$. The reaction was performed using 0.15 mL of enzyme extract and 0.25 mL of azocasein solution at room temperature (25 ± 1 °C). After 1 h, the reaction was stopped by addition of 1.0 mL of 10% trichloroacetic acid (TCA). Then, the samples were centrifuged at 8000× *g* for 20 min at 4 °C, 0.8 mL of supernatant were added to 0.2 mL of 1.8 M NaOH, and the absorbance of samples was measured at 420 nm with a spectrophotometer (Biochrom Libra S6, Cambridge, UK). Protease activity was defined as the enzyme amount responsible for a 0.1 increase in absorbance at 420 nm within 1 h. The total protein concentration was determined by the Bradford method [42] using bovine serum albumin (BSA) as a standard.

3.6. Effect of pH and Temperature on Protease Activity

The optimum pH of enzyme activity was determined through activity tests carried out at 25°C as described in the previous section using substrate solutions at pH values ranging from 5.0 to 12.0. For this purpose, the following buffers were used at 0.1 M concentration: citrate-phosphate (5.0–7.0), Tris-HCl (7.0–9.0) and glycine-NaOH (9.0–12.0). On the other hand, the influence of temperature on protease activity was investigated through activity tests carried out at different temperatures (30–80°C) and constant pH (7.2). The results were evaluated by the Tukey's test to verify statistically significant differences ($p < 0.05$) among different samples at different values of pH or temperature.

3.7. Kinetic and Thermodynamic Parameters of Azocasein Hydrolysis

The Michaelis constant (K_m) and maximum rate (V_{max}) of azocasein hydrolysis were estimated by the Lineweaver-Burk double reciprocal plot. For this purpose, protease activity assays were performed using different substrate concentrations (2.0–100 mg mL^{-1}) at pH 7.2 and 25 ± 1°C. The catalytic constant or turnover number (k_{cat}) was calculated as the ratio of V_{max} to the protein concentration. The activation energy (E^*_a) of substrate hydrolysis was estimated by the Arrhenius-type plot of $\ln A_0$ versus 1/T, being A_0 the starting activity and T the absolute temperature, using the results of enzyme activity tests. For a better understanding of temperature influence on the hydrolysis reaction, the temperature quotient (Q_{10}), which is the factor by which the enzyme activity increases due to a 10 °C temperature increase, was calculated by the following equation [43]:

$$Q10 = anti\,log\left(\frac{E*a\times 10}{RT2}\right) \quad (1)$$

where R is the gas constant (8.314 J mol^{-1} K^{-1}).

The thermodynamic parameters, namely the activation enthalpy (ΔH*), Gibbs free energy (ΔG*) and entropy (ΔS*) of azocasein hydrolysis, were calculated using the following equations:

$$\Delta H* = E*a - RT \quad (2)$$

$$\Delta G* = -RT\ln\left(\frac{k_{cat}h}{kbT}\right) \quad (3)$$

$$\Delta S* = \frac{\Delta H* - \Delta G*}{T} \quad (4)$$

where h (6.626 × 10^{-34} J s) is the Planck constant, k_b (1.381 × 10^{-23} J K^{-1}) the Boltzmann constant and *T* is the reference temperature (298.15 K).

3.8. Parameters Governing the Kinetics and Thermodynamics of Thermal Denaturation in Protease

As known, the increase in temperature not only accelerates the enzyme reaction but also promotes the reversible enzyme unfolding and the subsequent denaturation of its structure [31]. The enzyme thermal denaturation can then be kinetically described by the equation:

$$\frac{dA}{dt} = -kd.A \quad (5)$$

where k_d is the rate constant of denaturation, A the enzyme activity and t the time.

Accordingly, the k_d values were estimated through the linearized version of this equation:

$$In\left(\frac{A}{A_0}\right) = -K_d t \quad (6)$$

from the slopes of the straight lines obtained by plotting the experimental data of $\ln(A/A_0)$ versus time in the temperature range of 50 to 80 °C, where A_0 is the starting activity at t = 0.

From the k_d values, we also calculated the half-life ($t_{1/2}$), which is defined as the time after which the enzyme activity is reduced to one-half the initial value:

$$t1/2 = \frac{In\ 2}{kd} \quad (7)$$

and the decimal reduction time (D-value), i.e., the time required for a 10-fold activity reduction:

$$D\text{ - }value = \frac{In\ 10}{kd} \quad (8)$$

The D-value is often accompanied by the thermal resistance constant (Z-value) that corresponds to the temperature increase required to achieve a 10-fold reduction of the D-value. This parameter was estimated from the slope of the thermal death time plot of logD versus T (°C).

The thermodynamic parameters related to protease thermal denaturation were also estimated from an Arrhenius-type plot. Particularly, the activation energy (E^*_d) was estimated from the slope of the straight line of $\ln k_d$ vs. 1/T, while the activation enthalpy (ΔH^*_d), Gibbs free energy (ΔG^*_d) and entropy (ΔS^*_d) were calculated similarly to those of enzyme-catalyzed reaction using k_d instead of k_{cat} in Equations (2)–(4).

4. CONCLUSIONS

The sequential fermentation (SF) process proved to be an interesting approach for protease production, showing higher outcomes (180.17 U mL^{-1}) compared to the conventional submerged fermentation (20.0 U mL^{-1}). Several differences in the enzyme characteristics, especially an improved thermostability, were observed compared to the protease obtained in a previous study by solid-state fermentation using the same fungal strain and substrate, thereby indicating the need to carry out further studies on the protein structure for a better understanding of SF features. The enzyme produced has been classified as a neutral protease (optimum pH of 7.0), which would offer several advantages in typical food industry applications such as baking and brewing processes. Moreover, the kinetic and thermodynamic study pointed out a satisfactory thermostability of the protease at 50°C, a temperature commonly used in many industrial processes, evidenced especially by a half-life of 231.05 min and a decimal reduction time of 767.53 min. The results obtained in this study can contribute to the development of the SF technique to produce proteases and various other enzymes of industrial interest.

5. CONCISE SUMMARY

Microbial proteases are commonly produced by submerged (SmF) or solid-state fermentation (SSF), whose combination results in an unconventional method, called sequential fermentation (SF), which has already been used only to produce cellulolytic enzymes. In this context, the aim of the present study was the development of a novel SF method for protease production using wheat bran as a substrate. Moreover, the kinetic and thermodynamic parameters of azocasein hydrolysis were estimated, thus providing a greater understanding of the catalytic reaction. In SF, an approximately 9-fold increase in protease activity was observed compared to the conventional SmF method. Optimization of glucose concentration and medium volume by statistical means allowed us to achieve a maximum protease activity of 180.17 U mL^{-1}. The obtained enzyme had an optimum pH and temperature of 7.0 and 50°C, respectively. Kinetic and thermodynamic parameters highlighted that such a neutral protease is satisfactorily thermostable at 50°C, a temperature commonly used in many applications in the food industry. The results obtained suggested not only that SF could be a promising alternative to produce proteases, but also that it could be adapted to produce several other enzymes.

REFERENCES

1. Naveed, M.; Nadeem, F.; Mehmood, T.; Bilal, M.; Anwar, Z.; Amjad, F. A versatile and ecofriendly biocatalyst with multi-industrial applications: An updated review. *Catal. Lett.* **2020**, *151*, 307–323. [Google Scholar] [CrossRef]

2. Gurumallesh, P.; Alagu, K.; Ramakrishnan, B.; Muthusamy, S. A systematic reconsideration on proteases. *Int. J. Biol. Macromol.* **2019**, *128*, 254–267. [Google Scholar] [CrossRef]

3. Sharma, K.M.; Kumar, R.; Panwar, S.; Kumar, A. Microbial alkaline proteases: Optimization of production parameters and their properties. *J. Genet. Eng. Biotechnol.* **2017**, *15*, 115–126. [Google Scholar] [CrossRef] [PubMed]

4. Soccol, C.R.; da Costa, E.S.F.; Letti, L.A.J.; Karp, S.G.; Woiciechowski, A.L.; Vandenberghe, L.P.S. Recent developments and innovations in solid state fermentation. *Biotechnol. Res. Innov.* **2017**, *1*, 52–71. [Google Scholar] [CrossRef]

5. Farinas, C.S. Developments in solid-state fermentation for the production of biomass-degrading enzymes for the bioenergy sector. *Renew. Sustain. Energy Rev.* **2015**, *52*, 179–188. [Google Scholar] [CrossRef]

6. Yazid, N.A.; Barrena, R.; Komilis, D.; Sánchez, A. Solid-State Fermentation as a novel paradigm for organic waste valorization: A review. *Sustainability* **2017**, *9*, 224. [Google Scholar] [CrossRef][Green Version]

7. Cunha, F.M.; Esperança, M.N.; Zangirolami, T.C.; Badino, A.C.; Farinas, C.S. Sequential solid-state and submerged cultivation of *Aspergillus niger* on sugarcane bagasse for the production of cellulase. *Bioresour. Technol.* **2012**, *112*, 270–274. [Google Scholar] [CrossRef][Green Version]

8. Farinas, C.S.; Florencio, C.; Badino, A.C. On-site production of cellulolytic enzymes by the sequential cultivation method. In *Cellulases. Methods in Molecular Biology*; Lübeck, M., Ed.; Humana Press: New York, NY, USA, 2018; Volume 1796, pp. 273–282. [Google Scholar] [CrossRef]

9. Florencio, C.; Cunha, F.M.; Badino, A.C.; Farinas, C.S.; Ximenes, E.; Ladisch, M.R. Secretome analysis of *Trichoderma reesei* and *Aspergillus niger* cultivated by submerged and sequential fermentation processes: Enzyme production for sugarcane bagasse hydrolysis. *Enzyme Microb. Technol.* **2016**, *90*, 53–60. [Google Scholar] [CrossRef][Green Version]

10. Florencio, C.; Cunha, F.M.; Badino, A.C.; Farinas, C.S. Validation of a novel sequential cultivation method for the production of enzymatic cocktails from *Trichoderma* strains. *Appl. Biochem. Biotechnol.* **2014**, *175*, 1389–1402. [Google Scholar] [CrossRef]

11. An, Q.; Wu, X.; Han, M.; Chu, B.; He, S.; Dai, Y.; Si, J. Sequential Solid-State and Submerged Cultivation of the white rot fungus *Pleurotus ostreatus* on biomass and the activity of lignocellulolytic enzymes. *BioResources* **2016**, *11*, 8791–8805. [Google Scholar] [CrossRef][Green Version]

12. Cunha, F.M.; Vasconcellos, V.M.; Florencio, C.; Badino, A.C.; Farinas, C.S. On-site production of enzymatic cocktails using a non-conventional fermentation method with agro-industrial residues as renewable feedstocks. *Waste Biomass Valorization* **2017**, *8*, 517–526. [Google Scholar] [CrossRef]

13. Bankar, S.B.; Bule, M.V.; Singhal, R.S.; Ananthanarayan, L. Optimization of *Aspergillus niger* fermentation for the production of glucose oxidase. *Food Bioprocess Technol.* **2009**, *2*, 344–352. [Google Scholar] [CrossRef]

14. Ahmed, A.; Ejaz, U.; Sohail, M. Pectinase production from immobilized and free cells of *Geotrichum candidum* AA15 in galacturonic acid and sugars containing medium. *J. King Saud Univ. Sci.* **2020**, *32*, 952–954. [Google Scholar] [CrossRef]

15. Sun, H.; Ge, X.; Hao, Z.; Peng, M. Cellulase production by *Trichoderma* sp. on apple pomace under solid state fermentation. *Afr. J. Biotechnol.* **2010**, *9*, 163–166. [Google Scholar] [CrossRef]

16. Sandhya, C.; Sumantha, A.; Szakacs, G.; Pandey, A. Comparative evaluation of neutral protease production by *Aspergillus oryzae* in submerged and solid-state fermentation. *Process Biochem.* **2005**, *10*, 2689–2694. [Google Scholar] [CrossRef]

17. Robinson, P.K. Enzymes: Principles and biotechnological applications. *Essays Biochem.* **2015**, *59*, 1–41. [Google Scholar] [CrossRef]

18. Vitolo, M. Enzymes: The catalytic proteins. In *Pharmaceutical Biotechnology*, 1st ed.; Pessoa, A., Jr., Vitolo, M., Long, P.F., Eds.; CRC Press: Boca Raton, FL, USA, 2021; Volume 1, pp. 257–263. [Google Scholar] [CrossRef]

19. da Silva, O.S.; de Oliveira, R.L.; Souza-Motta, C.M.; Porto, A.L.F.; Porto, T.S. Novel protease from *Aspergillus tamarii* URM4634: Production and characterization using inexpensive agroindustrial substrates by Solid-State Fermentation. *Adv. Enzym. Res.* **2016**, *4*, 125–143. [Google Scholar] [CrossRef][Green Version]

20. Fernandes, L.M.G.; Carneiro-da-Cunha, M.N.; Silva, J.C.; Porto, A.L.F.; Porto, T.S. Purification and characterization of a novel *Aspergillus heteromorphus* URM 0269 protease extracted by aqueous two-phase systems PEG/citrate. *J. Mol. Liq.* **2020**, *317*, 113957. [Google Scholar] [CrossRef]

21. de Castro, R.J.S.; Nishide, T.G.; Sato, H.H. Production and biochemical properties of proteases secreted by *Aspergillus niger* under solid state fermentation in response to different agroindustrial substrates. *Biocatal. Agric. Biotechnol.* **2014**, *3*, 236–245. [Google Scholar] [CrossRef]

22. Salihi, A.; Asoodeh, A.; Aliabadian, M. Production and biochemical characterization of an alkaline protease from *Aspergillus oryzae* CH93. *Int. J. Biol. Macromol.* **2017**, *94*, 827–835. [Google Scholar] [CrossRef]

23. Razzaq, A.; Shamsi, S.; Ali, A.; Ali, Q.; Sajjad, M.; Malik, A.; Ashraf, M. Microbial proteases applications. *Front. Bioeng. Biotechnol.* **2019**, *7*, 110. [Google Scholar] [CrossRef] [PubMed][Green Version]

24. Tavano, O.L.; Berenguer-Murcia, A.; Secundo, F.; Fernandez-Lafuente, R. Biotechnological applications of proteases in food technology. *Compr. Rev. Food Sci. Food Saf.* **2018**, *17*, 412–436. [Google Scholar] [CrossRef] [PubMed] [Green Version]

25. Belmessikh, A.; Boukhalfa, H.; Mechakra-Maza, A.; Gheribi-Aoulmi, Z.; Amrane, A. Statistical optimization of culture medium for neutral protease production by *Aspergillus oryzae*. Comparative study between solid and submerged

fermentations on tomato pomace. *J. Taiwan Inst. Chem. Eng.* **2013**, *44*, 377–385. [Google Scholar] [CrossRef]

26. Alves, R.O.; de Oliveira, R.L.; da Silva, O.S.; Porto, A.L.F.; Porto, C.S. Extractive fermentation for process integration of protease production by *Aspergillus tamarii* Kita UCP1279 and purification by PEG-Citrate Aqueous Two-Phase System. *Prep. Biochem. Biotechnol.* **2021**, 1–8. [Google Scholar] [CrossRef]

27. Lee, S.K.; Hwang, J.Y.; Choi, S.H.; Kim, S.M. Purification and characterization of *Aspergillus oryzae* LK101- salt-tolerant acid protease isolated from soybean paste. *Food Sci. Biotechnol.* **2010**, *19*, 327–334. [Google Scholar] [CrossRef]

28. Hernández-Martínez, R.; Gutiérrez-Sánchez, G.; Bergmann, C.W.; Loera-Corral, O.; Rojo-Domínguez, A.; Huerta-Ochoa, S.; Regalado-González, C.; Prado-Barragán, L.A. Purification and characterization of a thermodynamic stable serine protease from *Aspergillus fumigatus*. *Process Biochem.* **2011**, *45*, 2001–2006. [Google Scholar] [CrossRef]

29. Melikoglu, M.; Lin, C.S.K.; Webb, C. Kinetic studies on the multi-enzyme solution produced via solid state fermentation of waste bread by *Aspergillus awamori*. *Biochem. Eng. J.* **2013**, *80*, 76–82. [Google Scholar] [CrossRef]

30. de Castro, R.J.S.; Ohara, A.; Nishide, T.G.; Albernaz, J.R.M.; Soares, M.H.; Sato, H.H. A new approach for proteases production by *Aspergillus niger* based on the kinetic and thermodynamic parameters of the enzymes obtained. *Biocatal. Agric. Biotechnol.* **2015**, *4*, 199–207. [Google Scholar] [CrossRef]

31. de Oliveira, R.L.; Silva, M.F.; Converti, A.; Porto, T.S. Biochemical characterization and kinetic/thermodynamic study of *Aspergillus tamarii* URM4634 β-fructofuranosidase with transfructosylating activity. *Biotechnol. Prog.* **2019**, *35*, 2879. [Google Scholar] [CrossRef]

32. Abdel-Naby, M.A.; Ahmed, S.A.; Wehaidy, H.R.; El-Mahdy, S.A. Catalytic, kinetic and thermodynamic properties of stabilized *Bacillus stearothermophilus* alkaline protease. *Int. J. Biol. Macromol.* **2017**, *96*, 265–271. [Google Scholar] [CrossRef] [PubMed]

33. Silva, J.C.; de França, P.R.L.; Converti, A.; Porto, T.S. Kinetic and thermodynamic characterization of a novel *Aspergillus aculeatus* URM4953 polygalacturonase. Comparison of free and calcium alginate-immobilized enzyme. *Process Biochem.* **2018**, *74*, 61–70. [Google Scholar] [CrossRef]

34. Mostafa, F.A.; Abdel, W.A.; Salah, H.A.; Nawwar, G.A.M.; Esawy, M.A. Kinetic and thermodynamic characteristic of *Aspergillus awamori* EM66 levansucrase. *Int. J. Biol. Macromol.* **2018**, *119*, 232–239. [Google Scholar] [CrossRef] [PubMed]

35. Converti, A.; Pessoa, A., Jr.; Silva, J.C.; de Oliveira, R.L.; Porto, T.S. Thermodynamics applied to biomolecules. In *Pharmaceutical Biotechnology*, 1st

ed.; Pessoa, A., Vitolo, M., Long, P.F., Eds.; CRC Press: Boca Raton, FL, USA, 2021; Volume 1, pp. 27–42. **[Google Scholar]** **[CrossRef]**

36. Abdel-Naby, M.A.; El-Wafa, W.M.A.; Salem, G.E.M. Molecular characterization, catalytic, kinetic and thermodynamic properties of protease produced by a mutant of *Bacillus cereus*-S6. *Int. J. Biol. Macromol* **2020**, *160*, 695–702. **[Google Scholar]** **[CrossRef]** **[PubMed]**
37. da Silva, O.S.; de Oliveira, R.L.; Silva, J.C.; Converti, A.; Porto, T.S. Thermodynamic investigation of an alkaline protease from *Aspergillus tamarii* URM4634: A comparative approach between crude extract and purified enzyme. *Int. J. Biol. Macromol.* **2018**, *109*, 1039–1044. **[Google Scholar]** **[CrossRef]** **[PubMed]**
38. Saqib, A.A.N.; Hassan, M.; Khan, N.F.; Baig, S. Thermostability of crude endoglucanase from *Aspergillus fumigatus* grown under solid state fermentation (SSF) and submerged fermentation (SmF). *Process Biochem.* **2010**, *45*, 641–646. **[Google Scholar]** **[CrossRef]**
39. de Oliveira, R.L.; da Silva, O.S.; Converti, A.; Porto, T.S. Thermodynamic and kinetic studies on pectinase extracted from *Aspergillus aculeatus*: Free and immobilized enzyme entrapped in alginate beads. *Int. J. Biol. Macromol.* **2018**, *115*, 1088–1093. **[Google Scholar]** **[CrossRef]** **[PubMed]**
40. Porto, A.L.F.; Campos-Takaki, G.M.; Lima Filho, J.L. Effects of culture conditions on protease production by *Streptomyces clavuligerus* growing on soybean flour medium. *Appl. Biochem. Biotechnol.* **1996**, *60*, 115–122. **[Google Scholar]** **[CrossRef]**
41. Ginther, C.L. Sporulation and the production of serine protease and cephamycin C by *Streptomyces lactamdurans*. *Antimicrob. Agents Chemother.* **1979**, *15*, 522–526. **[Google Scholar]** **[CrossRef]****[Green Version]**
42. Bradford, M.M. A rapid and sensitive method for the quantitation of microgram quantities of protein utilizing the principle of protein-dye binding. *Anal. Biochem.* **1976**, *72*, 248–254. **[Google Scholar]** **[CrossRef]**
43. Dixon, M.; Webb, C. Enzyme inhibition and inactivation. In *Enzymes*, 3rd ed.; Longman: London, UK, 1979; pp. 332–467. **[Google Scholar]**

CHAPTER-5

APPLICATION OF CLAY MATERIALS AS A SUBSTRATE FOR LIPASE FROM *ASPERGILLUS JAPONICUS*: AN ECO-FRIENDLY APPROACH

1. INTRODUCTION: AN OVERVIEW

In the pursuit of sustainable and eco-friendly enzymatic processes, the utilization of clay materials as a substrate for lipase from *Aspergillus japonicus* emerges as a promising avenue. This study delves into the innovative application of clay as a support medium, offering a greener alternative for enhancing the catalytic efficiency of *Aspergillus japonicus* lipase. Through a comprehensive exploration of this eco-friendly approach, the research aims to shed light on the potential advantages, synergies, and environmental benefits associated with incorporating clay materials in the context of lipase-based processes. As we navigate the landscape of biocatalysis, this investigation contributes to the ongoing dialogue on developing more sustainable practices in enzymatic applications.

Lipases (also triacylglycerol ester hydrolase) [EC 3.1.1.3] are enzymes that hydrolyze ester linkages of triglycerides [1,2]. Since these enzymes present large structural and functional versatility, they are of greatest importance in industrial sector to be applied in the food, pharmaceutical, detergent, textile, biodiesel production, cosmetic, among others [3,4]. Considering the high number of lipase applications, the microbial lipase market was valued at $400.6 million in 2017, with expectations to achieve a $590 million in 2023 at a Compound annual growth rate (CAGR) of 6.8% [5]. Over the last years, the growing interest in large-scale production of more natural enzymes, i.e., microbial lipase, has increased due to its outstanding advantages, *viz.*, high stability, high productivity yield, no seasonal restrictions, use of agro-waste/residues as feedstock and low-cost production as compared to other sources (animal or vegetable) [2,3,6].

In nature, a large number of microorganisms have been described as natural lipase producers, among these, filamentous fungi stand out due to their ability to biosynthesize extracellular lipases with improved catalytic characteristics in relation to stability and specificity [7,8,9,10]. The filamentous fungi from the genera *Aspergillus*, *Rhizopus*, *Penicillium*, *Mucor*, and *Fusarium* are commonly used for the commercial production of lipases [10], with special notoriety on *A. niger*, *Humicola lanuginosa*, *Mucor miehei*, *R. arrhizus*, *R. delemar*, *R. japonicus*, *R. niveus*, and *R. oryza* species [8]. The industrial needs for new microbial sources of lipases with different catalytic characteristics [11] encouraged the search of new strains, *viz*, *A. japonicus.* In this sense, following the selection of lipase-strain producer, the enzyme immobilization its necessary, as it allows easy control of reaction parameters (flow rate and substrates), reuse of immobilized catalysts and suitable chemical, mechanical and thermal stability [3,12].

A wide range of physical and chemical techniques have been used over the last few years for lipases immobilization purpose, cf., adsorption, encapsulation and entrapment, confinement, covalent binding and cross-linking [12], physical adsorption (lipases immobilized by hydrophobic interactions) [13,14,15] followed by covalent binding being the main methods used [16]. It should be noted that the choice of support for immobilization is crucial for the catalytic effectiveness of an enzyme, and in general, the support is considered ideal when it has good biocompatibility, high physical and chemical stability, presence of multiple enzyme-support binding points and low cost [17]. Inorganic clays are low-cost supports, with high adsorption capacity, environmentally friendly properties and renewable abundance [18,19]. Among those, clay minerals such as, diatomite, vermiculite, montmorillonite KSF and kaolinite are good examples of supports for enzyme immobilization [20], due to the unique physicochemical characteristics, viz, high thermal and chemical stability, mechanical strength, charge density (excellent ion exchange and adsorption capacity), relative hydrophobicity–hydrophilicity, high surface area, microbial resistance, environmental sustainability and economically viable [18,21,22,23,24,25,26,27,28,29,30].

The adsorption of enzymes to clay miners depends on a series of physical and chemical characteristics of the clays, as well as the structural characteristics of the enzymes, ionic strength of the adsorbent solution and the interaction between these factors [22,31]. Smectite clays such as montmorillonite have a large surface area and pore volume, useful characteristics in the adsorption process, the enzymes can be adsorbed on both the internal and external surface of the mineral [21,31,32]. These materials have been successful used to immobilize invertase [33,34], α-amylase [34], glucoamylase [34], lipase [22,35,36,37,38,39,40,41], inulinase [42], pectinase [25], phytase [21] and rhynchophorol [30].

Following these promising reports, the aim of this study was focused on the production by submerged culture and immobilization of *Aspergillus japonicus* (*A. japonicus*) lipase using different supports (diatomite, vermiculite, montmorillonite KSF and kaolinite) by physical adsorption procedure. As a final test to validate the industrial potential of *A. japonicus* lipase, the kinetic properties of enzymatic derivatives and its free form were evaluated and compared.

2. RESULTS AND DISCUSSION

2.1. Production of Lipase by A. japonicus

As aforementioned, microbial lipases, *cf.*, *Aspergillus*-based lipases present great potential to be used in many industrial fields, to produce detergents, biodiesel, bread, functional foods, among others [43]. There are few reports about the microbial production of lipase by *A. japonicus* in the literature [44,45,46,47,48,49,50]. However, among the works found, it is clear that the kinetic parameters (K_M and V_{max}) are directly affected by the composition of nutritional media, substrate used in the enzymatic assay, type of cultivation (solid or submerged) and several processual parameters, i.e., pH, temperature and agitation [44,45,46,47,48,49,50]. The ideal temperature to achieve high lipase production yields generally ranges from 30 to 50 °C [45,46,47,48,49,50] and pH 6.0 to 8.5 which are considered optimal for *A. japonicus* growth and therefore the successful production of lipase [45,46,47,48,49,50].

In this work, the production of extracellular lipase by *A. japonicus* using olive oil as substrate was evaluated, and the obtained enzymatic extract clearly demonstrate the ability of microorganism to hydrolyze triacylglycerols, achieving, 20.42 U ml^{-1} and 934.958 U mg^{-1} of lipase hydrolytic activity and specific activity, respectively. These results are in accordance with those (20.6 U ml^{-1}) obtained by Karanam et al. [44] in the hydrolysis of p-nitrophenyl palmitate (pNPP) by a lipase preparation using a genetically modified strain of *A. japonicus* MTCC 1975. Evaluating the lipase production by *A. japonicus* LAB01 in medium supplemented with sunflower oil (1% w w^{-1}), at pH 6.0, Souza et al. [47] achieved 28.04 U ml^{-1} on pNPP hydrolysis. Likewise, purified lipase obtained by *A. japonicus* produced in cultivation media containing malt extract, also showed specific activity of 36.83 U mg^{-1}, using *Jatropha* oil as substrate [45]. So, the filamentous fungi *A. japonicus* has the ability to metabolize different substrate to produce lipolytic enzymes with application in several biotechnological processes.

2.2. Characterization of Free A. japonicus Lipase

2.2.1. pH Effects on Enzyme Activity and Stability

The pH has great effect on the lipase activity, being essential to define this parameter for the efficient characterization of the *A. japonicus* lipase produced. The activity profiles and stability of the free lipases at pH 3.0 to 10.0 using different buffers at the same concentration (0.1 M) were measured (**Figure 1**a).

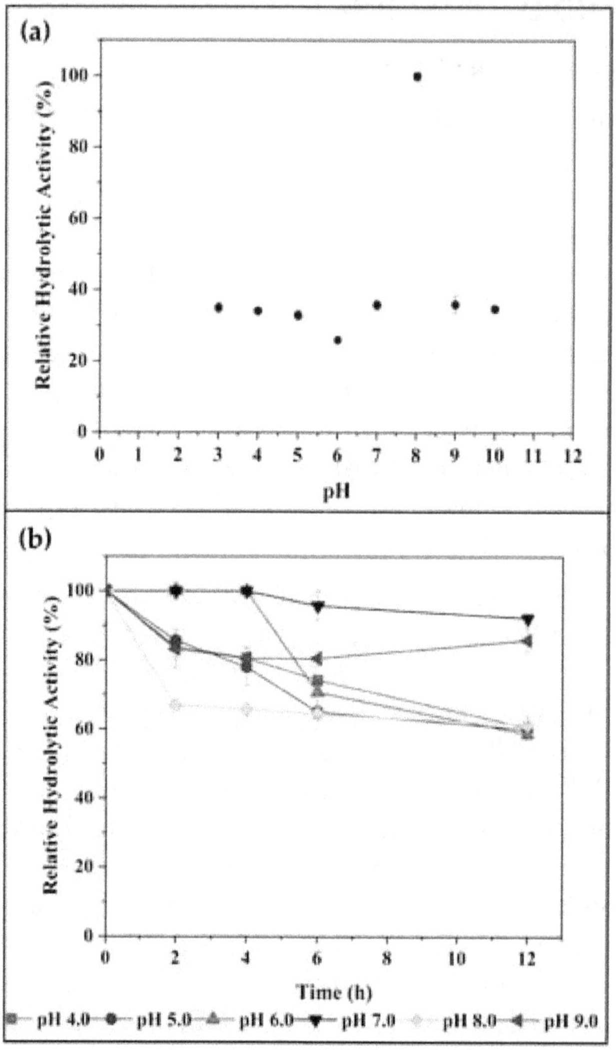

Figure 1. Characterization of lipase from *A. japonicus* in relation to optimal pH (a) and pH stability through time (until 12 h of incubation) (b). These experiments were performed at 30 °C. In the optimal pH, 100% is related to the highest hydrolytic activity achieved. In the pH stability studies, the initial activity obtained at time zero to each pH was considered 100% (different values for each pH values). The error bars represent the standard deviation of triplicates.

As depicted in Figure 1a the hydrolytic activity of the free lipase is highest in the pH 8.0. It is well-known that lipases show their activity at alkaline pH as result of a change in the ionization state of amino acid that varies with pH, and consequently affect the active site and the enzyme conformation [51,52,53]. The optimum pH found in this study for the activity of *A. japonicus* lipase are similar to results reported in literature using the same microbial specie, in which maximum lipase activities were obtained around pH 6.7 to 7.9 [45] and 8.5 [47].

The pH stability of lipases produced through submerged cultivation of *A. japonicus* was initially assessed at pH 4.0 to 9.0 (Figure 1b). In this particular case, the initial activity value (time zero) obtained for each pH (from 4.0 to 9.0) was defined as 100%. As shown in Figure 1b, at pH 7.0 lipase showed high stability value, preserving 92% of its original hydrolytic activity after 12 h of reaction. This result is in accordance with previous studies [54] in which lipase produced by *Aspergillus* sp., achieved high pH stability at 7.0 after 48 h of incubation. In this work, interesting results were also observed at pH 9.0 (Figure 1b), in which a slight decrease of ~20% after 4 h was detected maintaining the original activity (~80%) until the end of 12 h of incubation. The present results demonstrate that the microbial enzyme under study is an alkaline lipase, that maintain the stability under basic conditions and can be used in several industries for the production of detergents, for example [47].

2.2.2. Temperature Effects on Enzyme Activity and Stability

The effect of temperature (at pH 8.0, optimal pH conditions) on the lipase activity was also evaluated, and the enzyme demonstrate activity in a wide range of temperatures (20–60°C), with maximum activity value reached at 30°C (Figure 2a). The results were similar to those of *A. niger* strain MTCC 872 [55] and *A. niger* [56]. According to the literature, several microbial lipases produced by *Aspergillus* species exhibit optimal temperature close to 37°C [51,54,57,58].

The thermostability of lipase was studied by incubating it at different temperatures: 20°C, 30°C, 40°C and 50°C, at pH 8.0 for 12 h. Note that, the 100% is referred to the initial activity value obtained at time zero for each temperature evaluated. As depicted in Figure 2b, at 30°C, the enzyme was stable for the 12 h of study and retained 100% of its activity. However, when the enzyme was incubated at 20°C and 40°C the enzyme maintained 95.45% and 93.1% of its activity after 4 h of incubation. After this period, at 20°C a pronounced loss of enzyme activity was observed, reaching around of 50% of activity after 12 h of incubation. Interestingly, at 20°C, the activity loss initiated from 4 h of incubation, and after 12 h, it was achieved an activity loss around 75%. For 50°C, lipase lost about 40% of its relative activity after 2 h and at the end of 12 h of incubation time, a decrease in the hydrolytic activity (75–60%) was perceived, indicating a possible lipase denaturation caused by prolonged exposure to temperatures (20, 40 and 50°C) [54,59]. According to these results, 30°C in not only the temperature of optimal enzyme activity but also the one that keeps the lipase stability. From the literature, it is known that the formation of the enzyme-substrate complex is impacted by the temperature since it has influence in the number of collisions between the enzyme and substrate [54,59]. However, high temperatures, 40 and 50°C in the case of *A. japonicus* lipase, may promote the protein denaturation which explains the low activity measured at these conditions after 12 h of incubation.

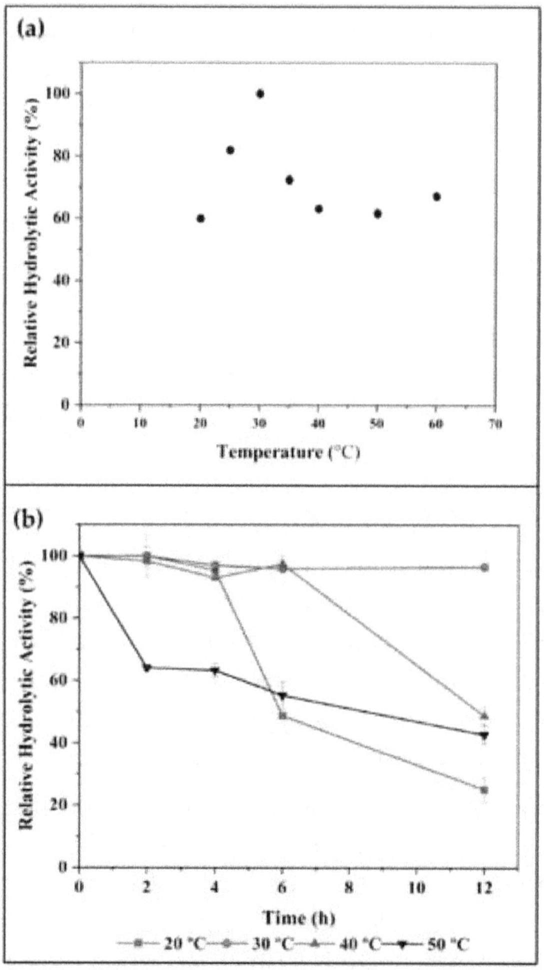

Figure 2. Characterization of *A. japonicus* lipase in relation to optimal temperature (a) and temperature stability through time (until 12 h of incubation) (b). These experiments were performed at pH 8.0. In the optimal temperature, 100% is related to the highest hydrolytic activity achieved. In the temperature stability studies, the initial activity obtained at time zero to each temperature was considered 100% (different values for each temperature values). The error bars represent the standard deviation of triplicates.

It was also performed the polyacrylamide gel electrophoretic of culture media containing the enzyme. According to Mala and Takeuchi [60], the molecular weight of microbial lipase is from 20 to 60 kDa. From the electrophoresis data, it can be seen that the culture media containing the *A. japonicus* lipase has a single band with a relative molecular weight of 25 kDa on SDS-PAGE gel (**Figure 3**), demonstrating that the obtained *A. japonicus* lipase has a single subunit. *A. Japonicus* lipase with molecular weight of 25 kDa was reported by Souza et. al [47] while the same specie produced an enzyme with molecular weight of 43 kDa using Jatropha oil as substrate [45].

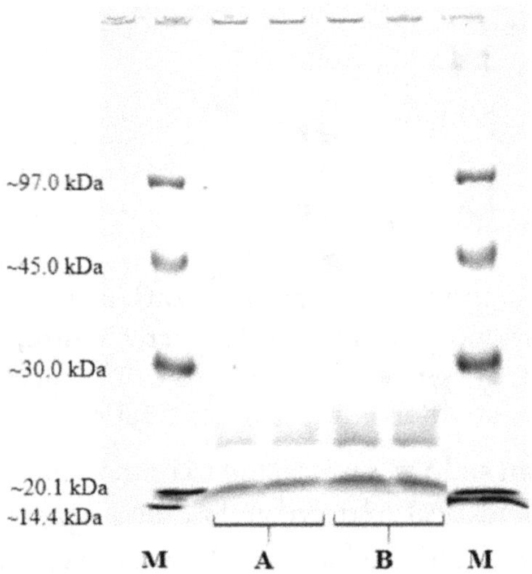

Figure 3. SDS-Page of *A. japonicus* lipase produced by submerged culture. The columns represent: M: molecular weight markers (14.4–97 kDa). A: culture media containing *A. japonicus* lipase at 155.2 mg mL^{-1}; B: culture media containing *A. japonicus* lipase at 257.2 mg mL^{-1}.

2.3. The immobilization of lipase from Aspergillus japonicus using various support materials

The immobilization of lipase was performed by physical adsorption using four different supports, namely, diatomite, vermiculite, montmorillonite KSF (MKSF) and kaolinite, and in all cases 1 g of support were suspended in 20 mg mL^{-1} of protein. The immobilization yield of each derivative was calculated by comparing the protein content of enzymatic solution in the initial and the final immobilization. The results in **Figure 4** exhibit the hydrolytic activity expressed in U g^{-1} and immobilization yield (%) of *A. japonicus* lipase immobilized in the above-mentioned supports.

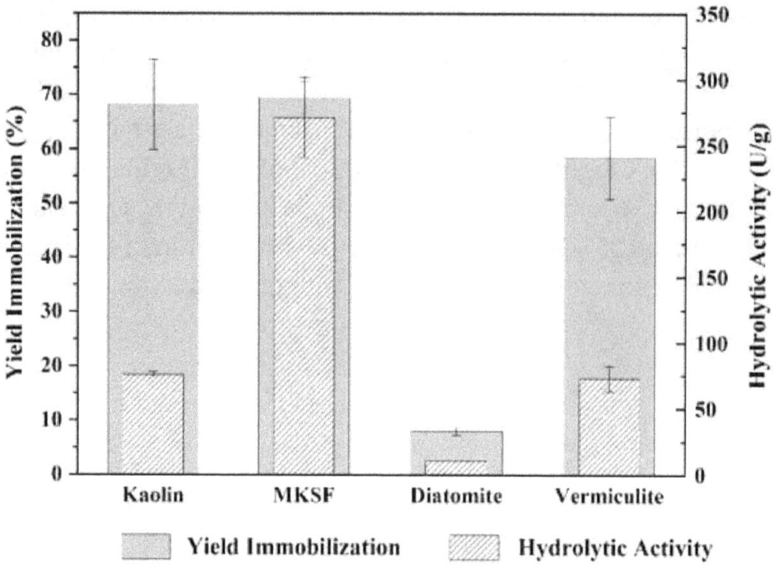

Figure 4. Enzymatic activity and immobilization yield of *A. japonicus* lipase immobilized on the supports: diatomite, vermiculite, montmorillonite KSF (MKSF) and kaolinite by physical adsorption. The error bars represent the standard deviation of triplicates.

As shown in **Figure 4** all supports were able to immobilize lipase from *A. japonicus*, the immobilization yields (%) and enzymatic activity (U g^{-1}) follow the trend: MKSF (69.47 and 270.7) > kaolinite (68.14 and 75.51) > vermiculite (58.45 and 72.89) > diatomite (7.97 and 10.5). So, the MKSF support presented the highest yield. The hydrolytic activity obtained for lipase immobilization using MKSF (270.7 U g^{-1}) were three-fold more than the obtained using kaolinite (75.51 U g^{-1}) with immobilizations yield around ~70%. The adsorption of enzymes in a different types of clay minerals, involves a wide variety of physical and/or chemical interactions, which mainly depend on the nature and type of both compounds. The surfaces of clay minerals may differ mainly due to the presence of the enzyme molecule [61], and even with high immobilization yields, i.e., kaolinite (68.14%) and vermiculite (58.45%) (**Figure 4**), low biocatalyst activity can occur. This behavior usually arises, when the adsorption occurs near to the enzyme active site, hindering the access of substrate molecules, or even due to the diffusion resistance [33].

Similar studies regarding lipases and MKSF, were also reported by Scherer et al. [38], in the immobilization of porcine pancreatic lipase by physical adsorption using MKSF. In this case the immobilization yield reached 38.2% and high esterification activity were also observed (1400 U g^{-1}) demonstrating the potential of MKSF effective support for use in the immobilization of lipases.

Proteins are usually adsorbed by several interactions such as, cation exchange, electrostatic interactions, van der Waals forces, hydrogen bonds, or by the hydrophobic/hydrophilic moiety present on the clay surface [31,62,63]. Often, in protein immobilization using clay supports, protein molecules spontaneously interact with hydrophobic regions [31] and in particular case of the increased activity of lipases in the presence of MKSF (hydrophobic interphases) is known as "interfacial activation" [35,64,65,66].

In the study developed by Sani et al. [67], the lipase from *Rhizopus oryzae* was immobilized by physical adsorption on a polypropylene support with additive CAVAMAX® W6 achieving an increase of 100% in relative activity after immobilization. According to the authors, the interactions between lipase and support were mostly hydrophobic, the hydrophobic effect resulted in open conformation and interfacial activation of lipase during adsorption, leading to an improvement in the enzymatic activity after immobilization procedure.

Most lipolytic enzymes present on their surface, close to the active site, a helical loop, referred in the literature as a "*lid*" [68,69]. When lipases come into contact with hydrophobic surfaces, the reaction is similar to the way they recognize natural substrates (generally lipids) and the "*lid*" opens, allowing exposure of the enzyme's active site, a phenomenon called interfacial activation [35,64,65,66]. In a hydrophobic or non-aqueous environment, the hydrophobic layer triggers the opening of the "*lid*", allowing the entry of the substrate into the active site, whereas in aqueous environments the lipase remains inactive [70,71]. Thus, the enzymatic activity of the lipases that contain the "*lid*" is controlled by the cap domain [71,72]. Thus, immobilization process on hydrophobic supports allows the immobilization of the open conformation of the lipase, improving its enzymatic activity, causing the enzyme to hyperactivate [66,73,74,75]. It is important to highlight that the lipase contains one or more caps in the helix form and that not all lipases undergo interfacial activation [71].

As a result of the interesting immobilization yield and hydrolytic activity obtained with lipase from *A. japonicus* immobilized with MKSF, this derivative was selected to evaluate the effect of pH and temperature on activity and stability and the determination of kinetic parameters in the next section.

2.4. Characterization of A. japonicus Lipase Immobilized on MKSF

In the previous section, it was shown that the great hydrolytic activity and immobilizations yields of lipase were obtained using MKSF as a support. Therefore, to obtain a better understanding of the impact of the support (MKSF) it is essential to understand the changes occurred in the kinetics parameters, pH and temperature stability during the lipase immobilization process, as discussed in the next subsections.

2.4.1. Effect of pH and Temperature on Immobilized Lipase Activity

The effect of pH in the immobilized lipase activity was studied in the range of pH from 3.0 to 10.0 using different buffers at 0.1 M (**Figure 5**). The immobilized lipase showed different behaviors of pH dependence than the free lipase, achieving the optimal pH at 7.0 and 8.0, with high relative activity of 95.5% and 100%, respectively (**Figure 5**a). As also depicted in **Figure 5**a, the immobilized lipase showed much higher activity than the free enzyme (**Figure 1**a), and improvements in relation to the hydrolytic activity in the range of pH 4.0 to 10.0 were observed. A similar observation was reported by Tacin et al. [76] using *Aspergillus* sp. lipase immobilized by octyl-sepharose. In this particular case, the authors suggested that the changes in the optimal pH of immobilized lipase is directly related to the conformation change of the enzyme molecules after immobilization, making the catalytic site more easily accessible to the ions (H^+ or OH^-).

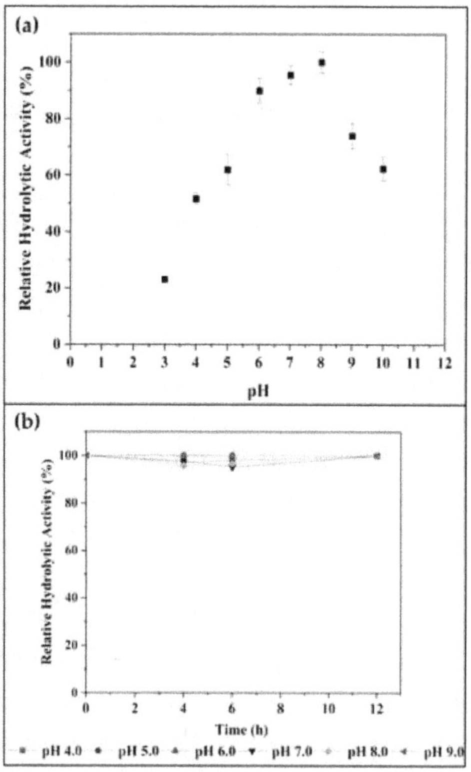

Figure 5. Characterization of *A. japonicus* lipase immobilized in MKSF support in relation to optimal pH (a) and pH stability through time (until 12 h of incubation) (b). These experiments were performed at 30 °C. In the optimal pH, 100% is related to the highest hydrolytic activity achieved. In the pH stability studies, the initial activity obtained at time zero to each pH was considered 100% (different values for each pH values). The error bars represent the standard deviation of triplicates.

In this set of trials, the initial activity (time zero) defined as 100%, was considered different to each pH as depicted in **Figure 5**b. In all evaluated pH (from 4.0 to 9.0) the immobilized lipase showed great pH stability, maintaining the catalytic activity (100%) for 12 h of incubation reaction, achieving an improvement (40%) in pH stability as compared with the free lipase (**Figure 1**b). A similar extended pH profile for lipase immobilized on montmorillonite K10 was also reported by Sanjay and Sugunan [33]. The shift in the pH optimum for lipase immobilized in usually, and as reported by Dong et. al [35], the maximum amount of H^+ ions in the support surfaces leads to the changes in pH value on the catalysis microenvironment.

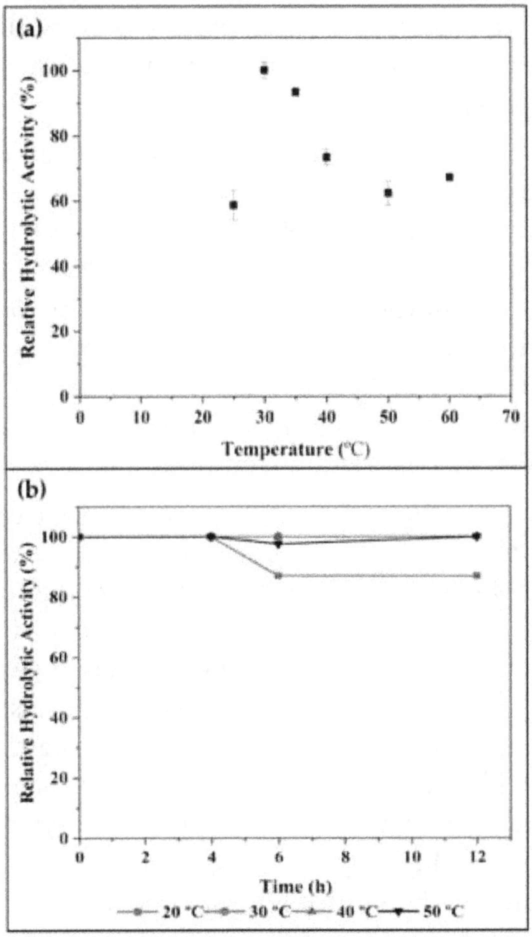

Figure 6. Characterization of *A. japonicus* lipase immobilized in MKSF support in relation to optimal temperature (a) and temperature stability through time (until 12 h of incubation) (b). These experiments were performed at pH 8 In the optimal pH, 100% is related to the highest hydrolytic activity achieved. In the pH stability studies, the initial activity obtained at time zero to each pH was considered 100% (different values for each temperature values). The error bars represent the standard deviation of triplicates.

In a study performed by Sanjay and Sugunan [33] to immobilize invertase enzyme using montmorillonite K10, this process changed the optimal pH from 5.0 to 6.0. According to the author, negatively charged supports, as montmorillonite K10, changes the optimal pH due to charge interaction changing the enzyme ionization.

Additionally, the effect of temperature on the immobilized *A. japonicus* lipase activities was also investigated in the range of 20 a 60°C (**Figure 6**a). The maximal activity of the immobilized lipase was as high as 30°C, the same value obtained by the free lipase (**Figure 2**a). Similar results were found by Zou et. al [41] and Sani et. al. [66] in which the enzymes kept their optimal temperature at 50°C. However, the immobilized enzymes were more stable in a wide range of temperature as compared with free-enzyme.

Figure 6b, demonstrates the effect of reaction temperature on the activities of the immobilized lipases within the range of 20, 40 and 50°C at optimal pH (8.0), and the initial activity (at time zero) was defined 100% for each temperature evaluated. As observed, the optimal reaction temperature (30, 40 and 50°C) of immobilized lipase achieved the 100% of the original catalytic activity for 12 h of incubation. On the contrary, with the reaction temperature constantly decreasing to 20°C, the relative activity of immobilized lipase only maintained 87.1% of its activity at the end of 12 h. In this set of experiments, it was observed that immobilized *A. japonicus* lipase present much wider temperature endurance range.

The temperature has two-main impacts on the *A. japonicus* lipase catalytic activity, namely, (i) the thermal energy of substrate (olive oil) is affected by the increasing of temperature that led to the collision of substrate molecules with enzymes and (ii) the increase of temperature can also lead the denaturation of lipase caused by the cleavage of non-covalent bonds [54,59]. In this way, the immobilization process can promote a protection against exposure to high temperature [37,77] and prevent the unfolding of the tertiary structure of the enzyme [78]. As cited before, high temperatures can promote the disruption of disulphide bonds and salt bridges (ionic bonds), interactions that maintain the spatial conformation of the enzymes [79,80]. Thus, the immobilized enzyme maintained these bonds in the studied temperature range.

Montmorillonite are 2:1 dioctahedral, and are the most commonly used clay mineral as outstanding supports with capability to adsorb enzymes on the external surface (hydrophobic interactions) and intercalated in interlayer space (cation exchange) [61]. The enzymes (e.g., lipases) immobilization by cation exchange, in the support (montmorillonite) interlayer space, provides to the enzyme, more stability, and protection against thermal denaturation [61,81,82,83]. Thus, the great results concerning the stability (in pH and temperature), of derivatives immobilized in MKSF may be related to the partial immobilization of *A. japonicus* lipase by intercalation.

Similar trends were observed by Benamia et. al [37] in a work focused on the identification of natural supports for immobilization of *Candida rugosa* lipase (CRL). In this work, the immobilized CRL on Maghnite-H exhibited good thermostability over a wide temperature range (30–90°C) compared to the free one. Tu et. al [40] studied the *Yarrawia lipolytica* lipase immobilization in chitosan modified with clay (Na-betonite) and the lipase immobilized were more thermostable than the enzyme immobilized in chitosan no-modified and the free one and showed more reusability (up to 7 reuses). According to the authors, the incorporation of clay provided a support with greater adsorption capacity and greater rigidity (mechanical resistance), improving its performance. In the studies performed by Morais et al. [84] in the immobilization of Endocellulase and β-glucosidase in magnetic iron oxide nanoparticles demonstrated that the enzymatic immobilization improved the thermal stability of the biocatalysts, achieving residual activities of 80% in 72 h of incubation at 60 °C.

2.4.2. Kinetic Parameters of Free and Immobilized Lipase

Apparent kinetic parameters are essential for the selection of enzyme regarding different industrial processes and applications. Kinetic parameters of the free and immobilized lipases were determined in the optimal temperature and pH conditions (30 °C, pH 8.0, 200 rpm at 5 min), varying only the olive oil (substrate) from 186 to 2604 mM in the reactional media. Through using software Origin 8.0, the Michaelis-Menten constant (K_M) and the initial maximum reaction velocity (V_{max}) of the lipase were calculated. The results for free and immobilized lipases are given in the **Figure 7**a and b, respectively.

As observed in **Figure 7**a the experimental data set fit with the calculated theoretical data, with an $R^2 = 0.9986$, demonstrating that the produced lipase from *A. japonicus* follows the Michaelis-Menten kinetics. The values of K_M and V_{max} obtained was 79.35 mM and 30.71 µmol min^{-1}, respectively. In study reported by Bharti et al. [45], lipase produced by other strain of *A. japonicus* also followed the Michaelis-Menten kinetics.

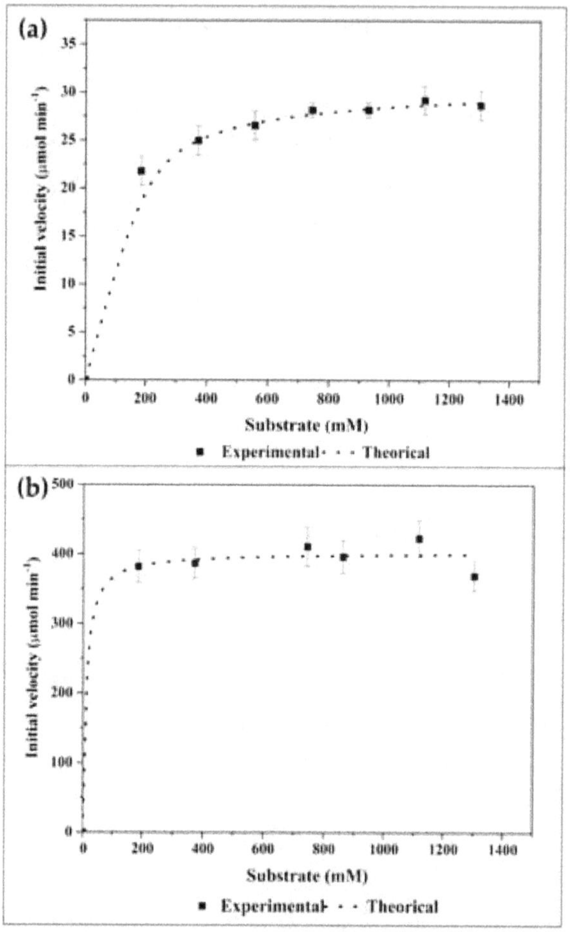

Figure 7. Hyperbolic Michaelis-Menten regression of *A. japonicus* lipase (a) free and (b) immobilized in MKSF support, at different concentration of olive oil substrate at 30 °C and pH 8.0. The error bars represent the standard deviation of triplicates.

As observed in **Figure 7**b, the predicted and experimentally K_M value illustrated a correlation of $R^2 = 0.9849$. The V_{max} and K_M values of immobilized lipase were 402.56 μmol min^{-1} and 9.87 mM, respectively. The K_M value obtained with the immobilized lipase was 4 times lower than those obtained with free lipase (79.35 mM) (**Figure 7**a). Note that, the higher K_M value suggested the lower affinity between the enzyme and substrate [**85**], and in this particular work, the low value of the K_M in the *A. japonicus* lipase immobilized MKSF suggested the higher affinity to the substrate in comparison to the free form. The V_{max} value of immobilized lipase (402.56 μmol min^{-1}) was 13 times greater than those obtained with the free one (30.71 μmol min^{-1}) (**Figure 7**a) proving that the enzyme activity improved. Moreover, these results suggest that the native lipase conformation was not altered by the immobilization process [**34**].

Actually, immobilized enzyme in different supports is applied in continuous process in food industry to produce corn syrup, cocoa butter analogues and galacto-oligosaccharides, synthesis of chiral molecules among others [86], in pharmaceutical industry to synthetize chiral molecules [86] and in the chemical industry to produce complex molecules such as chiral amines and herbicides [86]. So, there is a wide range of process that immobilized enzymes can be applied and the results reported in the present work demonstrate that the low cost MKSF is a promising support for *A. japonicus* lipase, and can encourage the scientific community to explore other clays-based substrates for enzymes immobilization, providing a natural and economical support to the industrial application of immobilized enzyme.

3. MATERIALS AND METHODS

3.1. Materials

Peptone (bacteriological) and Potato Dextrose Agar (PDA) were acquired from Acumedia®. Olive oil from Carbonell® (Córdoba, Spain) was obtained in a local market. Kaolinite was purchased from Sigma-Aldrich (St. Louis, MO, USA) and diatomite, vermiculite, montmorillonite KSF, were acquired from *Laboratório de Peneiras Moleculares* (LABPEMOL) from Federal University of Rio Grande do Norte (UFRN). All the other reagents were of analytical grade and used as received.

3.2. Microorganism

Aspergillus japonicus DPUA1727 was kindly provided by the Culture Collection of Federal University of Amazonas, DPUA, AM, Brazil. The *A. japonicus* preserved in distilled water was reactivated in PDA (39.0 g L^{-1} agar potato dextrose composed of dehydrated Potato Infusion and Dextrose) supplemented with yeast extract (0.5% w v^{-1}) and maintained at 30°C for 7 days. Afterward, the cultures were kept in the refrigerator at 4 °C and defined as a stock culture for the whole work.

3.3. Production of *Aspergillus japonicus* Lipase by Submerged Culture

Initially, lipase was produced using a methodology described by Tacin et al. [54]. Briefly, the inoculum was prepared in PDA plates, and the cultures were maintained at the same reactivation conditions. Following, 5 mycelial agar discs (8 mm diameter) of microorganism were inoculated in Erlenmeyer flasks (250 mL) containing 50 mL of culture medium composed of (g L^{-1}, in deionized water): peptone (40); olive oil (1.6); $MgSO_4 \cdot 7H_2O$ (1.2); KH_2PO_4 (2.0) and NH_4NO_3 (2.0). The experiments were performed in triplicate and the initial pH of the culture medium was adjusted to 7.0. Cells were grown for 72 h at 30 °C and 150 rpm in an orbital shaker New Ethics, model 521/2DE (Piracicaba, SP, Brazil). After the cultivation, *A. japonicus* biomass was separated from the fermented supernatant containing the enzyme by conventional filtration using a filter paper Whatman 80 g m^{-2} (Maidstone, Kent, England). The cell-free filtrate containing lipases, was used to measure the hydrolytic enzyme activity and total proteins, according methodologies described in **Section 3.4.1** and **Section 3.4.2**, respectively.

3.4. Analytical Methods

3.4.1. Determination of Hydrolytic Enzyme Activity

The hydrolytic enzyme activity was carried out using an emulsion containing olive oil 50% w w^{-1} and gum Arabic 7% w w^{-1} [87]. The unit of hydrolytic enzyme activity was calculated as the amount of enzyme required to hydrolyze 1 μmol of fatty acid per minute of reaction. The activity was expressed in μmol mL^{-1} min^{-1} (U mL^{-1}) for lipase in free form and in μmol g^{-1} min^{-1} (U g^{-1}) for immobilized derivatives.

3.4.2. Determination of Protein Content

The concentration of total proteins was determined by the method described by Bradford [88]. Briefly, 100 μL sample was mixed with 5 mL Bradford reagent Sigma-Aldrich (St. Louis, MO, USA). After 5 min, the absorbance was measured in the wavelength of 595 nm in a Spectrophotometer model Thermo Scientific™ GENESYS 10S UV-Vis (Waltham, MA, USA). Protein content was estimated by means of a calibration curve obtained using concentrations of 0.05 to 1 mg mL^{-1} of Bovine Serum Albumin from Sigma-Aldrich (St. Louis, MO, USA).

3.4.3. SDS-PAGE Analysis

The electrophoresis analyzes of *A. japonicus* extracts performed based on the methodology described by Laemmli [89]. The gel was prepared with 12% polyacrylamide by applying 30 μL of the sample ate the concentrations 155.2 mg mL^{-1} and 257.2 mg mL^{-1} and 14.4–97.0 kDa molecular weight marker BluEYE Sigma-Aldrich (St. Louis, MO, USA).

3.4.4. Lipase Immobilization Using Different Supports

The immobilizations of microbial lipase on different supports, namely diatomite, vermiculite, montmorillonite KSF and kaolinite, were performed based on the methodology described by Vescovi et al. [14]. For that, 1 g of each support was suspended in 25 mL of enzyme solution (20 mg mL^{-1} of protein prepared in 25 mM sodium phosphate buffer at pH 7.0). The suspension was kept under mild stirring in a roller Shaker Quimis (Diadema, SP, Brazil) for 24 h, at 25 °C. After that, the solution was centrifuged at 8161× *g* in centrifuge Eppendorf model 5415 D (Hamburg, Germany) for 5 min, to separate derivates, followed in the supernatant was determined the hydrolytic activity and the protein content, according methodologies described at **Section 3.4.1** and **Section 3.4.2**, respectively. The immobilization yield (n, %) was calculated according to the Equation (1):

$$N = 100 - ((P \times 100)/Po) \quad (1)$$

where: P = Protein content of immobilized derivative supernatant; Po = Protein content absorbance in time 0.

3.5. Characterization of Free and Immobilized Microbial Lipase

3.5.1. Determination of Optimum pH and Temperature

In order to evaluate the optimal pH of the enzyme, the hydrolytic lipase assay was carried out at different pH values (from 3.0 to 10.0) using different buffers at 0.1 M: Mcllvaine buffer (pH 3.0 to 6.0), phosphate buffer (pH 7.0 and 8.0) and carbonate bicarbonate Buffer (pH 9.0 to 10.0). These experiments were performed at 30 °C. The optimal temperature was determined in the optimum pH 8.0 and it was calculated using a hydrolytic lipase activity incubated in the temperature ranged from 20 to 60 °C, in a thermostatic bath New Ethics model 521/2DE (New Ethics, Piracicaba, Brazil).

3.5.2. Determination of Kinetic Parameters

The influence of olive oil substrate concentration on the initial hydrolysis activity of *A. japonicus* lipase (free or immobilized), was evaluated under the optimum pH and temperature conditions of the reaction (30 °C, pH8, 200 rpm at 5 min). Since olive oil concentration lower than 186 mmol did not generate a stable emulsion, the kinetic studies were performed varying the olive oil substrate from 186 to 2604 mmol, i.e, 5 to 70% w w^{-1}. For this relation (mmol in % w w^{-1}), the average molar mass of olive oil of 1344 g mol^{-1} was considered, based on its fatty acid composition. The kinetic constant: Michaelis-Menten constant (K_M) and specific enzyme activity (V_{max}) were calculated from the data obtained experimentally with the equation of Michaelis-Menten using software Origin 8.0 OriginLab (serial GA3S5-6089-7173339, USA).

3.5.3. Stability of the Free and Immobilized Microbial Lipase

For pH stability, the enzymatic extract (1 mL) or immobilized derivative (0.05 g) were incubated in a thermostatic bath at 30°C, pH from 4.0 to 9.0 for 12 h. In the case of temperature stability, the enzymatic extract (1 mL) or immobilized derivative (0.05 g) were also incubated in a thermostatic bath at 20°C, 30°C, 40°C and 50°C, in optimal pH for 12 h. In both cases, the residual hydrolytic activity of lipase was determined after 0, 2, 4, 6 and 12 h according to methodology described at **Section 3.4.1**.

For the stability assays, the relative activity (RA, %) of lipase from *A. japonicus* lipase immobilized by physical adsorption were calculated according to the Equation (2):

$$RA = (L_A \times 100)/L_{A0} \qquad (2)$$

where: L_A is the free/immobilized lipase activity after the incubation time and L_{A0} initial enzymatic activity.

4. CONCLUSIONS

In this work, the immobilization process of lipase produced by *A. japonicus* using clays as support was studied. The lipase immobilized in MSKF showed the more promising results comparing to the other supports evaluated achieving immobilization yield and hydrolytic activity of 69.47% and 270.7 U g^{-1}, respectively. Comparing the catalytic performance of enzyme immobilized and free, the immobilization process improved the lipase stability regarding temperature and pH and promoted an increment in the enzyme-substrate affinity since the value of V_{MAX} increased compared to the free one while the value of K_M decreased. These findings shows that the lipase immobilization on clay support creates a good synergetic able to improve the lipase catalytic performance being an interest purpose to apply in lipases with industrial interests.

5. CONCISE SUMMARY

Lipase is an important group of biocatalysts, which combines versatility and specificity, and can catalyze several reactions when applied in a high amount of industrial processes. In this study, the lipase produced by *Aspergillus japonicus* under submerged cultivation, was immobilized by physical adsorption, using clay supports, namely, diatomite, vermiculite, montmorillonite KSF (MKSF) and kaolinite. Besides, the immobilized and free enzyme was characterized, regarding pH, temperature and kinetic parameters. The most promising clay support was MKSF that presented 69.47% immobilization yield and hydrolytic activity higher than the other conditions studied (270.7 U g^{-1}). The derivative produced with MKSF showed high stability at pH and temperature, keeping 100% of its activity throughout 12 h of incubation in the pH ranges between 4.0 and 9.0 and at a temperature from 30 to 50°C. In addition, the immobilized lipase on MKSF support showed an improvement in the catalytic performance. The study shows the potential of using clays as support to immobilized lipolytic enzymes by adsorption method, which is a simple and cost-effective process.

REFERENCES

1. Seddigi, Z.S.; Malik, M.S.; Ahmed, S.A.; Babalghith, A.O.; Kamal, A. Lipases in asymmetric transformations: Recent advances in classical kinetic resolution and lipase–metal combinations for dynamic processes. *Coord. Chem. Rev.* **2017**, *348*, 54–70. [Google Scholar] [CrossRef]

2. Sankar, S.; Ponnuraj, K. Less explored plant lipases: Modeling and molecular dynamics simulations of plant lipases in different solvents and temperatures to understand structure-function relationship. *Int. J. Biol. Macromol.* **2020**, *164*, 3546–3558. [Google Scholar] [CrossRef]

3. Chandra, P.; Enespa; Singh, R.; Arora, P.K. Microbial lipases and their industrial applications: A comprehensive review. *Microb. Cell Fact.* **2020**, *19*, 169. [Google Scholar] [CrossRef]

4. Rafiee, F.; Rezaee, M. Different strategies for the lipase immobilization on the chitosan based supports and their applications. *Int. J. Biol. Macromol.* **2021**, *179*, 170–195. [Google Scholar] [CrossRef]

5. Microbial Lipase Market—Growth, Trends and Forecasts (2018–2023). Available online: **https://www.marketsandmarkets.com/Market-Reports/microbial-lipase-market-248464055.html** (accessed on 9 August 2021).

6. Geoffry, K.; Achur, R.N. Screening and production of lipase from fungal organisms. *Biocatal. Agric. Biotechnol.* **2018**, *14*, 241–253. [Google Scholar] [CrossRef]

7. Pandey, N.; Dhakar, K.; Jain, R.; Pandey, A. Temperature dependent lipase production from cold and pH tolerant species of *Penicillium*. *Mycosphere* **2016**, *7*, 1533–1545. [Google Scholar] [CrossRef]

8. Meghwanshi, G.K.; Vashishtha, A. Biotechnology of fungal lipases. In *Fungi and Their Role in Sustainable Development: Current Perspectives*; Gehlot, P., Singh, J., Eds.; Springer: Singapore, 2018; pp. 383–411. ISBN 9789811303937. [Google Scholar]

9. Basheer, S.M.; Chellappan, S.; Beena, P.; Sukumaran, R.K.; Elyas, K.; Chandrasekaran, M. Lipase from marine *Aspergillus awamori* BTMFW032: Production, partial purification and application in oil effluent treatment. *New Biotechnol.* **2011**, *28*, 627–638. [Google Scholar] [CrossRef]

10. Melani, N.; Tambourgi, E.B.; Silveira, E. Lipases: From Production to Applications. *Sep. Purif. Rev.* **2020**, *49*, 143–158. [Google Scholar] [CrossRef]

11. Bharathi, D.; Rajalakshmi, G. Microbial lipases: An overview of screening, production and purification. *Biocatal. Agric. Biotechnol.* **2019**, *22*, 101368. [Google Scholar] [CrossRef]

12. Contesini, F.J.; Calzado, F.; Madeira, J.V.; Rubio, M.V.; Zubieta, M.P.; de Melo, R.R.; Gonçalves, T.A. *Aspergillus* Lipases: Biotechnological and Industrial Application. In *Reference Series in Phytochemistry*; Springer: Cham, Switzerland, 2017; pp. 639–666. [**Google Scholar**]

13. Remonatto, D.; de Oliveira, J.V.; Guisan, J.M.; de Oliveira, D.; Ninow, J.; Fernandez-Lorente, G. Production of FAME and FAEE via Alcoholysis of Sunflower Oil by Eversa Lipases Immobilized on Hydrophobic Supports. *Appl. Biochem. Biotechnol.* **2018**, *185*, 705–716. [**Google Scholar**] [**CrossRef**]

14. Vescovi, V.; Kopp, W.; Guisán, J.M.; Giordano, R.L.; Mendes, A.A.; Tardioli, P.W. Improved catalytic properties of *Candida antarctica* lipase B multi-attached on tailor-made hydrophobic silica containing octyl and multifunctional amino-glutaraldehyde spacer arms. *Process. Biochem.* **2016**, *51*, 2055–2066. [**Google Scholar**] [**CrossRef**]

15. Virgen-Ortíz, J.J.; Pascacio, V.G.T.; Hirata, D.B.; Torrestiana-Sanchez, B.; Rosales-Quintero, A.; Fernandez-Lafuente, R. Relevance of substrates and products on the desorption of lipases physically adsorbed on hydrophobic supports. *Enzym. Microb. Technol.* **2017**, *96*, 30–35. [**Google Scholar**] [**CrossRef**] [**PubMed**]

16. Guisan, J.M.; López-Gallego, F.; Bolivar, J.M.; Rocha-Martín, J.; Fernandez-Lorente, G. The Science of Enzyme Immobilization. In *Methods in Molecular Biology*; Guisan, J., Bolivar, J., López-Gallego, F., Rocha-Martín, J., Eds.; Publisher: Humana, NY, USA, 2020; Volume 2100, pp. 1–26. [**Google Scholar**] [**CrossRef**]

17. Liu, J.; Ma, R.-T.; Shi, Y.-P. "Recent advances on support materials for lipase immobilization and applicability as biocatalysts in inhibitors screening methods"—A review. *Anal. Chim. Acta* **2020**, *1101*, 9–22. [**Google Scholar**] [**CrossRef**]

18. Etcheverry, M.; Cappa, V.; Trelles, J.; Zanini, G. Montmorillonite-alginate beads: Natural mineral and biopolymers based sorbent of paraquat herbicides. *J. Environ. Chem. Eng.* **2017**, *5*, 5868–5875. [**Google Scholar**] [**CrossRef**]

19. Ashkan, Z.; Hemmati, R.; Homaei, A.; Dinari, A.; Jamlidoost, M.; Tashakor, A. Immobilization of enzymes on nanoinorganic support materials: An update. *Int. J. Biol. Macromol.* **2021**, *168*, 708–721. [**Google Scholar**] [**CrossRef**] [**PubMed**]

20. Cecilia, J.A.; Jiménez-Gómez, C.P. Catalytic Applications of Clay Minerals and Hydrotalcites. *Catalysts* **2021**, *11*, 68. [**Google Scholar**] [**CrossRef**]

21. Naik, S.; Scholin, J.; Goss, B. Stabilization of phytase enzyme on montmorillonite clay. *J. Porous Mater.* **2016**, *23*, 401–406. [**Google Scholar**] [**CrossRef**]

22. Viana, A.C.; Ramos, I.G.; Mascarenhas, A.J.S.; dos Santos, E.L.; Sant'Ana, A.E.G.; Goulart, H.F.; Druzian, J.I. Release of aggregation pheromone rhynchophorol from clay minerals montmorillonite and kaolinite. *J. Therm. Anal. Calorim.* **2021**, *1*, 1–13. [Google Scholar] [CrossRef]

23. Tanasković, S.J.; Jokić, B.; Grbavčić, S.; Drvenica, I.; Prlainović, N.; Luković, N.; Knežević-Jugović, Z. Immobilization of *Candida antarctica* lipase B on kaolin and its application in synthesis of lipophilic antioxidants. *Appl. Clay Sci.* **2017**, *135*, 103–111. [Google Scholar] [CrossRef]

24. Pereira, F.A.; Sousa, K.S.; Cavalcanti, G.R.; França, D.B.; Queiroga, L.N.; dos Santos, I.M.G.; Fonseca, M.; Jaber, M. Green biosorbents based on chitosan-montmorillonite beads for anionic dye removal. *J. Environ. Chem. Eng.* **2017**, *5*, 3309–3318. [Google Scholar] [CrossRef][Green Version]

25. Długosz, O.; Banach, M. Kinetic, isotherm and thermodynamic investigations of the adsorption of Ag+ and Cu2+ on vermiculite. *J. Mol. Liq.* **2018**, *258*, 295–309. [Google Scholar] [CrossRef]

26. Mohammadi, M.; Heshmati, M.K.; Sarabandi, K.; Fathi, M.; Lim, L.-T.; Hamishehkar, H. Activated alginate-montmorillonite beads as an efficient carrier for pectinase immobilization. *Int. J. Biol. Macromol.* **2019**, *137*, 253–260. [Google Scholar] [CrossRef] [PubMed]

27. Zhao, Y.; Tian, G.; Duan, X.; Liang, X.; Meng, J.; Liang, J. Environmental Applications of Diatomite Minerals in Removing Heavy Metals from Water. *Ind. Eng. Chem. Res.* **2019**, *58*, 11638–11652. [Google Scholar] [CrossRef]

28. Brião, G.D.V.; da Silva, M.G.C.; Vieira, M.G.A. Efficient and Selective Adsorption of Neodymium on Expanded Vermiculite. *Ind. Eng. Chem. Res.* **2021**, *60*, 4962–4974. [Google Scholar] [CrossRef]

29. Song, X.; Li, C.; Chai, Z.; Zhu, Y.; Yang, Y.; Chen, M.; Ma, R.; Liang, X.; Wu, J. Application of diatomite for gallic acid removal from molasses wastewater. *Sci. Total Environ.* **2021**, *765*, 142711. [Google Scholar] [CrossRef]

30. Vallova, S.; Plevova, E.; Smutna, K.; Sokolova, B.; Vaculikova, L.; Valovicova, V.; Hundakova, M.; Praus, P. Removal of analgesics from aqueous solutions onto montmorillonite KSF. *J. Therm. Anal. Calorim.* **2021**, *1*, 1–9. [Google Scholar] [CrossRef]

31. Yu, W.H.; Li, N.; Tong, D.S.; Zhou, C.H.; Lin, C.X.; Xu, C.Y. Adsorption of proteins and nucleic acids on clay minerals and their interactions: A review. *Appl. Clay Sci.* **2013**, *80–81*, 443–452. [Google Scholar] [CrossRef]

32. Szabó, T.; Mitea, R.; Leeman, H.; Premachandra, G.S.; Johnston, C.T.; Szekeres, M.; Dékány, I.; Schoonheydt, R.A. Adsorption of protamine and papain proteins on saponite. *Clays Clay Miner.* **2008**, *56*, 494–504. [Google Scholar] [CrossRef]

33. Sanjay, G.; Sugunan, S. Enhanced pH and thermal stabilities of invertase immobilized on montmorillonite K-10. *Food Chem.* **2006**, *94*, 573–579. [Google Scholar] [CrossRef][Green Version]

34. Gopinath, S.; Sugunan, S. Enzymes immobilized on montmorillonite K 10: Effect of adsorption and grafting on the surface properties and the enzyme activity. *Appl. Clay Sci.* **2007**, *35*, 67–75. [Google Scholar] [CrossRef]

35. Dong, H.; Li, Y.; Li, J.; Sheng, G.; Chen, H. Comparative Study on Lipases Immobilized onto Bentonite and Modified Bentonites and Their Catalytic Properties. *Ind. Eng. Chem. Res.* **2013**, *52*, 9030–9037. [Google Scholar] [CrossRef]

36. Da Silva, V.C.F.; Contesini, F.; Carvalho, P.D.O. Enantioselective behavior of lipases from *Aspergillus niger* immobilized in different supports. *J. Ind. Microbiol. Biotechnol.* **2009**, *36*, 949–954. [Google Scholar] [CrossRef] [PubMed]

37. Benamia, F.; Benouis, S.; Belafriekh, A.; Semache, N.; Rebbani, N.; Djeghaba, Z. Efficient *Candida rugosa* lipase immobilization on Maghnite clay and application for the production of (1R)-(-)-Menthyl acetate. *Chem. Pap.* **2017**, *71*, 785–793. [Google Scholar] [CrossRef]

38. Scherer, R.P.; Dallago, R.L.; Penna, F.G.; Bertella, F.; de Oliveira, D.; de Oliveira, J.V.; Pergher, S.B. Influence of process parameters on the immobilization of commercial porcine pancreatic lipase using three low-cost supports. *Biocatal. Agric. Biotechnol.* **2012**, *1*, 290–294. [Google Scholar] [CrossRef]

39. Babavatan, E.O.; Yildirim, D.; Peksel, A.; Binay, B. Immobilization ofRhizomucor mieheilipase onto montmorillonite K-10 and polyvinyl alcohol gel. *Biocatal. Biotransform.* **2019**, *38*, 1–9. [Google Scholar] [CrossRef]

40. Tu, N.; Shou, J.; Dong, H.; Huang, J.; Li, Y. Improved Catalytic Performance of Lipase Supported on Clay/Chitosan Composite Beads. *Catalysts* **2017**, *7*, 302. [Google Scholar] [CrossRef][Green Version]

41. Zou, T.; Duan, Y.-D.; Wang, Q.-E.; Cheng, H.-M. Preparation of Immobilized Lipase on Silica Clay as a Potential Biocatalyst on Synthesis of Biodiesel. *Catalysts* **2020**, *10*, 1266. [Google Scholar] [CrossRef]

42. Coghetto, C.C.; Scherer, R.P.; Silva, M.F.; Golunski, S.; Pergher, S.; Oliveira, D.; Oliveira, J.V.; Treichel, H. Natural montmorillonite as support for the immobilization of inulinase from *Kluyveromyces marxianus* NRRL Y-7571. *Biocatal. Agric. Biotechnol.* **2012**, *1*, 284–289. [Google Scholar] [CrossRef]

43. Contesini, F.J.; Lopes, D.B.; Macedo, G.; Nascimento, M.D.G.; de Carvalho, P. Aspergillus sp. lipase: Potential biocatalyst for industrial use. *J. Mol. Catal. B Enzym.* **2010**, *67*, 163–171. [Google Scholar] [CrossRef]

44. Sita, K.K.; Narasimha, R.M.; Karanam, S.K.; Medicherla, N.R. Enhanced lipase production by mutation induced *Aspergillus japonicus*. *Afr. J. Biotechnol.* **2008**, *7*, 2064–2067. [Google Scholar] [CrossRef]

45. Bharti, M.K.; Khokhar, D.; Pandey, A.K.; Gaur, A.K. Purification and Characterization of Lipase from *Aspergillus japonicas*: A Potent Enzyme for Biodiesel Production. *Natl. Acad. Sci. Lett.* **2013**, *36*, 151–156. [Google Scholar] [CrossRef]

46. Sobha, K.; Lalitha, N.; Pavani, N.; Praharshini, R.; Pradeep, D. Statistical optimization of Factors influencing the activity and the kinetics of partially purified Lipase produced from *Aspergillus japonicus* (MTCC 1975) cultured in protein enriched medium. *Int. J. Chem. Tech. Res.* **2014**, *6*, 4817–4831. [Google Scholar]

47. Souza, L.T.A.; Oliveira, J.S.; Dos Santos, V.L.; Régis, W.C.B.; Santoro, M.M.; Resende, R.R. Lipolytic Potential of *Aspergillus japonicus* LAB01: Production, Partial Purification, and Characterisation of an Extracellular Lipase. *BioMed Res. Int.* **2014**, 1–11. [Google Scholar] [CrossRef][Green Version]

48. Jayaprakash, A.; Ebenezer, P. Optimization of *Aspergillus japonicus* lipase production by Response Surface Methodology. *J. Acad. Ind. Res.* **2012**, *1*, 23–30. [Google Scholar]

49. Jayaprakash, A.; Ebenezer, P. Investigation on extracellular lipase production by *Aspergillus japonicus* isolated from the paper nest of Ropalidia marginata. *Indian J. Sci. Technol.* **2010**, *3*, 113–117. [Google Scholar] [CrossRef]

50. Rani, N.V.K.M.E.; Kannan, R.R.G.N.D. Lipase Production using *Aspergillus japonicus* MF-1 through Biotransformation of Agro-Waste and Medicinal Oil Effluent. *Int. J. Curr. Microbiol. Appl. Sci.* **2017**, *6*, 2005–2020. [Google Scholar] [CrossRef][Green Version]

51. Colla, L.M.; Ficanha, A.M.M.; Rizzardi, J.; Bertolin, T.E.; Reinehr, C.; Costa, J.A.V. Production and Characterization of Lipases by Two New Isolates of *Aspergillus* through Solid-State and Submerged Fermentation. *BioMed Res. Int.* **2015**, 1–9. [Google Scholar] [CrossRef][Green Version]

52. Jyoti, G.; Keshav, A.; Anandkumar, J. Review on Pervaporation: Theory, Membrane Performance, and Application to Intensification of Esterification Reaction. *J. Eng.* **2015**, *2015*, 1–24. [Google Scholar] [CrossRef][Green Version]

53. Sharma,R.;Soni,S.;Vohra,R.;Gupta,L.;Gupta,J.Purificationandcharacterisation of a thermostable alkaline lipase from a new thermophilic *Bacillus* sp. RSJ-1. *Process. Biochem.* **2002**, *37*, 1075–1084. [Google Scholar] [CrossRef]

54. Tacin, M.V.; Massi, F.P.; Fungaro, M.H.P.; Teixeira, M.F.S.; de Paula, A.V.; Santos-Ebinuma, V. Biotechnological valorization of oils from agro-industrial wastes to produce lipase using *Aspergillus* sp. from Amazon. *Biocatal. Agric. Biotechnol.* **2019**, *17*, 369–378. **[Google Scholar]** **[CrossRef]**

55. Mandari, V.; Nema, A.; Devarai, S.K. Sequential optimization and large scale production of lipase using tri-substrate mixture from *Aspergillus niger* MTCC 872 by solid state fermentation. *Process. Biochem.* **2020**, *89*, 46–54. **[Google Scholar]** **[CrossRef]**

56. Utami, T.S.; Hariyani, I.; Alamsyah, G.; Hermansyah, H. Production of dry extract extracellular lipase from *Aspergillus niger* by solid state fermentation method to catalyze biodiesel synthesis. *Energy Procedia* **2017**, *136*, 41–46. **[Google Scholar]** **[CrossRef]**

57. Das, A.; Shivakumar, S.; Bhattacharya, S.; Shakya, S.; Swathi, S.S. Purification and characterization of a surfactant-compatible lipase from *Aspergillus tamarii* JGIF06 exhibiting energy-efficient removal of oil stains from polycotton fabric. *3 Biotech* **2016**, *6*, 131. **[Google Scholar]** **[CrossRef][Green Version]**

58. Saxena, R.; Davidson, W.; Sheoran, A.; Giri, B. Purification and characterization of an alkaline thermostable lipase from *Aspergillus carneus*. *Process. Biochem.* **2003**, *39*, 239–247. **[Google Scholar]** **[CrossRef]**

59. Gomes, F.M.; De Paula, A.V.; Silva, G.D.S.; De Castro, H.F. Determinação das propriedades catalíticas em meio aquoso e orgânico da lipase de *Candida rugosa* imobilizada em celulignina quimicamente modificada por carbonildiimidazol. *Química Nova* **2006**, *29*, 710–718. **[Google Scholar]** **[CrossRef]**

60. Mala, J.G.S.; Takeuchi, S. Understanding Structural Features of Microbial Lipases—An Overview. *Anal. Chem. Insights* **2008**, *3*, ACI.S551-19. **[Google Scholar]** **[CrossRef]**

61. An, N.; Zhou, C.H.; Zhuang, X.Y.; Tong, D.S.; Yu, W.H. Immobilization of enzymes on clay minerals for biocatalysts and biosensors. *Appl. Clay Sci.* **2015**, *114*, 283–296. **[Google Scholar]** **[CrossRef]**

62. Quiquampoix, H.; Staunton, S.; Baron, M.-H.; Ratcliffe, R.G. Interpretation of the pH dependence of protein adsorption on clay mineral surfaces and its relevance to the understanding of extracellular enzyme activity in soil. *Colloids Surf. A Physicochem. Eng. Asp.* **1993**, *75*, 85–93. **[Google Scholar]** **[CrossRef]**

63. De Oliveira, M.F.; Johnston, C.T.; Premachandra, G.S.; Teppen, B.J.; Li, H.; Laird, D.A.; Zhu, D.; Boyd, S.A. Spectroscopic Study of Carbaryl Sorption on Smectite from Aqueous Suspension. *Environ. Sci. Technol.* **2005**, *39*, 9123–9129. **[Google Scholar]** **[CrossRef]** **[PubMed]**

64. Basso, A.; Froment, L.; Hesseler, M.; Serban, S. New highly robust divinyl benzene/acrylate polymer for immobilization of lipase CALB. *Eur. J. Lipid Sci. Technol.* **2013**, *115*, 468–472. [Google Scholar] [CrossRef]

65. Basso, A.; Hesseler, M.; Serban, S. Hydrophobic microenvironment optimization for efficient immobilization of lipases on octadecyl functionalised resins. *Tetrahedron* **2016**, *72*, 7323–7328. [Google Scholar] [CrossRef]

66. Lorente, F.; Cabrera, Z.; Godoy, C.; Fernandez-Lafuente, R.; Palomo, J.M.; Guisan, J.M. Interfacially activated lipases against hydrophobic supports: Effect of the support nature on the biocatalytic properties. *Process. Biochem.* **2008**, *43*, 1061–1067. [Google Scholar] [CrossRef]

67. Sani, F.; Mokhtar, N.F.; Shukuri, M.; Ali, M.; Noor, R.; Raja, Z.; Rahman, A. Enhanced Performance of Immobilized *Rhizopus oryzae* Lipase on Coated Porous Polypropylene Support with Additives. *Catalysts* **2021**, *11*, 303. [Google Scholar] [CrossRef]

68. Mulinari, J.; Oliveira, J.V.; Hotza, D. Lipase immobilization on ceramic supports: An overview on techniques and materials. *Biotechnol. Adv.* **2020**, *42*, 107581. [Google Scholar] [CrossRef] [PubMed]

69. Jaeger, K.E.; Reetz, M.T. Microbial lipases form versatile tools for biotechnology. *Trends Biotechnol.* **1998**, *16*, 396–403. [Google Scholar] [CrossRef]

70. Khan, F.I.; Lan, D.; Durrani, R.; Huan, W.; Zhao, Z.; Wang, Y. The Lid Domain in Lipases: Structural and Functional Determinant of Enzymatic Properties. *Front. Bioeng. Biotechnol.* **2017**, *5*, 16. [Google Scholar] [CrossRef] [PubMed][Green Version]

71. Lai, O.-M.; Lee, Y.Y.; Phuah, E.-T.; Akoh, C.C. Lipase/Esterase: Properties and Industrial Applications. In *Encyclopedia of Food Chemistry*; Melton, L., Shahidi, F., Varelis, P., Eds.; Academia Press: Waltham, MA, USA, 2019; pp. 158–167. [Google Scholar]

72. Nascimento, P.A.M.; Picheli, F.P.; Lopes, A.; Pereira, J.F.B.; Santos-Ebinuma, V.D.C. Effects of cholinium-based ionic liquids on *Aspergillus niger* lipase: Stabilizers or inhibitors. *Biotechnol. Prog.* **2019**, *35*, e2838. [Google Scholar] [CrossRef] [PubMed]

73. Palomo, J.M.; Muñoz, G.; Lorente, F.; Mateo, C.; Fernández-Lafuente, R.; Guisán, J.M. Interfacial adsorption of lipases on very hydrophobic support (octadecyl–Sepabeads): Immobilization, hyperactivation and stabilization of the open form of lipases. *J. Mol. Catal. B Enzym.* **2002**, *19-20*, 279–286. [Google Scholar] [CrossRef]

74. Fernandez-Lafuente, R.; Armisén, P.; Sabuquillo, P.; Lorente, F.; Guisán, J.M. Immobilization of lipases by selective adsorption on hydrophobic supports. *Chem. Phys. Lipids* **1998**, *93*, 185–197. [Google Scholar] [CrossRef]

75. Bastida, A.; Sabuquillo, P.; Armisen, P.; Fernández-Lafuente, R.; Huguet, J.; Guisán, J.M. A Single Step Purification, Immobi-lization, and Hyperactivation of Lipases via Interfacial Adsorption on Strongly Hydrophobic Supports. *Biotechnol. Bioeng.* **1998**, *58*, 486–493. [Google Scholar] [CrossRef]

76. Tacin, M.V.; Costa-Silva, T.A.; de Paula, A.V.; Palomo, J.M.; Santos-Ebinuma, V.D.C. Microbial lipase: A new approach for a heterogeneous biocatalyst. *Prep. Biochem. Biotechnol.* **2021**, *51*, 749–760. [Google Scholar] [CrossRef]

77. Wu, C.; Zhou, G.; Jiang, X.; Ma, J.; Zhang, H.; Song, H. Active biocatalysts based on *Candida rugosa* lipase immobilized in vesicular silica. *Process. Biochem.* **2012**, *47*, 953–959. [Google Scholar] [CrossRef]

78. Balcão, V.; Paiva, A.L.; Malcata, F. Bioreactors with immobilized lipases: State of the art. *Enzym. Microb. Technol.* **1996**, *18*, 392–416. [Google Scholar] [CrossRef]

79. Tang, A.; Zhang, Y.; Wei, T.; Wu, J.; Li, Q.; Liu, Y. Immobilization of *Candida cylindracea* Lipase by Covalent Attachment on Glu-Modified Bentonite. *Appl. Biochem. Biotechnol.* **2018**, *187*, 870–883. [Google Scholar] [CrossRef]

80. Zhao, Z.; Hou, S.; Lan, D.; Wang, X.; Liu, J.; Khan, F.I.; Wang, Y. Crystal structure of a lipase from *Streptomyces* sp. strain W007—Implications for thermostability and regiospecificity. *FEBS J.* **2017**, *284*, 3506–3519. [Google Scholar] [CrossRef] [Green Version]

81. Lozzi, I.; Calamai, L.; Fusi, P.; Bosetto, M.; Stotzky, G. Interaction of horseradish peroxidase with montmorillonite homoionic to Na^+ and Ca^{2+}: Effects on enzymatic activity and microbial degradation. *Soil Biol. Biochem.* **2001**, *33*, 1021–1028. [Google Scholar] [CrossRef]

82. Sanjay, G.; Sugunan, S. Acid activated montmorillonite: An efficient immobilization support for improving reusability, storage stability and operational stability of enzymes. *J. Porous Mater.* **2008**, *15*, 359–367. [Google Scholar] [CrossRef]

83. Wang, Q.; Peng, L.; Li, G.; Zhang, P.; Li, D.; Huang, F.; Wei, Q. Activity of Laccase Immobilized on TiO2-Montmorillonite Complexes. *Int. J. Mol. Sci.* **2013**, *14*, 12520–12532. [Google Scholar] [CrossRef] [PubMed][Green Version]

84. Morais, W.G., Jr.; Pacheco, T.F.; Gao, S.; Martins, P.A.; Guisán, J.M.; Caetano, N.S. 1 Sugarcane Bagasse Saccharification by Enzymatic Hydrolysis. *Catalysts* **2021**, *11*, 340. [Google Scholar] [CrossRef]

85. Carvalho, T.; Pereira, A.D.S.; Bonomo, R.C.F.; Franco, M.; Finotelli, P.V.; Amaral, P.F. Simple physical adsorption technique to immobilize *Yarrowia lipolytica* lipase purified by different methods on magnetic nanoparticles: Adsorption isotherms and thermodynamic approach. *Int. J. Biol. Macromol.* **2020**, *160*, 889–902. [Google Scholar] [CrossRef] [PubMed]

86. Basso, A.; Serban, S. Industrial applications of immobilized enzymes—A review. *Mol. Catal.* **2019**, *479*, 110607. [Google Scholar] [CrossRef]

87. Soares, C.M.F.; De Castro, H.F.; De Moraes, F.F.; Zanin, G.M. Characterization and utilization of *Candida rugosa* lipase immobilized on controlled pore silica. *Appl. Biochem. Biotechnol. Part A Enzym. Eng. Biotechnol.* **1999**, *77–79*, 745–757. [Google Scholar] [CrossRef]

88. Bradford, M.M. A rapid and sensitive method for the quantitation of microgram quantities of protein utilizing the principle of protein-dye binding. *Anal. Biochem.* **1976**, *72*, 248–254. [Google Scholar] [CrossRef]

89. Laemmli, U.K. Cleavage of Structural Proteins during the Assembly of the Head of Bacteriophage T4. *Nature* **1970**, *227*, 680–685. [Google Scholar] [CrossRef]

CHAPTER-6

PHENYLALANINE, TYROSINE, AND DOPA ARE *BONA FIDE* SUBSTRATES FOR *BAMBUSA OLDHAMII* BOPAL4

1. INTRODUCTION: AN OVERVIEW

Phenylalanine ammonia-lyase (PAL, EC 4.3.1.24) is the enzyme that catalyzes the first committed step of general phenylpropanoid pathways in plants, catalyzing the conversion of L-phenylalanine (L-Phe) to *trans*-cinnamic acid via a non-oxidative deamination reaction (**Scheme 1**) [1,2,3]. *Trans*-cinnamic acid served as the start point of secondary metabolism is then hydroxylated by a membrane-bounded cinnamate 4-hydroxylase (C4H, EC 1.14.14.91) to yield *p*-coumaric acid [4] for the synthesis of thousands of phenylpropanoids [5,6]. Plant hormone, salicylic acid, is also synthesized via the PAL-mediated pathway [7]. PAL activity is elevated by light treatment, maintaining phenolic compound levels in fresh-cut sweet peppers [8]. *Trans*-cinnamic acid and *p*-coumaric acid have been shown to exert multifarious health benefits for humans, including anti-diabetic, anti-obesity, anti-oxidant, and anti-microbial activities [9,10]. *P-coumaric* acid is further hydroxylated by 4-coumarate 3-hydroxylase (4CMH or C3H, EC 1.14.14.9) to yield caffeic acid, an intermediate involved in the early stage of lignin biosynthesis [11]. In humans, caffeic acid provides numerous beneficial biological activities, including anti-oxidant and anti-microbial activities [12,13,14].

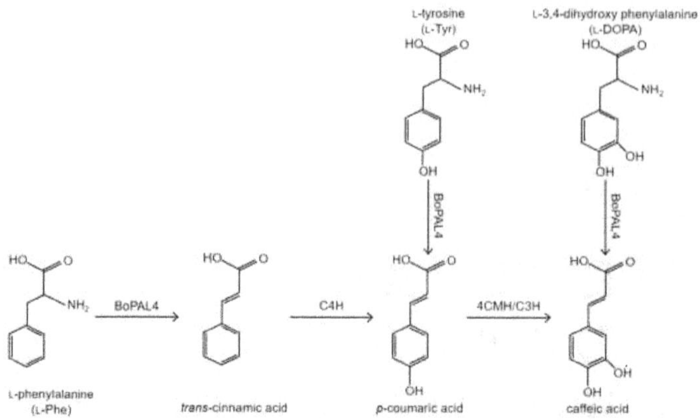

Scheme 1. The enzyme reaction catalyzed by the bamboo BoPAL4 phenyl alanine/tyrosine ammonia-lyase (PTAL). BoPAL4 PTAL catalyzes the non-oxidative deamination of L-phenylalanine to yield *trans*-cinnamic acid for the synthesis of *p*-coumaric acid via the enzyme reaction catalyzed by the C4H cinnamate 4-hydroxylase. BoPAL4 PTAL can bypass the BoC4H requirement to produce *p*-coumaric acid by the deamination of L-tyrosine. In plants, *p*-coumaric acid is hydroxylated at C-3 position to yield 3,4,-dihydroxy *trans*-cinnamic acid, also known as caffeic acid, by the 4CMH/C3H 4-coumarate 3-hydroxylase. BoPAL4 PTAL can also bypass the 4CMH requirement to produce caffeic acid by the deamination of L-3,4,-dihydroxy phenylalanine (L-DOPA).

Histidine ammonia-lyase (HAL, EC 4.3.1.3) [15], PAL, and tyrosine ammonia-lyase (TAL, EC 4.3.1.24) share the conserved active site motif, Ala-Ser-Gly, which can spontaneously merge into a 4-methylidine-imidazol-5-one (MIO) group [16,17]. PAL is one of the highly post-translational modified enzymes in plant secondary metabolism [18]. The degradation of PAL is mediated by ubiquitin proteasome system [19,20]. PAL is a phosphoprotein [21,22,23], and the phosphorylation site identified in French bean is the Thr-545 residue in a conserved R/K-X-X-S/T motif [24]. PAL proteins generally function as tetrameric enzymes [25,26,27].

In dicot plants, PAL exclusively uses L-Phe as substrate [28,29,30]; however, several PAL isoforms discovered in monocot plants exhibit dual functions using both L-Phe and L-tyrosine (L-Tyr) as substrates for the synthesis of *trans*-cinnamic acid and *p*-coumaric acid, respectively (**Scheme 1**) [27,31,32]. Therefore, the bifunctional PAL is renamed as phenylalanine/tyrosine ammonia-lyase (PTAL, EC 4.3.1.25) [32,33,34]. Unlike PALs, specific tyrosine ammonia-lyase (TAL, EC 4.3.1.24) solely occurs in some microorganisms [35,36,37,38,39]. Hence, the detectable TAL activity in monocot plants is contributed by the PTAL bifunctional enzyme [29,40], accounting for half lignin biosynthesis in grass plants [33,41].

Rhodobacter sphaeroides TAL (RsTAL) enzyme is firstly utilized as a model to investigate the substrate specificity between L-Phe and L-Tyr; two independent studies reveal that His-89 of RsTAL is the major substrate specificity switch site [36,42]. *Bambusa oldhamii* BoPAL1 is only specific to L-Phe but not to L-Tyr; site-directed mutagenesis studies have demonstrated that TAL activities are detectable in the BoPAL1-F133H and BoPAL1-V197I mutants [27,32]. *Sorghum bicolor* SbPAL1 is also a PTAL enzyme, utilizing L-Phe and L-Tyr as substrates (**Scheme 1**) [34]. In addition, SbPAL1 exhibits L-3,4-dihydroxy phenylalanine (L-DOPA) ammonia-lyase (DAL) activity to convert L-DOPA to caffeic acid [34]. TAL activities are completely eliminated in RsTAL-H89F [36] and SbPAL1-H123F mutants [34]. Taken together, Phe or His residue at the substrate specificity switch site is critical for PAL or TAL activity, respectively.

Bamboo PAL proteins are encoded by a multi-gene family, namely, BoPAL1 [43], BoPAL2 [26], BoPAL3, and BoPAL4 [27]. The substrate specificity site of BoPAL1, BoPAL2, or BoPAL3 is a Phe residue, whereas an His residue is present in the BoPAL4 protein (**Figure 1**A) [27]. BoPAL4 exhibits both PAL and TAL activities [27], making it a good candidate for studying substrate specificity. In the present study, a site-directed mutagenesis combined with PAL/TAL/DAL activities analysis toward diverse substrates, L-Phe/L-Tyr/L-DOPA, was carried out to better understand the catalytic functions in the BoPAL4. Furthermore, the optimum reaction conditions of BoPAL4 using three substrates were examined.

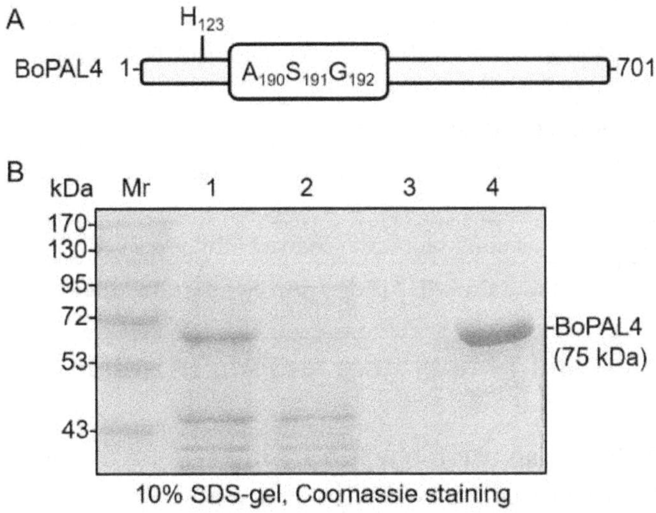

Figure 1. Expression and purification of recombinant BoPAL4 in *Escherichia coli*. (A) The primary structure of the BoPAL4 is illustrated. BoPAL4 is a 701-a.a protein with the conserved Ala-Ser-Gly catalytic motif as well as the substrate specificity site at His-123. (B) Recombinant BoPAL4 protein was purified using Ni-NTA affinity chromatography and separated using 10% SDS–PAGE and then stained with Coomassie Blue. Mr, molecular weight SDS–PAGE marker; lane 1, crude extract; lane 2, unbound protein; lane 3, flow-through; lane 4, 125 mM imidazole-buffer-eluted BoPAL4.

2. RESULTS

2.1. Expression and Purification of Recombinant BoPAL4 in Escherichia coli

BoPAL4 contained a 2106 bp open-reading frame (ORF) and encoded a 701- amino acids polypeptide (**Figure 1**A). The Ala-Ser-Gly catalytic motifs of BoPAL4 and SbPAL1 [**34**] are from 190 to 192 and from 189 to 191, respectively. Coincidentally, the substrate specificity switch site of both proteins is located at His-123 (**Figure 1**A). To better understand the catalytic function of BoPAL4, the coding region of *BoPAL4* was inserted into pTrcHisA plasmid and expressed heterologously in the *E. coli* Top10 strain (**Table 1**). N-terminal His_6-tag was fused to the recombinant BoPAL4 protein for facilitating affinity purification by the Ni-NTA resin (**Figure 1**B). The recombinant BoPAL4 protein was highly expressed and mainly eluted at 125 mM imidazole buffer (**Figure 1**B, lane 4). On the SDS-gel, a nearly homogeneous protein band with a molecular mass of 75 kDa corresponded to the predicted molecular mass of BoPAL4 protein. Thus, *E. coli* expression system was useful for producing recombinant BoPAL4 protein.

Table 1. Bacterial strain and plasmids used for recombinant protein expressions.

Strain	Relevant Characteristics	Source or Ref.
E. coli Top10	F^- *mcr*A Δ(*mrr-hsd*RMS-*mcr*BC) φ80*lac*ZΔM15 Δ*lac*X74 *rec*A1 *ara*D139 Δ(*ara-leu*)7697 *gal*U *gal*K λ^- *rps*L(Str^R) *end*A1 *nup*G	Invitrogen
Plasmid	**Relevant Characteristics**	**Source or Ref.**
pTrcHisA	*E. coli* expression vector with N-terminal His_6-tag fusion	Invitrogen
pTrcHisA-BoPAL4	*BoPAL4* coding sequence inserted into pTrcHisA	[27]
pTrcHisA-BoPAL4-H123F	Point mutation H123F derivative of pTrcHisA-BoPAL4	This study

2.2. Optimum pH and Temperature for PAL, TAL and DAL Activities of BoPAL4

To examine if BoPAL4 can use L-Phe, L-Tyr, and L-DOPA as substrates, PAL, TAL, and DAL activities were performed with 12.1 mM L-Phe, 1.9 mM L-Tyr, and 10 mM L-DOPA in a range of temperatures between 25 and 80 °C (**Figure 2**A). The maximum/optimum PAL, TAL, and DAL activities of BoPAL4 were measured at 50, 60, and 40 °C, similar to the optimum temperatures of BoPAL1 and BoPAL2 (**Table 2**) [**26,43**]. Accordingly, PAL, TAL, and DAL activities were performed with standard assay conditions in a range of pH between 5 and 11 (**Figure 2**B). The maximum/optimum PAL, TAL, and DAL activities of BoPAL4 were measured at pH 9.0, 8.5, and 9.0 °C, also similar to the optimum temperatures of BoPAL1 and BoPAL2 (**Table 2**)

[26,43]. Therefore, L-Phe, L-Tyr, and L-DOPA were bona fide substrates for BoPAL4 PTAL.

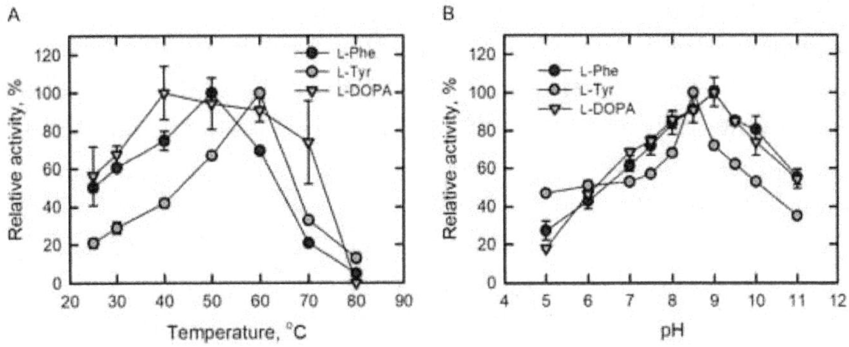

Figure 2. Optimum pH and temperature of PAL, TAL, and DAL activities for BoPAL4 PTAL. (A) Optimum temperatures of BoPAL4 using L-Phe (•), L-Tyr (•), or L-DOPA ( ) as substrate. Activities were measured under standard assay conditions in a range of temperatures from 25 to 80 °C. (B) Optimum pH of BoPAL4 using L-Phe (•), L-Tyr (•), and L-DOPA ( ) as substrates. Activities were measured under standard assay conditions in a range of pH from 5 to 11. All experiments were performed in triplicate and expressed as average ± standard deviation (S.D., error bars).

Table 2. Comparison of biochemical properties and kinetic parameters of E. coli expressed BoPAL proteins.

Protein	Substrate [1]	Optimum pH	Optimum Temp (°C)	k_{cat} (s^{-1})	K_m (μM)	k_{cat}/K_m (s^{-1} $μM^{-1}$)
BoPAL4	L-Phe	9.0	50	1.44	2084	6.9×10^{-4}
	L-Tyr	8.5	60	0.18	98	18.4×10^{-4}
	L-DOPA	9.0	40	0.06	956	0.6×10^{-4}
BoPAL4-H123F	L-Phe	9.0	50	1.87	640	29.2×10^{-4}

[1] L-Phe, phenylalanine; L-Tyr, tyrosine; and L-DOPA, 3,4-dihydroxy-phenylalanine.

2.3. Kinetic Parameters for PAL, TAL, and DAL Activities of BoPAL4

The kinetic parameters of BoPAL4 using L-Phe as substrate were measured with its PAL activity. Hyperbolic saturation curve (Figure 3A) and double reciprocal plot (Figure 3B) were obtained to calculate the kinetic parameters. The K_m value of BoPAL4 for L-Phe was estimated as 2084 μM, higher than the values of BoPAL1 (230 μM) [43], BoPAL2 (333 μM) [27], and SbPAL1 (340 μM) [34]. The kinetic parameters of BoPAL4 using L-Tyr as substrate were measured with its TAL activity. Hyperbolic saturation curve (Figure 3C) and double reciprocal plot (Figure 3D) were obtained to calculate the kinetic parameters. The K_m and k_{cat} values of BoPAL4 for L-Tyr were estimated as 98 μM and 0.18 s^{-1}, respectively. The kinetic parameters of BoPAL4 using L-DOPA as substrate were measured with its DAL activity. Hyperbolic saturation curve (**Figure 3**E) and double reciprocal plot (**Figure 3**F) were obtained to calculate the kinetic parameters. The K_m value of BoPAL4 for L-DOPA was estimated as 956 μM, which was 2.4-fold higher than SbPAL1 (0.40 mM) [**34**]. Taken together, BoPAL proteins were highly active at about 50 °C in alkaline reaction conditions.

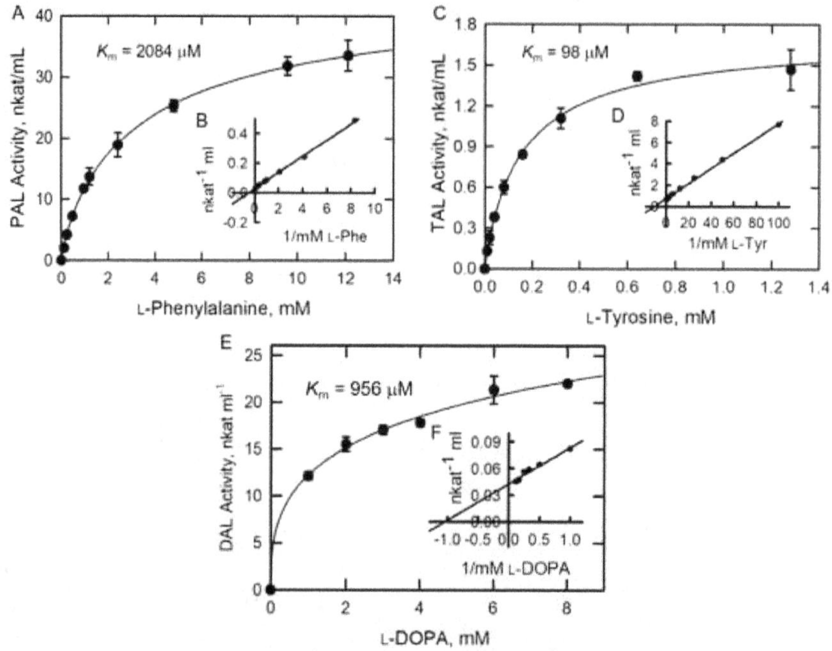

Figure 3. Kinetic parameters of BoPAL4 PTAL. To determine kinetic parameters using L-Phe as substrate, substrate saturation curve (A) and Lineweaver–Burk double reciprocal plot (B) of the initial rate result of BoPAL4 were used. To determine kinetic parameters using L-Tyr as substrate, substrate saturation curve (C) and Lineweaver–Burk double reciprocal plot (D) of the initial rate result of BoPAL4 were used. To determine kinetic parameters using L-DOPA as substrate, substrate saturation curve (E) and Lineweaver–Burk double reciprocal plot (F) of the initial rate result of BoPAL4 were used. All experiments were performed in triplicate and expressed as average ± standard deviation (S.D., error bars).

2.4. Kinetic Parameters for PAL Activity of BoPAL4-H123F

His-123 is the predicted substrate specificity switch site in the BoPAL4 (**Figure 1**A). Accordingly, site-directed mutant protein BoPAL4-H123F (**Table 1**) was also expressed and purified in the *E. coli* Top10 strain under the same procedure of BoPAL4 expression. After affinity purification, the purities of the 125 mM imidazole-buffer-eluted wild-type (WT) BoPAL4 and BoPAL4-H123F proteins were migrated at the same position on the SDS-gel (**Figure 4**A). The expression level of

BoPAL4-H123F was comparable to that of BoPAL4, indicating that *E. coli* expression system is also useful for producing BoPAL4-H123F protein.

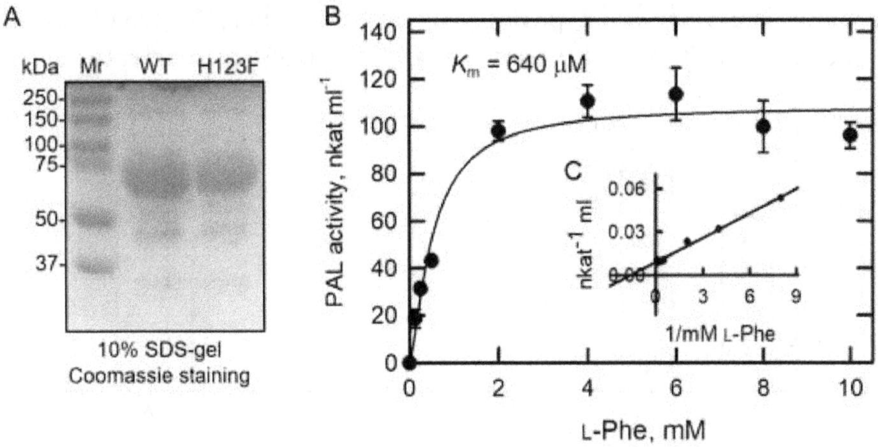

Figure 4. Kinetic parameters of BoPAL4-H123F. (A) BoPAL4-WT and BoPAL4-H123F were expressed and purified in *E. coli*. The 125 mM imidazole eluted fractions were separated using 10% SDS–PAGE and then stained with Coomassie Blue. Mr, molecular weight SDS–PAGE marker. To determine kinetic parameters using L-Phe as substrate, substrate saturation curve (B) and Lineweaver–Burk double reciprocal plot (C) of the initial rate result of BoPAL4-H123F were used. All experiments were performed in triplicate and expressed as average ± standard deviation (S.D., error bars).

The kinetic parameters of BoPAL4-H123F using L-Phe as substrate were measured its PAL activity. Hyperbolic saturation curve (Figure 4B) and double reciprocal plot (Figure 4C) were obtained to calculate the kinetic parameters. The K_m and k_{cat} values of BoPAL4-H123F for L-Phe were estimated as 640 μM and 1.87 s^{-1}, respectively (Table 2). The overall catalytic properties (k_{cat}/K_m value) of BoPAL4-H123F using L-Phe as substrate were 4.2-fold higher than that of wild-type BoPAL4.

TAL and DAL activities were significantly decreased in the BoPAL4-H123F. By using L-Tyr and L-DOPA as substrates, kinetic parameters of BoPAL4-H123F are not readily obtained, presumably due to their miniature TAL and DAL activities. Therefore, PAL-, TAL-, and DAL-specific activities were compared instead between WT and mutant proteins.

2.5. Comparison of Specific Activities in Wild-Type BoPAL4 and BoPAL4-H123F Mutants

The K_m value of BoPAL4-H123F using L-Phe (640 μM) was lower than that of BoPAL4 (2084 μM), indicating that BoPAL4-H123F mutant protein increases the binding affinity toward its substrate L-Phe. Likewise, the specific PAL activity of BoPAL4-

H123F was slightly higher (1.3-fold) than that of BoPAL4 (**Figure 5**A). Unlike this, the specific TAL and DAL activities only retained 7.5% (**Figure 5**B) and 17.8% (**Figure 5**C) in comparison with wild-type BoPAL4. Taken together, His-123 of BoPAL4 is the major specificity switch site for using L-Tyr and L-DOPA as substrates.

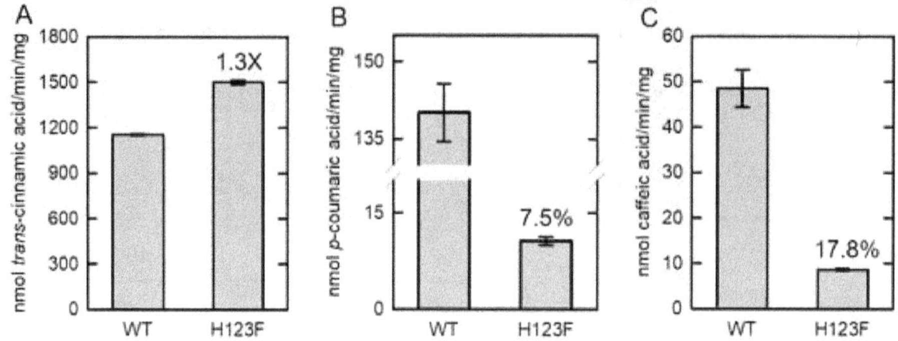

Figure 5. Comparison of specific activities in the BoPAL4 and BoPAL4-H123F proteins. PAL (A)-, TAL (B)-, and DAL (C)-specific activities of BoPAL4 and BoPAL4-H123F were measured under standard assay conditions. All experiments were performed in triplicate and expressed as average ± standard deviation (S.D., error bars).

3. DISCUSSION

Green bamboo BoPAL4 is a multifunctional enzyme, catalyzing the nonoxidative deamination of L-Phe, L-Tyr, and L-DOPA to yield *trans*-cinnamic acid, *p*-coumaric acid, and caffeic acid, respectively. The K_m value of L-Phe in the BoPAL4-H123F significantly decreases, indicating that His-123 is the major substrate specificity switch site. Phe-123 to His mutation of BoPAL4 exhibits higher L-Phe binding affinity than wild-type BoPAL4 as well as an elevated k_{cat} value of 1.87 s^{-1}. On the contrary, the TAL and DAL activities are significantly reduced in BoPAL4-H123F, which further confirms that the His-123 of BoPAL4 is essential for utilizing L-Tyr and L-DOPA as substrates. Two studies have shown that TAL activities are completely abolished in both RsTAL-H89F and SbPAL1-H123F mutants [**34,36**]. To our knowledge, L-DOPA utilization has never been studied in those mutants. Our results show that BoPAL4-H123F still can monitor TAL and DAL activities to some degree. Site-directed mutagenesis study can be performed in the BoPAL4-H123F to further identify more substrate specificity site (s).

Trans-cinnamic acid derivatives are shown to have useful effects in human health, such as anti-obesity and anti-microbial activities [**9,10**]. Recently, several studies have successfully immobilized PAL proteins on different carriers [**44,45,46,47**]. Electrospun nanofibers combined with dextran polyaldehyde crosslinker are suitable for BoPAL1 and BoPAL2 immobilization [**47**]. In this study, we showed that BoPAL4 had broad substrate specificities for synthesizing at least three products. In the meantime, other putative aromatic substrates are being tested to see if they can be utilized by the BoPAL4 enzyme. Furthermore, immobilization of BoPAL4 on electrospun nanofibers has great potential for the synthesis of aromatic compounds.

4. MATERIALS AND METHODS

4.1. Reagents

Bio-Rad protein assay dye reagent concentrate [**48**], Nuvia™ IMAC resin, and general chemicals for protein electrophoresis were purchased from Bio-Rad, Hercules, CA, USA. Restriction endonuclease enzymes, DNA proofreading polymerase, PrimeStar, and T4 DNA ligase for molecular biology manipulations were obtained from Takara, Kusatsu, Shiga, Japan. Plasmid mini-preparation kit and gel extraction/PCR cleanup kit were from Geneaid Biotech, New Taipei City, Taiwan. Oligonucleotide synthesis and DNA sequencing were serviced by Tri-I Biotech, New Taipei City, Taiwan. L-phenylalanine, L-tyrosine, L-DOPA, *trans*-cinnamic acid, *p*-coumaric acid, and caffeic acid were purchased from MilliporeSigma, Burlington, MA, USA. All other chemicals and reagents were ACS grade or higher.

4.2. Bacterial Strains, Plasmids Construction, Site-Directed Mutagenesis, and Bacterial Growth Conditions

E. coli strain and plasmids used in this study are listed in **Table 1**. Top10 strain was utilized for propagation of plasmids and for the induction of recombinant proteins. Plasmid pTrcHisA-BoPAL4 and pTrcHisA-BoPAL4-H123F directed the expression of N-terminal His_6-tagged wild-type BoPAL4 [27] and BoPAL4-H123F mutant protein, respectively. The pTrcHisA-BoPAL4-H123F plasmid containing an His to Phe point mutation was generated by QuikChange® site-directed mutagenesis (Strategene, Agilent, Santa Clara, CA, USA) as described by Hsieh et al. 2020 [**32**]. PCR product was obtained by PrimeStar DNA polymerase, and mixture was digested by *Dpn* I restriction endonuclease overnight and transformed chemically into Top10 component cells followed by DNA sequencing.

E. coli Top10 cells carrying BoPAL4 wild-type and mutant plasmids were grown at 37°C in Luria-Bertani medium (1% tryptone, 0.5% yeast extract, 1% NaCl, and pH 7.0) supplemented with 100 mg/mL ampicillin. The expression of BoPAL4 proteins was induced at 30°C with 1 mM isopropyl-β-D-thiogalactoside (IPTG) for 6–8 h [27]. Cells were centrifuged at 6000× *g* for 10 min, and pellets were stored at -20°C freezer before use.

4.3. Preparations of BoPAL4 Enzymes

E. coli cells that expressed His_6-tagged BoPAL proteins were freshly collected or resuspended from -20°C freezer, disrupting by sonication using a Branson cell disruptor [32,49,50]. Proteins were purified from cell lysates by affinity chromatography with Nuvia™ Ni-NTA resin. Samples were fractioned and eluted by varied concentrations of imidazole. Protein content of each step of the preparation was measured at $A_{595\ nm}$ using a microplate reader (BioTek 800TS, Winooski, VT, USA), and bovine serum albumin (BSA) was served as reference. All purification procedures were conducted at 4°C cold room.

4.4. Sodium Dodecyl Sulfate-Polyacrylamide Gel Electrophoresis

Purified BoPAL4 proteins were analyzed by polyacrylamide gel electrophoresis (PAGE) using a Mini-PROTEAN Tetra Cell system (Bio-Rad, Hercules, CA, USA). Proteins were stained by Coomassie Brilliant Blue R-250 and destained by 20% methanol solution. Gel image was obtained by a Gel Doc XR+ Imaging System (Bio-Rad, Hercules, CA, USA)

4.5. PAL Activity Assay

Phenylalanine ammonia-lyase activity was measured by following the formation of *trans*-cinnamic acid as the absorbance varied at 290 nm [32] using *trans*-cinnamic acid as standard. The reaction mixture for PAL contained 50 mM Tris-HCl (pH 8.5), 12.1 mM L-Phe, and an aliquot amount of BoPAL4 enzyme in a total volume of 1.0 mL. The reaction was incubated at 37°C for 30 min and ceased by adding 100 μL 6N HCl. PAL activity was defined as nkat (nanomole of *trans*-cinnamic acid formed per second).

4.6. TAL Activity Assay

Tyrosine ammonia-lyase activity was measured by following the formation of *p*-coumaric acid as the absorbance varied at 310 nm [32] using *p*-coumaric acid as standard. The reaction mixture for TAL contained 50 mM Tris-HCl (pH 8.5), 1.9 mM L-Tyr, and an aliquot amount of BoPAL enzyme in a total volume of 1.0 mL. The reaction was incubated at 37°C for 30 min and ceased when 100 μL 6N HCl was added. TAL activity was defined as nkat (nanomole of *p*-coumaric acid formed per second).

4.7. DAL Activity Assay

DOPA ammonia-lyase activity was measured by following the formation of caffeic acid as the absorbance varied at 350 nm [34] using caffeic acid as standard. The reaction mixture for DAL contained 50 mM Tris-HCl (pH 8.5), 10 mM L-DOPA, and an aliquot amount of BoPAL enzyme in a total volume of 1.0 mL. The reaction was incubated at 37°C for 30 min and ceased when 100 μL 6N HCl was added. DAL activity was defined as nkat (nanomole of caffeic acid formed per second).

4.8. Biochemical Properties and Enzyme Kinetic

To determine the optimum reaction pH, activity assays were performed at standard reaction mixture using various pH universal buffers. To obtain the optimum reaction temperature, activity assays were carried out at standard reaction mixture using in various temperatures.

To determine the kinetic parameter for PAL, TAL, and DAL, a range of concentration of L-Phe, L-Tyr, and L-DOPA was varied from 0 to 12.1 mM, from 0 to 1.28 mM, and from 0 to 10 mM, respectively. Substrate saturation curves were carried out after 10 min incubation [32,51] based on Michaelis–Menten equation [52], adapting to double reciprocal plot [53] for the calculation of the K_m and k_{cat} values.

5. CONCISE SUMMARY

Phenylalanine ammonia-lyase (PAL) links the plant primary and secondary metabolisms, and its product, *trans*-cinnamic acid, is derived into thousands of diverse phenylpropanoids. *Bambusa oldhamii* BoPAL4 has broad substrate specificity using L-phenylalanine, L-tyrosine, and L-3,4-dihydroxy phenylalanine (L-DOPA) as substrates to yield *trans*-cinnamic acid, *p*-coumaric acid, and caffeic acid, respectively. The optimum reaction pH of BoPAL4 for three substrates was measured at 9.0, 8.5, and 9.0, respectively. The optimum reaction temperatures of BoPAL4 for three substrates were obtained at 50, 60, and 40°C, respectively. The *Km* values of BoPAL4 for three substrates were 2084, 98, and 956 μM, respectively. The k_{cat} values of BoPAL4 for three substrates were 1.44, 0.18, and 0.06 σ^{-1}, respectively. The major substrate specificity site mutant, BoPAL4-H123F, showed better affinity toward L-phenylalanine by decreasing its K_m value to 640 μM and increasing its k_{cat} value to 1.87 s^{-1}. In comparison to wild-type BoPAL4, the specific activities of BoPAL4-H123F using L-tyrosine and L-DOPA as substrates retained 5.4% and 17.8% residual activities. Therefore, L-phenylalanine, L-tyrosine, and L-DOPA are bona fide substrates for BoPAL4.

REFERENCES

1. Koukol, J.; Conn, E.E. Metabolism of aromatic compounds in higher plants. IV. Purification and properties of phenylalanine deaminase of *Hordeum vulgare*. *J. Biol. Chem.* **1961**, *236*, 2692–2698. [Google Scholar] [CrossRef]
2. Zhang, X.; Liu, C.J. Multifaceted regulations of gateway enzyme phenylalanine ammonia-lyase in the biosynthesis of phenylpropanoids. *Mol. Plant* **2015**, *8*, 17–27. [Google Scholar] [CrossRef] [PubMed][Green Version]
3. Deng, Y.; Shanfa, L. Biosynthesis and regulation of phenylpropanoids in plants. *Crit. Rev. Plant Sci.* **2017**, *36*, 257–290. [Google Scholar] [CrossRef]

4. Bell-Lelong, D.A.; Cusumano, J.C.; Meyer, K.; Chapple, C. Cinnamate 4-hydroxylase expression in *Arabidopsis*. Regulation in response to development and the environment. *Plant Physiol.* **1997**, *113*, 729–738. [Google Scholar] [CrossRef][Green Version]

5. Achnine, L.; Blancaflor, E.B.; Rasmussen, S.; Dixon, R.A. Colocalization of L-phenylalanine ammonia-lyase and cinnamate 4-hydroxylase for metabolic channeling in phenylpropanoids biosynthesis. *Plant Cell* **2004**, *16*, 3098–3109. [Google Scholar] [CrossRef] [PubMed]

6. Chen, O.; Deng, L.; Ruan, C.; Yi, L.; Zeng, K. *Pichia galeiformis* induced resistance in postharvest citrus by activating the phenylpropanoid biosynthesis pathway. *J. Agri. Food Chem.* **2021**, *69*, 2619–2631. [Google Scholar] [CrossRef] [PubMed]

7. Zhang, K.; Lu, J.; Li, J.; Yu, Y.; Zhang, J.; He, Z.; Ismail, O.M.; Wu, J.; Xie, X.; Li, X.; et al. Efficiency of chitosan application against *Phytophthora infestans* and the activation of defence mechanisms in potato. *Int. J. Biol. Macromol.* **2021**, *182*, 1670–1680. [Google Scholar] [CrossRef]

8. Maroga, G.M.; Soundy, P.; Sivakumar, D. Different postharvest responses of fresh-cut sweet peppers related to quality and antioxidant and phenylalanine ammonia lyase activities during exposure to light-emitting diode treatment. *Foods* **2019**, *8*, 359. [Google Scholar] [CrossRef][Green Version]

9. Mitani, T.; Ota, K.; Inaba, N.; Kishida, K.; Koyama, H.A. Antimicrobial activity of the phenolic compounds of *Prunus mume* against enterobacteria. *Biol. Pharm. Bull.* **2018**, *41*, 208–212. [Google Scholar] [CrossRef][Green Version]

10. Wang, Z.; Ge, S.; Li, S.; Lin, H.; Lin, S. Anti-obesity effect of *trans*-cinnamic acid on HepG2 cells and HFD-fed mice. *Food Chem. Toxicol.* **2020**, *137*, 111148. [Google Scholar] [CrossRef]

11. Barros, J.; Escamilla-Trevino, L.; Song, L.; Rao, X.; Serrani-Yarce, J.C.; Palacios, M.D.; Engle, N.; Choudhury, F.K.; Tschalinski, T.J.; Venables, B.J.; et al. 4-Coumarate 3-hydroxylase in the lignin biosynthesis pathway is a cytosolic ascorbate peroxidase. *Nat. Commun.* **2019**, *10*, 1994. [Google Scholar] [CrossRef][Green Version]

12. Abdallah, F.B.; Fetoui, H.; Fakhfakh, F.; Keskes, L. Caffeic acid and quercetin protect erythrocytes against the oxidative stress and the genotoxic effects of lambda-cyhalothrin in vitro. *Hum. Exp. Toxicol.* **2012**, *31*, 92–100. [Google Scholar] [CrossRef]

13. Lima, V.N.; Oliveira-Tintino, C.D.M.; Santos, E.S.; Morais, L.P.; Tintino, S.R.; Freitas, T.S.; Geraldo, Y.S.; Pereira, R.L.S.; Cruz, R.P.; Menezes, I.R.A.; et al. Antimicrobial and enhancement of the antibiotic activity by phenolic

compounds: Gallic acid, caffeic acid and pyrogallol. *Microb. Pathog.* **2016**, *99*, 56–61. [Google Scholar] [CrossRef] [PubMed]

14. Shull, T.E.; Kurepa, J.; Miller, R.D.; Martinez-Ochoa, N.; Smalle, J.A. Inhibition of *Fusarium oxysporum* f. sp. nicotianae growth by phenylpropanoid pathway intermediates. *Plant Pathol. J.* **2020**, *36*, 637–642. [Google Scholar] [CrossRef] [PubMed]
15. Schwede, T.F.; Rétey, J.; Schulz, G.E. Crystal structure of histidine ammonia-lyase revealing a novel polypeptide modification as the catalytic electrophile. *Biochemistry* **1999**, *38*, 5355–5361. [Google Scholar] [CrossRef] [PubMed]
16. Rétey, J. Discovery and role of methylidene imidazolone, a highly electrophilic prosthetic group. *Biochim. Biophys. Acta.* **2003**, *1647*, 179–184. [Google Scholar] [CrossRef]
17. Ritter, H.; Schulz, G.E. Structural basis for the entrance into the phenylpropanoid metabolism catalyzes by phenylalanine ammonia-lyase. *Plant Cell* **2004**, *16*, 3426–3436. [Google Scholar] [CrossRef] [PubMed]
18. Sulis, D.; Wang, J.P. Regulation of lignin biosynthesis by post-translational protein modifications. *Front Plant Sci.* **2020**, *11*, 914. [Google Scholar] [CrossRef] [PubMed]
19. Zhang, X.; Gou, M.; Liu, C.-J. *Arabidopsis* Kelch repeat F-box proteins regulate phenylpropanoid biosynthesis via controlling the turnover of phenylalanine ammonia-lyase. *Plant Cell* **2013**, *25*, 4994–5010. [Google Scholar] [CrossRef] [Green Version]
20. Yu, H.; Li, D.; Yang, D.; Xue, Z.; Li, J.; Xing, B.; Yan, K.; Han, R.; Liang, Z. SmKFB5 protein regulates phenolic acid biosynthesis by controlling the degradation the degradation of phenylalanine ammonia-lyase in *Salvia miltiorrhiza*. *J. Exp. Bot.* **2021**, *72*, 4915–4929. [Google Scholar] [CrossRef]
21. Bolwell, G.P. A role for phosphorylation in the down-regulation of phenylalanine ammonia-lyase in suspension-cultured cells of French bean. *Phytochemistry* **1992**, *31*, 4081–4086. [Google Scholar] [CrossRef]
22. Cheng, S.H.; Sheen, J.; Gerrish, C.; Bolwell, G.P. Molecular identification of phenylalanine ammonia-lyase as a substrate of a specific constitutively active *Arabidopsis* CDPK expressed in maize protoplasts. *FEBS Lett.* **2001**, *503*, 185–188. [Google Scholar] [CrossRef][Green Version]
23. Allwood, E.G.; Davies, D.R.; Gerrish, C.; Bolwell, G.P. Regulation of CDPKs, including identification of PAL kinase, in biotically stressed cells of French bean. *Plant Mol. Biol.* **2002**, *49*, 533–544. [Google Scholar] [CrossRef]

24. Allwood, E.G.; Davies, D.R.; Gerrish, C.; Ellis, B.E.; Bolwell, G.P. Phosphorylation of phenylalanine ammonia-lyase: Evidence for a novel protein kinase and identification of the phosphorylated residue. *FEBS Lett.* **1999**, *457*, 47–52. [Google Scholar] [CrossRef][Green Version]

25. Reichert, A.I.; He, X.Z.; Dixon, R.A. Phenylalanine ammonia-lyase (PAL) from tobacco (*Nicotiana tabacum*): Characterization of the four tobacco PAL genes and active heterotetrameric enzymes. *Biochem. J.* **2009**, *424*, 233–242. [Google Scholar] [CrossRef][Green Version]

26. Hsieh, L.-S.; Yeh, C.-S.; Cheng, C.-Y.; Yang, C.-C.; Lee, P.-D. Cloning and expression of a phenylalanine ammonia-lyase gene (*BoPAL2*) from *Bambusa oldhamii* in *Escherichia coli* and *Pichia pastoris*. *Protein Expr. Purif.* **2010**, *71*, 224–230. [Google Scholar] [CrossRef]

27. Hsieh, L.-S.; Ma, G.-J.; Yang, C.-C.; Lee, P.-D. Cloning, expression, site-directed mutagenesis and immunolocalization of phenylalanine ammonia-lyase in *Bambusa oldhamii*. *Phytochemistry* **2010**, *71*, 1999–2009. [Google Scholar] [CrossRef]

28. Appert, C.; Logemann, E.; Hahlbrock, K.; Schmid, J.; Amrhein, N. Structural and catalytic properties of the four phenylalanine ammonia-lyases from parsley (*Petroselinum crispum* Nym). *Eur. J. Biochem.* **1994**, *225*, 2177–2185. [Google Scholar] [CrossRef] [PubMed]

29. Barros, J.; Dixon, R.A. Plant phenylalanine/tyrosine ammonia-lyases. *Trends Plant Sci.* **2020**, *25*, 66–79. [Google Scholar] [CrossRef] [PubMed]

30. Moisă, M.E.; Amariei, D.A.; Nagy, E.Z.A.; Szarvas, N.; To□a, M.I.; Paizs, C.; Bencze, L.C. Fluorescent enzyme-coupled activity assay for phenylalanine ammonia-lyases. *Sci. Rep.* **2020**, *10*, 18418. [Google Scholar] [CrossRef] [PubMed]

31. Rösler, J.; Krekel, F.; Amrhein, N.; Schmid, J. Maize phenylalanine ammonia lyase has tyrosine ammonia-lyase activity. *Plant Physiol.* **1997**, *113*, 175–179. [Google Scholar] [CrossRef][Green Version]

32. Hsieh, C.-Y.; Huang, Y.-H.; Lin, Z.-Y.; Hsieh, L.-S. Insights into the substrate selectivity of *Bambusa oldhamii* phenylalanine ammonia-lyase 1 and 2 through mutational analysis. *Phytochem. Lett.* **2020**, *38*, 140–143. [Google Scholar] [CrossRef]

33. Barros, J.; Serrani-Yarce, J.C.; Chen, F.; Baxter, D.; Venables, B.J.; Dixon, R.A. Role of bifunctional ammonia-lyase in grass cell wall biosynthesis. *Nat. Plants* **2016**, *2*, 16050. [Google Scholar] [CrossRef]

34. Jun, S.Y.; Sattler, S.A.; Cortez, G.S.; Vermerris, W.; Sattler, S.E.; Kang, C. Biochemical and structural analysis of substrate specificity of a phenylalanine ammonia-lyase. *Plant Physiol.* **2018**, *176*, 1452–1468. [Google Scholar] [CrossRef] [PubMed][Green Version]

35. Kyndt, J.A.; Meyer, T.E.; Cusanovich, M.A.; Van Beeumen, J.J. Characterization of a bacterial tyrosine ammonia lyase, a biosynthetic enzyme for the photoactive yellow protein. *FEBS Lett.* **2002**, *512*, 240–244. [Google Scholar] [CrossRef][Green Version]

36. Louie, G.V.; Bowman, M.E.; Moffitt, M.C.; Baiga, T.J.; Moore, B.S.; Noel, N.P. Structural determinants and modulation of substrate specificity in phenylalanine–tyrosine ammonia-lyase. *Chem. Biol.* **2006**, *13*, 1327–1338. [Google Scholar] [CrossRef] [PubMed][Green Version]

37. Moffitt, M.C.; Louie, G.V.; Bowman, M.E.; Pence, J.; Noel, J.P.; Moore, B.S. Discovery of two cyanobacterial phenylalanine ammonia-lyases: Kinetic and structural characterization. *Biochemistry* **2008**, *46*, 1004–1012. [Google Scholar] [CrossRef][Green Version]

38. Jendresen, C.B.; Stahlhut, S.G.; Li, M.; Gasper, P.; Siedler, S.; Förster, J.; Maury, J.; Borodina, I.; Nielsen, A.T. Highly active and specific tyrosine ammonia-lyases from diverse origins enable enhanced production of aromatic compounds in bacteria and *Saccharomyces cerevisiae*. *Appl. Environ. Microbiol.* **2015**, *81*, 4458–4476. [Google Scholar] [CrossRef] [PubMed][Green Version]

39. MacDonald, M.C.; Arivalagan, P.; Barre, D.E.; Maclnnis, J.A.; D'Cunha, G.B. *Rhodotorula glutinis* phenylalanine/tyrosine ammonia lyase enzyme catalyzed synthesis of the methyl ester of para-hydroxycinnamic acid and its potential antibacterial activity. *Front Microbiol.* **2016**, *7*, 281. [Google Scholar] [CrossRef][Green Version]

40. Feduraev, P.; Skrypnik, L.; Riabova, A.; Pungin, A.; Tokupova, E.; Maslennikov, P.; Chupakhina, G. Phenylalanine and tyrosine as exogenous precursors of wheat (*Triticum aestivum* L.) secondary metabolism through PAL-associate pathways. *Plants* **2020**, *9*, 476. [Google Scholar] [CrossRef][Green Version]

41. Maeda, E.A. Lignin biosynthesis: Tyrosine shortcut in grasses. *Nat. Plants* **2016**, *2*, 16080. [Google Scholar] [CrossRef] [PubMed]

42. Watts, K.T.; Mijts, B.N.; Lee, P.C.; Manning, A.J.; Schmidt-Dannert, C. Discovery of a substrate selectivity switch in tyrosine ammonia-lyase, a member of the aromatic amino acid lyase family. *Chem. Biol.* **2006**, *13*, 1317–1326. [Google Scholar] [CrossRef] [PubMed][Green Version]

43. Hsieh, L.-S.; Hsieh, Y.-L.; Yeh, C.-S.; Cheng, C.-Y.; Yang, C.-C.; Lee, P.-D. Molecular characterization of a phenylalanine ammonia-lyase gene (*BoPAL1*) from *Bambusa oldhamii*. *Mol. Biol. Rep.* **2011**, *38*, 283–290. [Google Scholar] [CrossRef] [PubMed]

44. Cui, J.; Zhao, Y.; Feng, Y.; Lin, T.; Zhong, C.; Tan, Z.; Jia, S. Encapsulation of spherical cross-linked phenylalanine ammonia lyase aggregates in mesoporous biosilica. *J. Agri. Food Chem.* **2017**, *65*, 618–625. [Google Scholar] [CrossRef] [PubMed]

45. Cui, J.; Zhao, Y.; Tan, Z.; Zhong, C.; Han, P.; Jia, S. Mesoporous phenylalanine ammonia lyase microspheres with improved stability through calcium carbonate templating. *Int. J. Biol. Macromol.* **2017**, *98*, 887–896. [Google Scholar] [CrossRef]

46. Boros, K.; Moisă, M.E.; Nagy, C.L.; Paizs, C.; Toşa, M.I.; Bencze, L.C. Robust, site-specificity immobilized phenylalanine ammonia-lyases for the enantioselective ammonia addition of cinnamic acids. *Catal. Sci. Technol.* **2021**, *11*, 5553. [Google Scholar] [CrossRef]

47. Hong, P.-Y.; Huang, Y.-H.; Lim, G.C.W.; Chen, Y.-P.; Hsiao, C.-J.; Chen, L.-H.; Ciou, J.-Y.; Hsieh, L.-S. Production of trans-cinnamic acid by immobilization of the Bambusa oldhamii BoPAL1 and BoPAL2 phenylalanine ammonia-lyases on electrospun nanofibers. *Int. J. Mol. Sci.* **2021**, *22*, 11184. [Google Scholar] [CrossRef]

48. Bradford, M.M. A rapid and sensitive method for the quantitation of microgram quantities of protein utilizing the principle of protein dye-binding. *Anal. Biochem.* **1976**, *72*, 248–254. [Google Scholar] [CrossRef]

49. Hsiao, C.-J.; Hsieh, C.-Y.; Hsieh, L.-S. Cloning and characterization of the *Bambusa oldhamii BoMDH*-encoded malate dehydrogenase. *Protein Expr. Purif.* **2020**, *174*, 105665. [Google Scholar] [CrossRef]

50. Hsu, W.-H.; Huang, Y.-H.; Chen, P.-R.; Hsieh, L.-S. NLIP and HAD-like domains of Pah1 and Lipin 1 phosphatidate phosphatases are essential for their catalytic activities. *Molecules* **2021**, *26*, 5470. [Google Scholar] [CrossRef]

51. De Jong, F.; Hanley, S.J.; Beale, M.H.; Karp, A. Characterization of the willow phenylalanine ammonia-lyase (PAL) gene family reveals expression differences compared with poplar. *Phytochemistry* **2015**, *117*, 90–97. [Google Scholar] [CrossRef] [PubMed][Green Version]

52. Michaelis, L.; Menten, M.L. Die kinetik der invertinwirkung. *Biochem. Z.* **1913**, *49*, 333–369. [Google Scholar]

53. Lineweavwer, H.; Burk, D. The determination of enzyme dissociation constants. *J. Am. Chem. Soc.* **1934**, *56*, 658–666. [Google Scholar] [CrossRef]

CHAPTER-7

UTILIZATION OF NITRILE HYDRATASE FROM *GORDONIA HYDROPHOBICA* FOR THE SYNTHESIS OF AMIDES FROM NITRILES

1. INTRODUCTION: AN OVERVIEW

In the realm of enzymatic catalysis, the application of nitrile hydratase from *Gordonia hydrophobica* represents a noteworthy avenue for the synthesis of amides from nitriles. This study delves into the utilization of this enzymatic catalyst, exploring its potential in facilitating the conversion of nitriles into amides. As amides play a crucial role in diverse chemical processes, the investigation of enzymatic pathways for their preparation holds significant promise for sustainable and efficient synthetic methodologies. This research aims to contribute insights into the utilization of nitrile hydratase from *Gordonia hydrophobica*, shedding light on its catalytic capabilities and its potential application in the synthesis of valuable amide compounds.

Nitrile hydratases (NHase; EC 4.2.1.84) catalyze the hydration of nitriles to the corresponding amides (**Scheme 1**A). Two types of NHases can be distinguished, the non-corrin cobalt-containing and non-heme iron-containing NHases. Both types are composed of an α- and a β-subunit, and the functional heterologous expression depends on the action of an accessory protein [1]. NHases are valuable biocatalysts for atom-economic amide synthesis [2,3]. Few examples of NHases with excellent enantioselectivity have been reported; instead, there are many examples of NHases with a poor to moderate ability to distinguish between enantiomers [4,5].

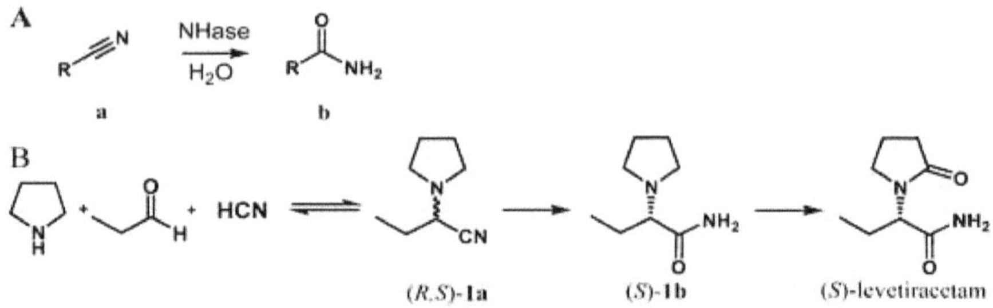

Scheme 1. (A) NHase-mediated nitrile hydration; (B) synthetic route to levetiracetam.

Levetiracetam is an active pharmaceutical ingredient used for the treatment and prevention of hypoxic- and ischemic-type aggressions of the central nervous system. This compound, also known as Keppra, is used as medication for epilepsy [6]. Levetiracetam is somewhat different from other antiepileptic drugs. It has, for example, a distinctive pharmacological profile in animal models of seizures and epilepsy, and it can inhibit neuronal hypersynchronization when epileptiform activity is evoked in rat hippocampal slices [7]. Its exact mechanism of action remains elusive, even after many years of clinical application and related research [8]. One observation indicates the partial blockade of N-type calcium currents [9], another the inhibition of AMPA-mediated currents in cortical neurons [10]. Levetiracetam was shown to bind to the synaptic vesicle protein SV2A in brain membranes [7], but the precise physiological role of SV2A and how levetiracetam binding influences SV2A are not yet fully understood yet [8]. Since the release of both glutamate and GABA is reduced upon levetiracetam binding, the mechanism of action is perhaps best described as a modulation of neurotransmitter release [11].

The chemical structure of levetiracetam is characterized by a heterocyclic pyrrolidinone in which the ring nitrogen resembles the amine of the (*S*)-enantiomer of 2-amino-butanoic acid amide. The respective (*R*) enantiomer lacks the above described biological activity [12]. Various routes to this molecule have been proposed in the literature. They typically involve multi-step reactions with either chiral auxiliaries or the final resolution of the desired enantiomer from the racemic mixture [13,14]. Analyzing levetiracetam from the retrosynthetic viewpoint, the non-oxidized precursor molecule (*S*)-2-(pyrrolidine-1-yl)butanamide [(*S*)-**1b**] would be accessible by stereoselective hydration of the respective amino nitrile 2-(pyrrolidine-1-yl)butanenitrile (**1a**), which can readily be synthesized from the three simple precursors pyrrolidine, propanal and hydrocyanic acid in a non-stereoselective Strecker-type reaction (**Scheme 1**B). The reversibility of this step enables the set-up of a dynamic kinetic resolution (DKR). The stereoselective conversion of the racemic aminonitrile to the aminoamide can be accomplished biocatalytically using an (*S*)-selective nitrile hydratase [15].

Nitrile hydratases are metallo proteins that activate a nitrile via a non-corrin cobalt or a non-heme iron-catalytic center in their active sites and incorporate a water molecule in a perfectly atom-economic step [16]. We recently investigated the synthetic sequence from racemic **1a** to levetiracetam via enantioenriched amide **1b** by combining enzymatic kinetic resolution with ex-cell electrochemical oxidation and investigated a non-corrin cobalt-type nitrile hydratase from *Comamonas testosteroni* in this context [14,17]. Cobalt-dependent NHases are abundant in contrast to the few iron-dependent NHases that have been discovered and investigated. Strikingly, however, iron-type NHases appear to be particularly capable of producing (*S*)-**1b**: in a preliminary screening, all four tested iron type NHases catalyzed the desired reaction, whereas only 3 of 15 cobalt-type NHases accepted racemic **1a** as a substrate [14,18]. A novel NHase originating from *Gordonia hydrophobica* showed the most promising combination of high (*S*)-selectivity and activity. The aim of this study was to investigate the substrate scope and biophysical characteristics of this iron-dependent NHase (*Gh*NHase).

2. RESULTS AND DISCUSSION

In an aqueous solution, aminonitrile **1a** is present in equilibrium with propanal, pyrrolidine and cyanide. A highly (*S*)-selective NHase can convert (*S*)-**1a** to (*S*)-**1b**, leaving the undesired (*R*)-enantiomer to disintegration and reassembly to (*R*,*S*)-**1a**, which enables a dynamic kinetic resolution to theoretically yield 100% of (*S*)-**1b**. The operational window of this reaction is determined by several factors. To name a few, the chemical equilibrium is dependent on pH, temperature and concentration of the single components, and so is the performance of the biocatalyst. Therefore, a thorough understanding of the system by systematic characterization of the reaction conditions was required to assess the scope and limitations of this key step in levetiracetam synthesis.

2.1. Influence of Reaction Temperature, Substrate Load and Catalyst Amount on Amide Production and Enantiomeric Excess (ee)

Gordonia hydrophobica nitrile hydratase (*Gh*NHase) was produced in *Escherichia coli* and applied as a cell free extract (CFE). In a first round of exploration of reaction conditions, reactions for the formation of aminoamide (*S*)-**1b** were performed at different temperatures. At a substrate concentration of 10 mM, the reactions were incubated for 16 h. The reactions were analyzed by reversed-phase chiral chromatography, using chemically-synthesized (*R*,*S*)-**1b** to generate a calibration curve. By linear interpolation, both the amount of the product and its enantiomeric excess (*ee*) were determined in a single chromatography run per reaction. There was 14% (*ee* 78%) of amide (*S*)-**1b** detected at 25°C, 13% (*ee* 78%) at 37°C and 16% (*ee* 77%) at 50°C (**Table 1**, Entry 1–3, respectively). A comparison of reaction progress at 5 °C and 25 °C, also gives highly similar product amounts and *ee* at modified reaction conditions (**Figure S1**). Under various sets of conditions, little effect by the reaction temperature on product formation and enantio-selectivity was observed. The enantio-selectivity of wild-type *Gh*NHase is in the typical range of other NHases [**4,19,20,21**]. In iterative rounds of reactions, several reaction parameters were varied to determine combinations that would lead to both high product titers and *ee*. For example, increasing substrate concentration from 10 to 100 mM increased product titers (**Table 1**, Entry 1, 4–6), but not in a linear fashion. An increase beyond 100 mM **1a** impeded product formation when **1a** was added as a single portion at the start of the reaction (**Figure S2**). Doubling the amount of biocatalyst led to a three-fold higher product concentration (**Table 1**, entry 5 versus entry 7); however, it also led to a slight decrease in the enantiomeric excess of (*S*)-**1b**. Neither the product titer nor the *ee* increased by applying only 5°C as the reaction temperature (**Table 1**, entry 7 versus entry 8). On the contrary, the *ee* decreased slightly. This result indicates that the chemical racemization might be impaired at very low temperatures.

Table 1. Effects of reaction temperature, amount of racemic substrate (*R,S*)-1a and amount of *Gh*NHase on product titer and enantiomeric excess. Data selected from iterative reaction engineering study.

Entry	*Gh*NHase [1] [mg/mL]	T [°C]	(*R,S*)-1a [mM]	(*S*)-1b [mM]	*ee* (*S*)-1b [%]
1	0.34	25	10	1.4	78
2	0.54	37	10	1.3	78
3	0.54	50	10	1.6	77
4	0.34	25	20	4.2	79
5	0.34	25	50	5.0	78
6	0.34	25	100	6.1	77
7	0.69	25	50	13.5	65
8	0.69	5	50	11.8	61

[1] Applied as cell free extract in 50/40 mM Tris-butyrate buffer, pH 7.2.

As indicated in **Table 1**, entry 7, elevated biocatalyst amounts significantly increased the formation of the desired amide. **Figure 1** shows the dependence of the reaction on various biocatalyst amounts. The highest enantioselectivity was observed between 300 and 600 µg/mL of *Gh*NHase and is associated with 85–90% of the analytical yield at 50 mM of the substrate load. A further increase in the catalyst amount resulted in a decrease in the product *ee*. At the same time, the total amount of product could not be increased further. The result was remarkably different at the 200 mM substrate load. Small amounts of catalysts could not cope with the high substrate load and produced much less amide with moderate *ee*. By increasing the biocatalyst load, both the product amount and its *ee* approached high levels. We assume that this phenomenon is due to the inhibiting effects of cyanide. During the course of the racemization of (*R*)-**1a**, the reaction solution contains propanal and cyanide. At a neutral pH, the formation of a racemic cyanohydrin occurs. Since cyanohydrins have been reported to be substrates of nitrile hydratases [22], the formation of a hydroxyamide is a possible side reaction, which would pull two essential reaction components from the system in an irreversible manner. This side reaction was elucidated in the case of the cobalt-dependent *Ct*NHase by HPLC/MS and explained why (*S*)-**1b** titers remained well below the theoretical 100% [14]. We therefore hypothesized that the remaining 10–15% required for full conversion by *Gh*NHase was also associated with the formation of 2-hydroxybutane amide. Increasing the substrate load to 200 mM indicates inhibition by substrate or one of the emerging components during DKR, because less product is formed in the same timeframe as compared to the 50 mM reactions. These results emphasize the importance of balancing the velocity of the *Gh*NHase-mediated hydration of the desired enantiomer with the reassembly velocity of racemic **1a**.

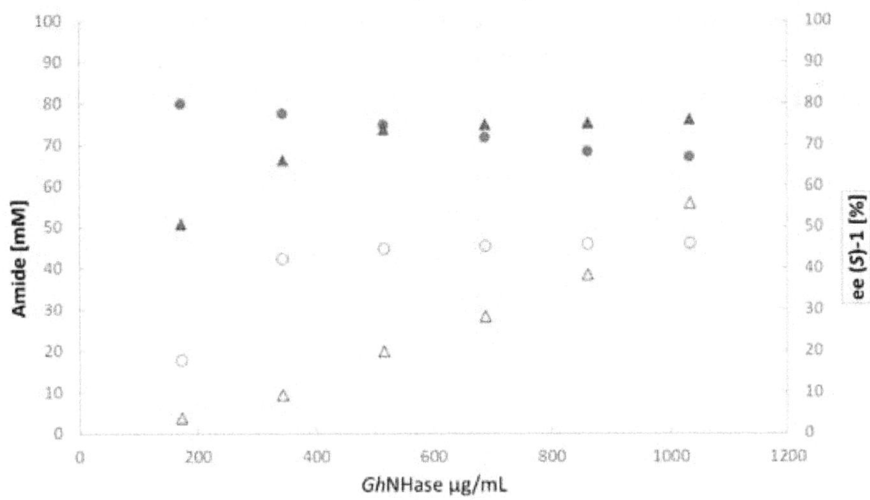

Figure 1. Effect of catalyst amount for the formation of (*S*)-1b by *Gh*NHase-CFE. The hydrations of 50 mM (dots) or 200 mM (triangles) of (*R,S*)-1a were performed in Tris-HCl buffer (300 mM, pH 7.5) at 25 °C and 500 rpm for 30 min. Filled symbols (top) indicate the *ee* (right y-axis); empty symbols (bottom) indicate product concentrations (left y-axis).

2.2. Enzyme Activity in the Presence of Additives

Iron-dependent NHases are characterized by a low-spin Fe^{III} ion in the active site [16]. The inhibition caused by cyanide as well as product inhibition have been reported [23]. Metal ions can form cyanide complexes. A strategy to alleviate cyanide inhibition might be the addition of extra metal ions to dynamic kinetic resolution reactions. We therefore tested the effect of metal ion addition. For this purpose, the hydration of methacrylonitrile (MAN) was monitored in a plate reader [24]. Notably, the activity of *Gh*NHase for MAN hydration was an order of magnitude lower than that of other NHases [18]. Externally added Fe^{III}, Fe^{II} and Mn^{II} decreased the activity of *Gh*NHase in the photometric assay, with the exception of 1 mM Fe^{III} .

2.3. Effect of Reaction pHs

The pH value is a critical factor in biocatalytic amide synthesis in general, and particularly in (*S*)-**1b** synthesis. The substrate is a labile α-aminonitrile. It is in equilibrium with its three building blocks pyrrolidine, propanal and cyanide [14]. When **1a** is incubated at different pH values, the formation of hydrocyanic acid can be detected in the headspace of the reactions [25]. At pH 10, **1a** disintegrates slowly, whereas low pH (pH 5.0) favours substrate dissociation within seconds. Nitrile hydratases show the highest activity from pH 7 to 8 [18], and the activity may also depend on the nature and strength of buffer components [26,27]. Taken together, these interdependent parameters require experimental exploitation, and we observed the same trend in two buffers as depicted in **Figure 2**: The lower the pH, the more product formed. Confirming the observations made in **Figure 1**, the analytical yield and the *ee* followed a reciprocal dependency. A similar trend can be observed when *Gh*NHase is applied as a whole-cell biocatalyst. The most promising combination of the product titer and *ee* was obtained in Tris-HCl at pH 7.5.

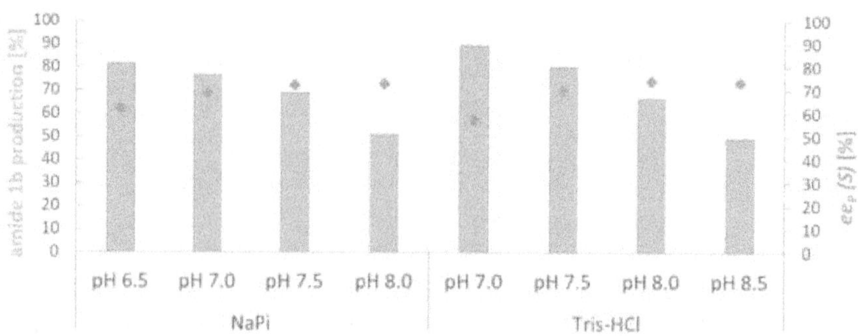

Figure 2. Formation of (*S*)-1b by *Gh*NHase-CFE at different pH. The orange-colored diamonds indicate the enantiomeric excess. (*R*,*S*)-1a (50 mM) was treated with 0.64 mg/mL *Gh*NHase in sodium phosphate or Tris-HCl buffer (100 mM) at 25°C and 500 rpm.

2.4. Synthesis Reaction from Single Components

The experiments up to this point have been performed via the addition of synthesized and purified racemic **1a**. Ultimately, the isolation of (*R*,*S*)-**1a** should be obsolete, since compound (*R*,*S*)-**1a** forms spontaneously under the applied conditions. To demonstrate the proof of concept, KCN was solubilized in a buffer and mixed with pyrrolidine and propanal prior to the addition of the biocatalyst (344 µg/mL *Gh*NHase added as CFE in finally 10% of the reaction volume). It needs to be noted that the strong basic character of KCN and pyrrolidine required a high buffer capacity and posed limitations with respect to substrate concentrations on this analytical level. Hence, the theoretical maximal yield was 27 mM of (*S*)-**1b**. KCN was chosen as a limiting component and propanal was applied in excess on the basis of experience with *Ct*NHase [14]. The *Ct*NHase wildtype was tested under identical conditions for comparison. The excess of propanal had a pronounced effect on product formation (**Table 2**, entry 1 versus 3). In the direct comparison, *Gh*NHase delivered significantly more of the desired amide (*S*)-**1b** as compared to the Co-dependent wild-type *Ct*NHase, albeit in lower enantiomeric purity. Product yields could be improved even further by carrying out the reaction in reactors that control pH and substrate feed. Protein engineering is a promising means of improving the enantio-selectivity of *Gh*NHase and suppressing by-product formation as shown in the example of *Ct*NHase [14,28]. The enantio-selectivity of the low molecular weight NHase from *Rhodococcus rhodochrous* J1, for example, was improved by using steered molecular dynamics simulations to identify key residues, followed by site saturation mutagenesis [29].

Table 2. One-pot enzymatic dynamic kinetic resolution by Strecker reaction from three precursors. Conditions: pH 7.3, 25 °C, 344 µg/mL of *Gh*NHase, 30 min reaction time.

Entry	NHase	KCN [mM]	Pyrrolidine [mM]	Propanal [mM]	(*S*)-1b [mM]	*ee* (*S*)-1b [%]
1	*Gh*HNase	26.9	29.4	approx. 30 [1]	2.3 ± 0.0	75.9 ± 0.2
2	*Ct*NHase	26.9	29.4	approx. 30 [1]	0.1 ± 0.0	n.d. [2]
3	*Gh*HNase	26.9	29.4	approx. 90 [1]	8.4 ± 0.4	73.9 ± 0.1
4	*Ct*NHase	26.9	29.4	approx. 90 [1]	4.9 ± 0.1	82.1 ± 0.1

[1] Propanal volatility impairs correct pipetting of small volumes. [2] Peak area of (*R*)-**1b** below detection limit.

2.5. Exploration of Substrate Scope

*Gh*NHase proved to be a promising enzyme for levetiracetam synthesis. Many other APIs that harbor amide groups and nitriles are frequently chosen as the precursor group to introduce the amide moiety [30]. In general, enzymes are frequently incorporated in API synthesis due to their unique selectiveness [31,32,33]. In this light, the ability of *Gh*NHase to produce a variety of structurally diverse compounds was studied (Figure 3). Benzonitrile (**2a**) was chosen as a model substrate for aromatic amide precursors. Cinnamonitrile (**3a**), [34] mandelonitrile (**4a**) [20] and 3-amino-3-tolyl-propanenitrile (**8a**) [35] are representatives of aryl-aliphatic nitriles. Methacrylonitrile (**5a**) is a small, aliphatic nitrile that is often used as a test substrate for NHase activity [18]. The nitrogen-containing aromatic heterocyclic compound 3-cyanopyridine (**6a**) is the precursor of nicotinamide (**6b**) [36]. Pyrazine-2-carbonitrile (**7a**), a heteroaromatic compound with two nitrogen atoms, gives pyrazine-2-carboxamide upon hydration, which is known to be an antitubercular agent [37]. The levetiracetam precursor **1a** represents a non-aromatic heterocyclic compound. Consistent with our results in the photometric assay that was used to study inhibition phenomena (data not shown), **5a** was a poor substrate of *Gh*NHase [18], and so was **4a**. About 50% of **8a** was converted to the respective amide under the tested conditions, indicating enantio-selectivity for one of the enantiomers. *Gh*NHase showed full conversion of the nitriles **2a**, **6a** and **7a**, indicating a preference for substrates with the cyano group directly attached on an aromatic system. Under modified conditions, the nicotinamide precursor **6a** appears to be the preferred substrate of *Gh*NHase among the three substrates with full conversion: **2a**, **6a**, and **7a**.

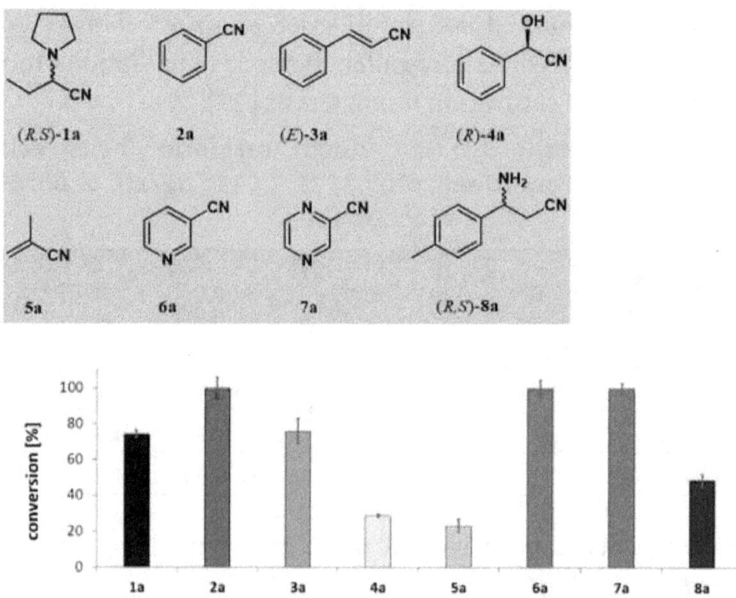

Figure 3. Substrate scope of *Gh*NHase. Top: chemical structures of investigated substrates. Bottom: nitriles a (50 mM) were converted by *Gh*NHase CFE (344 μg/mL NHase) at 25 °C and 500 rpm for 30 min. Each reaction was carried out in triplicate.

3. MATERIALS AND METHODS

3.1. General

Tris was purchased from Carl Roth, Karlsruhe; IPTG from Serva Electrophoresis, Heidelberg; MAN from Fluka, Buchs; and $FeSO_4*7H_2O$ from Merck, Darmstadt. 2-(Pyrrolidine-1-yl)butanenitrile was prepared according to the procedure of Orejarena Pacheco et al. [**38**]. Pyrrolidine, propanal, KCN and all other chemicals were obtained from Sigma–Aldrich, St. Louis, MO, USA, and used without further purification. *Gh*NHase (accession numbers WP_066163464.1, WP_066163466.1 and NZ_BCWU01000002.1, for the α-subunit, β-subunit and accessory protein, respectively) was previously prepared in *Escherichia coli* [**18**]. A Series 1100 HPLC system equipped with a diode array detector (DAD) was used for chiral analyses (Hewlett Packard, Palo Alto, CA, USA).

3.2. Biocatalyst Preparation

E. coli BL21 Gold (DE3) cells expressing NHases were grown at 37°C in 400 mL of LB-ampicillin (100 µg/mL) media up to an OD_{600} of 0.8–1.0. Protein expression was induced with 0.1 mM IPTG and 1 mM or 2.5 mM $FeSO_4*7H_2O$, 0.1 mM or 1 mM $CoCl_2$ at 20°C for 24 h. Cells were pelleted by centrifugation at 5000× *g* and 4°C for 15 min. An amount of 25 mL of 50/40 mM Tris-butyrate buffer, pH 7.2, was used for resuspension of the cell pellet (1.2–3.0 g cell wet weight) and disrupted on ice by sonication for 6 min at 70–80% duty cycle and 7–8 output control with a Sonifier 250 (Branson, Danbury). After centrifugation for 1 h at 4°C and 48,250× *g*, cell-free extracts were filtered using 0.45 µm syringe filters before being applied as biocatalyst. The Pierce™ BCA Protein Assay Kit (ThermoFisher Scientific, Waltham, MA, USA) was used for determining protein concentrations.

3.3. Conversions of (R,S)-1a to (S)-1b

Cell-free extract containing *Gh*NHase (162 µg/mL–1.08 mg/mL) was mixed with substrate solution (*R*,*S*)-**1a** in buffer, (40 Mm–300 mM, pH 6.5–8.5) at a total volume of 500 µL and incubated for up to 16 h at 5–50 °C and 300 rpm in Eppendorf thermomixers. The reactions were stopped by the addition of 1 mL ethanol and vortexing. After centrifugation of precipitated protein, the supernatants were analyzed by chiral HPLC on a Chiralpak AD-RH (150 × 4.6 mM, 5 µm) using Na-borate buffer, (20 mM, pH 8.5) and acetonitrile in a ratio of 70:30 as the mobile phase. The flow rate was 0.5 mL/min for 15 min. The compounds were detected at 210 nm (DAD). A calibration curve of racemic **1b** (external standard) was used for quantification by linear interpolation. The enantiomeric excess was calculated with the following formula, where [R] is the peak area of the (*R*)-**1b** and [S] the peak area of (*S*)-**1b**. Retention times of (*R*)- and (*S*)-**1b** were 5.8 min and 6.4 min, respectively. Nitrile **1a** cannot be analyzed in the same chromatography run, as it decomposes under these conditions.

$$ee\ [\%] = \frac{[S] - [R]}{[R] + [S]}\ \%$$

Analytical yield % of **1b** was defined as the concentration of both enantiomers of **1b** in relation to the concentration of the added starting material racemic **1a**. The term production is used as synonym for yield.

$$Analytical\ yield\ [\%] = \frac{\{[(S) - \mathbf{1b}] + [(R) - \mathbf{1b}]\}}{[(R,S) - \mathbf{1a}]_0}\ \%$$

Conversion % refers to substrate consumption and is calculated with the substrate concentration [S] at a given time and the initial substrate concentration $[S]_0$.

$$Conversion\ [\%] = 1 - \frac{[S]}{[S]_0}\ \%$$

3.4. Enzyme Activity in the Presence of Additives

The hydration of methacrylonitrile (MAN, **5a**) was monitored as described previously [18]. Metals were added as solutions of $MnCl_2$, $FeCl_2$ or $FeCl_3$.

3.5. One-Pot Enzymatic Dynamic Kinetic Resolution by Strecker Reaction from Three Precursors

KCN (3.65 mg/mL) was dissolved in Tris-HCl buffer (300 mM, pH 7.5) and mixed with pyrrolidine (3.22 mg/mL) and propanal (4.2 µL/mL or 12.4 µL/mL). KH_2PO_4 (200 mM) was used to adjust the pH to 7.3. Final buffer concentrations were 150 mM Tris HCl and 100 mM potassium phosphate. Per reaction, 50 µL of cell-free extract was mixed with 450 µL of this reaction mixture. Each reaction was carried out in triplicate. The reactions proceeded at 25 °C and 500 rpm in an Eppendorf thermomixer. Reactions were terminated and analyzed as described in **Section 3.2**.

3.6. Chromatographic Assay for Substrate Scope Determination

The assay to explore the substrate scope of *Gh*NHase was carried out as follows: 50 µL of cell-free extract containing *Gh*NHase (approximately 611 µg) was mixed with 450 µL of substrate solution (final concentration 50 mM in potassium phosphate buffer, pH 7.2, and 5% of DMSO). The samples were incubated for 30 min at 25°C and 500 rpm. The reactions were stopped by the addition of 500 µL EtOH. Precipitated protein was removed by centrifugation, and the supernatants were analyzed by HPLC-UV. Agilent Technologies 1100 Series equipped with a G1379B degasser, 1200

Series Binary pump G1312B, G1367A autosampler, G1330B autosampler thermostat, G1316A thermostated column compartment and a G1315B diode array detector (DAD) was used. The analysis was carried out with a Kinetex 2.6µ Biphenyl 100A HPLC column (Phenomenex). The mobile phases were ammonium acetate (5 mM) with 0.5% (*v/v*) acetic acid in water and ACN at a flow rate of 0.26 mL min^{-1}. The following stepwise gradient was used: 5–35% ACN (5 min), 35–60% ACN (5.0–7.2 min) and 60–90% ACN (7.2–7.5 min). For quantification of compounds, calibration curves were determined at 254 nm (nitriles) or 230 nm (amides), and linear interpolation was used.

4. CONCLUSIONS

In conclusion, a virtually unexplored nitrile hydratase from *Gordonia hydrophobica* (*Gh*NHase) was evaluated as a biocatalyst for the preparation of API building blocks and, in particular, for the preparation of the API levetiracetam. Focusing on a dynamic kinetic resolution strategy, the non-oxidized levetiracetam precursor molecule 2-(pyrrolidine-1-yl)butane amide was synthesized enzymatically, either starting from the racemic mixture of 2-(pyrrolidine-1-yl)butanenitrile or from an aqueous solution of pyrrolidine, propanal and KCN. Whereas analytical yields were significantly better than those described for *Comamonas testosteroni* nitrile hydratase (up to 60% versus up to 90%), enantiomeric excess values reached lower levels (>90% *ee* versus 70–80% *ee*). Due to higher product titers, the formation of the putative byproduct 2-hydroxybutanamide was less pronounced for *Gh*NHase as compared to the case of the cobalt-dependent *Ct*NHase. Conversions were strongly determined by substrate concentrations, the reaction pH and the biocatalyst amount, and these parameters also influenced the degree of selective formation of the desired (*S*)-configured amide. The promising activity with this heterocyclic API building block motivated us to explore the substrate scope by providing structurally diverse nitriles as substrates to *Gh*NHase, which revealed a preference for heterocyclic nitriles. *Gh*NHase-mediated hydration of 3-cyano-pyridine to nicotinamide was found to be the most efficient reaction, but the broad substrate tolerance of *Gh*NHase indicates that this enzyme is a valuable addition to the limited toolbox of iron-dependent nitrile hydratases.

5. CONCISE SUMMARY

The active pharmaceutical ingredient levetiracetam has anticonvulsant properties and is used to treat epilepsies. Herein, we describe the enantioselective preparation of the levetiracetam precursor 2-(pyrrolidine-1-yl) butanamide by enzymatic dynamic kinetic resolution with a nitrile hydratase enzyme. A rare representative of the family of iron-dependent nitrile hydratases from *Gordonia hydrophobica* (*Gh*NHase) was evaluated for its potential to form 2-(pyrrolidine-1-yl) butanamide in enantioenriched form from the three small, simple molecules, namely, propanal, pyrrolidine and cyanide. The yield and the enantiomeric excess (*ee*) of the product are determined most significantly by the substrate concentrations, the reaction pH and the biocatalyst amount. *Gh*NHase is also active for the hydration of other nitriles, in particular for the formation of *N*-heterocyclic amides such as nicotinamide, and may therefore be a tool for the preparation of various APIs.

REFERENCES

1. Yukl, E.T.; Wilmot, C.M. Cofactor biosynthesis through protein post-translational modification. *Curr. Opin. Chem. Biol.* **2012**, *16*, 54–59. **[Google Scholar]** **[CrossRef]****[Green Version]**
2. Asano, Y.; Yasuda, T.; Tani, Y.; Yamada, H. A New Enzymatic Method of Acrylamide Production. *Agric. Biol. Chem.* **1982**, *46*, 1183–1189. **[Google Scholar]** **[CrossRef]**
3. Martinkova, L.; Mylerova, V. Synthetic Applications of Nitrile-Converting Enzymes. *Curr. Org. Chem.* **2003**, *7*, 1279–1295. **[Google Scholar]** **[CrossRef]**
4. van Pelt, S.; Zhang, M.; Otten, L.G.; Holt, J.; Sorokin, D.Y.; van Rantwijk, F.; Black, G.W.; Perry, J.J.; Sheldon, R.A. Probing the enantioselectivity of a diverse group of purified cobalt-centred nitrile hydratases. *Org. Biomol. Chem.* **2011**, *9*, 3011. **[Google Scholar]** **[CrossRef]** **[PubMed]**
5. Gotor, V.; Gotor-Fernández, V.; Busto, E. 7.6 Hydrolysis and Reverse Hydrolysis: Hydrolysis and Formation of Amides. In *Comprehensive Chirality*; Elsevier Ltd.: Amsterdam, The Netherlands, 2012; Volume 7, pp. 101–121. ISBN 9780080951683. **[Google Scholar]**
6. Lin Lin Lee, V.; Kar Meng Choo, B.; Chung, Y.-S.; Kundap, U.P.; Kumari, Y.; Shaikh, M. Treatment, Therapy and Management of Metabolic Epilepsy: A Systematic Review. *Int. J. Mol. Sci.* **2018**, *19*, 871. **[Google Scholar]** **[CrossRef]** **[Green Version]**
7. Lynch, B.A.; Lambeng, N.; Nocka, K.; Kensel-Hammes, P.; Bajjalieh, S.M.; Matagne, A.; Fuks, B. The synaptic vesicle is the protein SV2A is the binding site for the antiepileptic drug levetiracetam. *Proc. Natl. Acad. Sci. USA* **2004**, *101*, 9861–9866. **[Google Scholar]** **[CrossRef]** **[PubMed]****[Green Version]**
8. Sills, G.J.; Rogawski, M.A. Mechanisms of action of currently used antiseizure drugs. *Neuropharmacology* **2020**, *168*, 107966. **[Google Scholar]** **[CrossRef]**
9. Lukyanetz, E.A.; Shkryl, V.M.; Kostyuk, P.G. Selective Blockade of N-Type Calcium Channels by Levetiracetam. *Epilepsia* **2002**, *43*, 9–18. **[Google Scholar]** **[CrossRef]** **[PubMed]**

10. Carunchio, I.; Pieri, M.; Ciotti, M.T.; Albo, F.; Zona, C. Modulation of AMPA Receptors in Cultured Cortical Neurons Induced by the Antiepileptic Drug Levetiracetam. *Epilepsia* **2007**, *48*, 654–662. [Google Scholar] [CrossRef] [PubMed]

11. Meehan, A.L.; Yang, X.; Yuan, L.L.; Rothman, S.M. Levetiracetam has an activity-dependent effect on inhibitory transmission. *Epilepsia* **2012**, *53*, 469–476. [Google Scholar] [CrossRef]

12. Noyer, M.; Gillard, M.; Matagne, A.; Hénichart, J.P.; Wülfert, E. The novel antiepileptic drug levetiracetam (ucb L059) appears to act via a specific binding site in CNS membranes. *Eur. J. Pharmacol.* **1995**, *286*, 137–146. [Google Scholar] [CrossRef]

13. Kotkar, S.P.; Sudalai, A. A short enantioselective synthesis of the antiepileptic agent, levetiracetam based on proline-catalyzed asymmetric α-aminooxylation. *Tetrahedron Lett.* **2006**, *47*, 6813–6815. [Google Scholar] [CrossRef]

14. Arndt, S.; Grill, B.; Schwab, H.; Steinkellner, G.; Pogorevčnik, U.; Weis, D.; Nauth, A.M.; Gruber, K.; Opatz, T.; Donsbach, K.; et al. The sustainable synthesis of levetiracetam by an enzymatic dynamic kinetic resolution and an: Ex-cell anodic oxidation. *Green Chem.* **2021**, *23*, 388–395. [Google Scholar] [CrossRef]

15. Prasad, S.; Bhalla, T.C. Nitrile hydratases (NHases): At the interface of academia and industry. *Biotechnol. Adv.* **2010**, *28*, 725–741. [Google Scholar] [CrossRef]

16. Hopmann, K.H. Full reaction mechanism of nitrile hydratase: A cyclic intermediate and an unexpected disulfide switch. *Inorg. Chem.* **2014**, *53*, 2760–2762. [Google Scholar] [CrossRef]

17. Petrillo, K.L.; Wu, S.; Hann, E.C.; Cooling, F.B.; Ben-Bassat, A.; Gavagan, J.E.; DiCosimo, R.; Payne, M.S. Over-expression in Escherichia coli of a thermally stable and regio-selective nitrile hydratase from *Comamonas testosteroni* -5MGAM4-D. *Appl. Microbiol. Biotechnol.* **2005**, *67*, 664–670. [Google Scholar] [CrossRef] [PubMed]

18. Grill, B.; Glänzer, M.; Schwab, H.; Steiner, K.; Pienaar, D.; Brady, D.; Donsbach, K.; Winkler, M. Functional Expression and Characterization of a Panel of Cobalt and Iron-Dependent Nitrile Hydratases. *Molecules* **2020**, *25*, 2521. [Google Scholar] [CrossRef]

19. Přepechalová, I.; Martínková, L.; Stolz, A.; Ovesná, M.; Bezouška, K.; Kopecký, J.; Křen, V. Purification and characterization of the enantioselective nitrile hydratase from *Rhodococcus equi* A4. *Appl. Microbiol. Biotechnol.* **2001**, *55*, 150–156. [Google Scholar] [CrossRef] [PubMed]

20. Pawar, S.V.; Yadav, G.D. Enantioselective Enzymatic Hydrolysis of *rac*- Mandelonitrile to *R*-Mandelamide by Nitrile Hydratase Immobilized on Poly(vinyl alcohol)/Chitosan–Glutaraldehyde Support. *Ind. Eng. Chem. Res.* **2014**, *53*, 7986–7991. [Google Scholar] [CrossRef]

21. D'Antona, N.; Morrone, R.; Gambera, G.; Pedotti, S. Enantiorecognition of planar "metallocenic" chirality by a nitrile hydratase/amidase bienzymatic system. *Org. Biomol. Chem.* **2016**, *14*, 4393–4399. [Google Scholar] [CrossRef] [PubMed]

22. Reisinger, C.; Osprian, I.; Glieder, A.; Schoemaker, H.E.; Griengl, H.; Schwab, H. Enzymatic hydrolysis of cyanohydrins with recombinant nitrile hydratase and amidase from *Rhodococcus erythropolis*. *Biotechnol. Lett.* **2004**, *26*, 1675–1680. [Google Scholar] [CrossRef]

23. Bui, K.; Maestracci, M.; Thiery, A.; Arnaud, A.; Galzy, P. A note on the enzymic action and biosynthesis of a nitrile-hydratase from a *Brevibacterium* sp. *J. Appl. Bacteriol.* **1984**, *57*, 183–190. [Google Scholar] [CrossRef]

24. Murakami, T.; Nojiri, M.; Nakayama, H.; Dohmae, N.; Takio, K.; Odaka, M.; Endo, I.; Nagamune, T.; Yohda, M. Post-translational modification is essential for catalytic activity of nitrile hydratase. *Protein Sci.* **2000**, *9*, 1024–1030. [Google Scholar] [CrossRef][Green Version]

25. Krammer, B.; Rumbold, K.; Tschemmernegg, M.; Pöchlauer, P.; Schwab, H. A novel screening assay for hydroxynitrile lyases suitable for high-throughput screening. *J. Biotechnol.* **2007**, *129*, 151–161. [Google Scholar] [CrossRef] [PubMed]

26. Huang, W.; Jia, J.; Cummings, J.; Nelson, M.; Schneider, G.; Lindqvist, Y. Crystal structure of nitrile hydratase reveals a novel iron centre in a novel fold. *Structure* **1997**, *5*, 691–699. [Google Scholar] [CrossRef][Green Version]

27. Nagasawa, T.; Nanba, H.; Ryuno, K.; TakeuchiI, K.; Yamada, H. Nitrile hydratase of *Pseudomonas chlororaphis* B23. Purification and characterization. *Eur. J. Biochem.* **1987**, *162*, 691–698. [Google Scholar] [CrossRef]

28. Tucker, J.L.; Xu, L.; Yu, W.; Scott, R.W.; Zhao, L.; Ran, N. Chemoenzymatic processes for preparation of levetiracetam. *PCT Int. Appl.* **2009**, *9*, A3. [Google Scholar]

29. Cheng, Z.; Peplowski, L.; Cui, W.; Xia, Y.; Liu, Z.; Zhang, J.; Kobayashi, M.; Zhou, Z. Identification of key residues modulating the stereoselectivity of nitrile hydratase toward rac-mandelonitrile by semi-rational engineering. *Biotechnol. Bioeng.* **2018**, *115*, 524–535. [Google Scholar] [CrossRef]

30. Mashweu, A.R.; Chhiba-Govindjee, V.P.; Bode, M.L.; Brady, D. Substrate Profiling of the Cobalt Nitrile Hydratase from *Rhodococcus rhodochrous* ATCC BAA 870. *Molecules* **2020**, *25*, 238. [Google Scholar] [CrossRef][Green Version]

31. Patel, R.N. Pharmaceutical Intermediates by Biocatalysis: From Fundamental Science to Industrial Applications. In *Applied Biocatalysis: From Fundamental Science to Industrial Applications*; Wiley-VCH Verlag GmbH & Co. KGaA: Weinheim, Germany, 2016; pp. 367–403. [**Google Scholar**]

32. Arroyo, M.; de la Mata, I.; García, J.L.; Barredo, J.L. Biocatalysis for Industrial Production of Active Pharmaceutical Ingredients (APIs). In *Biotechnology of Microbial Enzymes: Production, Biocatalysis and Industrial Applications*; Elsevier Inc.: Amsterdam, The Netherlands, 2017; pp. 451–473. ISBN 9780128037461. [**Google Scholar**]

33. Santi, M.; Sancineto, L.; Nascimento, V.; Braun Azeredo, J.; Orozco, E.V.M.; Andrade, L.H.; Gröger, H.; Santi, C. Flow Biocatalysis: A Challenging Alternative for the Synthesis of APIs and Natural Compounds. *Int. J. Mol. Sci.* **2021**, *22*, 990. [**Google Scholar**] [**CrossRef**]

34. Cowan, D.; Cramp, R.; Pereira, R.; Graham, D.; Almatawah, Q. Biochemistry and biotechnology of mesophilic and thermophilic nitrile metabolizing enzymes. *Extremophiles* **1998**, *2*, 207–216. [**Google Scholar**] [**CrossRef**] [**PubMed**]

35. Chhiba, V.; Bode, M.L.; Mathiba, K.; Kwezi, W.; Brady, D. Enantioselective biocatalytic hydrolysis of β-aminonitriles to β-amino-amides using *Rhodococcus rhodochrous* ATCC BAA870-. *J. Mol. Catal. B Enzym.* **2012**, *76*, 68–74. [**Google Scholar**] [**CrossRef**]

36. Nagasawa, T.; Mathew, C.D.; Mauger, J.; Yamada, H. Nitrile Hydratase-Catalyzed Production of Nicotinamide from 3-Cyanopyridine in *Rhodococcus rhodochrous* J1. *Appl. Environ. Microbiol.* **1988**, *54*, 1766–1769. [**Google Scholar**] [**CrossRef**] [**PubMed**][**Green Version**]

37. Zhou, S.; Yang, S.; Huang, G. Design, synthesis and biological activity of pyrazinamide derivatives for anti-*Mycobacterium tuberculosis*. *J. Enzyme Inhib. Med. Chem.* **2017**, *32*, 1183–1186. [**Google Scholar**] [**CrossRef**] [**PubMed**] [**Green Version**]

38. Orejanrena Pacheco, J.C.; Opatz, T. NEXT Ring Expansion of 1,2,3,4-Tetrahydroisoquinolines to Dibenzo[c,f]azonines. An Unexpected [1,4]-Sigmatropic Rearrangement of Nitrile-Stabilized Ammonium Ylides. *J. Org. Chem.* **2014**, *79*, 5182–5192. [**Google Scholar**] [**CrossRef**] [**PubMed**]

CHAPTER-8

EFFECTS OF LOWER TEMPERATURE ON EXPRESSION AND BIOCHEMICAL CHARACTERISTICS OF HCV NS3 ANTIGEN RECOMBINANT PROTEIN

1. INTRODUCTION: AN OVERVIEW

Exploring the impact of reduced temperatures on the expression and biochemical attributes of the Hepatitis C Virus (HCV) NS3 antigen recombinant protein is the focal point of this study. As temperature intricately influences protein folding, stability, and overall expression, investigating its effects on the NS3 antigen becomes crucial for optimizing production processes and understanding the protein's functional properties. This research aims to unravel the intricate relationship between lower temperatures and the recombinant NS3 antigen, providing valuable insights into potential strategies for enhancing protein yield and maintaining its biochemical integrity. Such insights hold significance in the context of both biotechnological applications and advancing our understanding of HCV-related antigenic proteins.

The hepatitis C virus (HCV), which infects approximately 3% of the world population annually, is a major etiology of the blood transfusion-associated non-A and non-B hepatitis [1,2]. The reduction in post-transfusion HCV incidence for blood donors largely depends on proper execution and lab practice. The most common screening method currently used employs ELISA (enzyme-linked immunosorbent assay) or NAT (nucleic acid amplification technology) to detect anti-HCV antibodies or HCV RNA in the serum sample. Although NAT could identify an extremely low level of virus at a very early stage of the infection, it is time-consuming, and the contamination potential limits its clinical application. ELISA is hence still a more favourable choice for its relatively cheap and rapid output [3].

HCV, which belongs to the Flaviviridae family, is a small-enveloped virus with a single-stranded sense (positive) RNA genome [4]. The viral genome encodes a single polyprotein, which is composed of about 3000 amino acids [5]. This polyprotein is cleaved by viral and cellular proteases into several mature proteins, which includes three structural proteins (core protein, envelopes 1 and 2) and six nonstructural proteins (NS2, NS3, NS4a, NS4b, NS5a, NS5b). Among these mature proteins, only a few are immunogenic [6,7], whereas among nonstructural proteins, protein 3 (NS3) has been demonstrated to possess potential antigenic efficacy and be able to induce high levels of antibodies in the early infection stage [8]. It also possesses enzymatic activities, including protease [9] and helicase [10], which are required for viral replication. Mainly due to its conserved sequence, apart from its immunogenic property, NS3 has been commonly used as an antigen of several commercial ELISA diagnostic products (MUREX, MP, ORTHO, INNOTEST, GBC) [11,12,13]. Several studies have been carried out to produce soluble NS3 fractions in high quantities through the cloning approach, including the refolding of over-expressed bulk fractions as inclusion bodies of NS3 recombinant proteins [14,15], and to produce an active domain such as the protease domain [16] or the helicase domain [17]. Nevertheless, due to its auto-protease activity and its aggregation in the *Escherichia coli* expression system, the industrial-scale production of this recombinant protein is still a major challenge.

Therefore, in the current study, a clone specially designed to synthesize a truncated NS3 recombinant protein (without the protease domain) when over-expressed in the *E. coli* expression system was developed. The aim of this research is to overcome the refolding challenge of the recombinant protein (as inclusion body: lacking proper folding and activity) when expressed under a strong promoter in the *E. coli* system growing at 37°C. The refolding of the inclusion body undergoes several time-consuming tedious steps of high salt solubilization and salt removing purification. Alternatively, to reduce the rate of protein expression (to avoid inclusion body formation) and improve its folding (soluble fraction), a lower temperature (25°C) culture incubation was opted for its expression besides the usual temperature incubation at 37°C. Temperature effects on recombinant protein production, protein stability, and antigenicity were characterized, and a comparative account was presented between both recombinant proteins obtained from the 25°C and 37°C incubations. The comparison of recombinant HCV ELISA with two commercial kits, the Abbott HCV EIA 2.0 and the Ortho HCV EIA 3.0, were also carried out and thoroughly discussed.

2. RESULTS AND DISCUSSIONS

2.1. Preparation of Recombinant HCV NS3

2.1.1. Expression and Purification of the Recombinant NS3 Proteins

To prevent auto-degradation, a recombinant clone encoding a truncated NS3 with partial protease domain deletion was generated. The recombinant clone was over expressed in the *E. coli* expression system by IPTG induction at 37°C or 25°C, and then the *E. coli* cell was collected. After cell disruption, about 15% of the NS3 protein was present in the cell debris (data not shown) of the total culture biomass, and only the soluble NS3 protein present in the supernatant fraction was used for purification. As shown in **Figure 1**, the bands for the NS3 protein in the supernatant fraction were more abundant at the 25°C than at the 37°C incubations. When chromatography was applied, both 37°C and 25°C incubated NS3 could be isolated efficiently by a single Talon affinity column, as shown in **Figure 2**A. Finally, as listed in **Table 1**, the NS3 production yield of the 25°C incubation was three-fold higher than that of the 37°C incubation. Even though the incubation at lower temperatures was designed to extend the incubation time from 4 to 18 h, the production yield was elevated dramatically from 4.15 to 11.1 mg/L. It was determined that when the incubation temperature was reduced from 37 to 25°C, the growth rate of bacteria was also reduced; however, the lower temperature conditions significantly increased the 3D folding degree of NS3 towards proper protein structure and further promoted the protein's solubility and productivity.

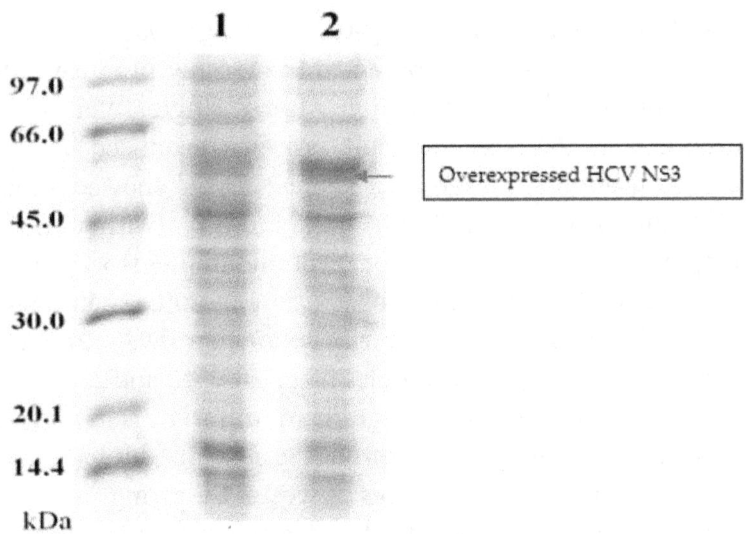

Figure 1. SDS-PAGE (12%) analysis of a crude extract. The supernatant fraction of cell lysate was incubated at 37°C (lane 1) or 25°C (lane 2). A low molecular weight protein marker was loaded and labelled their molecular weights.

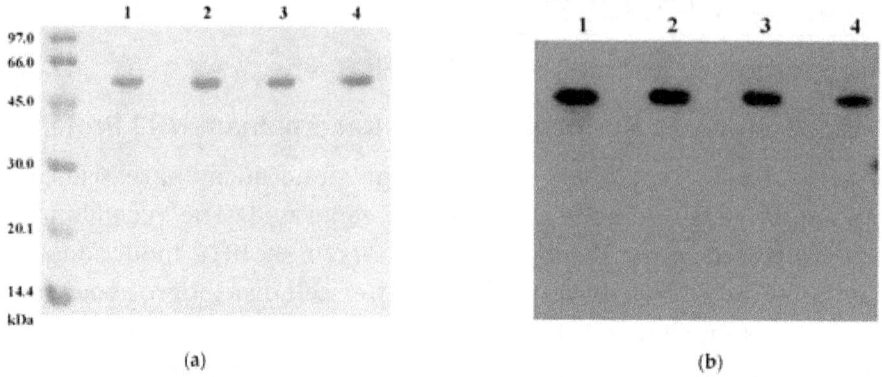

Figure 2. Purity and specificity of purified HCV NS3 antigens on SDS-PAGE (a) and western blotting (b). Lane 1 and lane 2 are long-term-stored and fresh NS3 antigens produced at 37°C, respectively. Lane 3 and lane 4 are long-term stored and fresh NS3 antigens produced at 25°C, respectively.

Table 1. Production yield of HCV NS3 in a 40 L reaction, with incubation temperatures of 37°C and 25°C.

Temperature	Crude Extract	HCV NS3	Production Yield
(°C)	(mg)	(mg)	(mg/L)
37	1030 (±5%)	166 (±3%)	4.15 (±3%)
25	910 (±4%)	445 (±1%)	11.1 (±1%)

2.1.2. Property and Stability of Fresh and Stored NS3 Proteins

The protein stability of NS3 after long-term storage was evaluated for their homogeneity and properties as compared to the freshly produced NS3 fractions in both proteins produced at 25 and 37°C. After a single Talon affinity column purification, both fresh NS3 recombinant proteins produced at an incubation temperature of 37°C and 25°C (**Figure 2**A: lane 2 and 4) showed a major band in their respective lanes of SDS-PAGE. Similar band intensity (without the smearing of degraded oligopeptides) also obtained from protein samples stored over six months (**Figure 2**A: lane 1 and 3) in a 10% glycerol solution at -20°C. Their degree of purity was above 90% by HPLC (**Figure 3**). The purity obtained was 92%, 97%, 92%, and 97%, respectively, from fresh and stored samples from the 37°C and 25°C incubations. In these two experiments, there were no significant differences in antigenic efficacy between the fresh and the long-term-stored NS3 protein antigens. The SDS-PAGE was further analyzed for their specificities by western blotting, as shown in **Figure 2**B. Besides the major anti-HCV specified band, a smear fraction was found for the long-term-stored NS3 proteins produced at 37°C (**Figure 2**B, lane 1). In contrast, no smear fraction (no protein degradation) was observed for the recombinant NS3 protein sample produced at 25°C even after six months of storage. This finding illustrates that the protein stability of NS3 antigens possessed in these fractions depended on the applied incubation temperature for growth, although less on the storage time. Moreover, the degradation of long-term-stored NS3 incubated at 37°C might be due to protease activity from the highly homogenous stock enzyme solution. Although the cloned HCV NS3 was not full-length, it still retains a part of the protease fragment region, therefore it still retains the protease hydrolysis properties. [18]. However, the recombinant protein NS3 degradation was still very low compared to other proteins reported [19,20]. Such property improves HCV resistance and for which two mutations have been reported: D168N and L153I [20]. Their molecular mechanisms are discussed in detail by recent studies covering the destabilization of receptor–ligand hydrogen bonds [18,20].

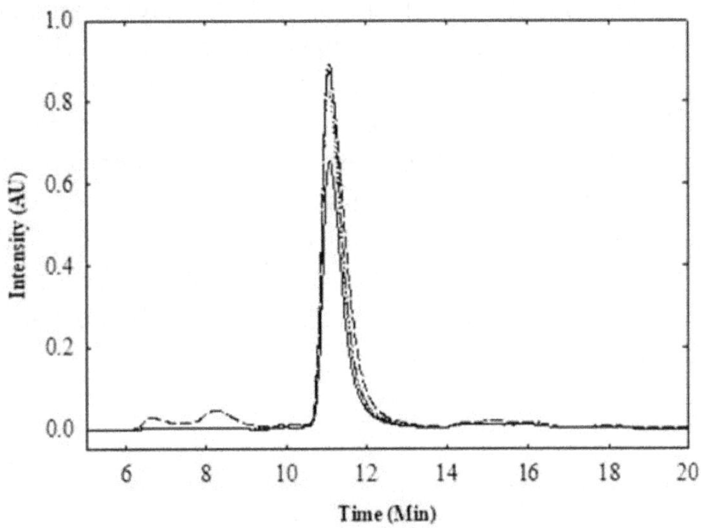

Figure 3. Size exclusion HPLC of purified HCV NS3. This assay was performed in a 0.1 M phosphate buffer at pH 6.7 by a flow rate of 0.3 mL/min using SEC column (BioSuite™ 125, 4 µm HR SEC, 7.8 mm × 300 mm, Waters Corporation, Milford, MA, USA). All samples were eluted as a major 280 nm absorption peak (retention time near 12 min). The purified recombinant protein elution peaks of fresh 37°C produced (-), fresh 25°C produced (...), long-term-stored 37°C produced (---), and long-term-stored 25°C produced (-..), obtained in order.

2.1.3. Electrospray Ionization Mass Spectrometric (ESI-MS) Analysis

The deconvolutions of the mass spectra of fresh NS3 proteins produced at 25°C or 37°C gave the same molar mass of 54,540 atomic mass units. However, cluster peaks at around 1600–2100 *m*/*z* were detected only in 25°C produced NS3 proteins (**Figure 4**A). When calculating the charge states of the monomeric NS3 (as the numbers labeled above peaks in **Figure 4**), an additional peak in the NS3 produced at 25°C was observed. These cluster peaks were isolated and deconvoluted as 109,080 amu belonging to the dimeric form of NS3. Even though the dimerization of NS3 was expected for its helicase activity [21], however, the monomeric form was still a major part of the 25°C produced protein. The truncated recombinant HCV (for protease) is well-reported for its helicase activity [22,23]. The theoretical molecular mass of the above-mentioned protein was equal to 54,656.36 [24]; however, the obtained Mw from the mass spectrum was 54,540. In general, high *m/z* values signify either a higher molecular weight or a lower charge state for the protein. In most cases, unfolded proteins require more protons and would thus appear at lower *m/z* values than those of the folded forms [25]. Thus, based on these findings, it can be concluded that the NS3 protein produced at 25°C was either more compact or with more dimeric form than those produced at 37°C. According to the amino acid sequence, this protein has 15 cysteine units; thus, it can form a maximum of up to 6–7 disulfide bonds. Nevertheless, studies have not yet been carried out to confirm the number of disulfide bonds that are in place and their effects on its properties during the ELISA test. The disulfide bond is highly important for immunogenic and antigenic performance as well as in diagnosis. It was reported from mutagenic studies that the replacement of even a single cysteine out of six cysteine residues in herpes simplex virus type 1 glycoprotein resulted in either a great reduction or a complete loss of binding with those monoclonal antibodies recognizing irregular epitopes. However, there was no effect on its binding with monoclonal antibodies recognizing continuous epitopes [26].

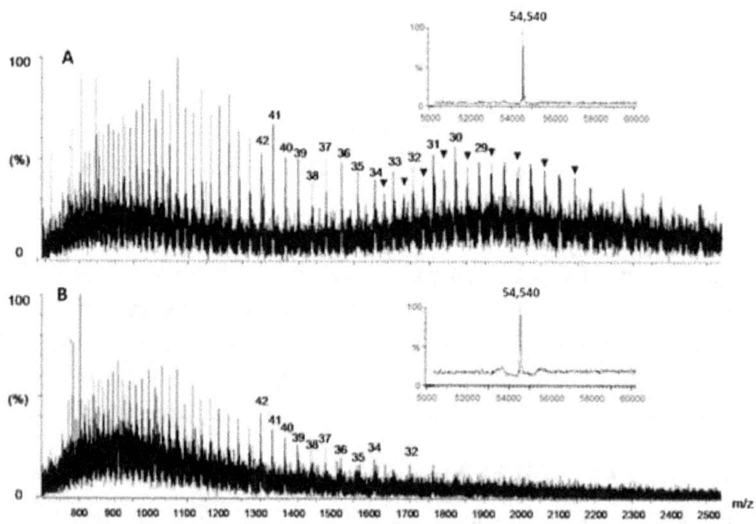

Figure 4. The mass spectra of HCV NS3 proteins produced at 25 °C (A) and 37°C (B). They are plots of ion intensity vs. *m/z* (mass-to-charge ratio). The deconvolution of the spectra (inset) give the same molar mass of 54,540 atomic mass units. Numbers above the peaks stand for the z value (charge state) of the monomeric protein, and triangle labels are the signals specific to the dimeric protein.

2.1.4. Protein Structure Stability

The protein structure was analyzed by CD spectrometry. As shown in **Figure 5**, the far-UV CD spectra showing a negative peak at 222 nm and 208 nm were a CD spectrum typical for an α-helix conformation. However, at 210–220 nm, the elliptical values of long-term-stored NS3 proteins produced at 37°C were significantly smaller than the other freshly prepared protein samples. It can be concluded that after long-term storage, the protein structure of the NS3 protein produced at 37°C was slightly changed, followed by aggregation and precipitation in the solution. The same phenomenon was described by a previous study carried out in 1999 [19]. This was also explained by the less compact 37°C produced NS3 protein structure.

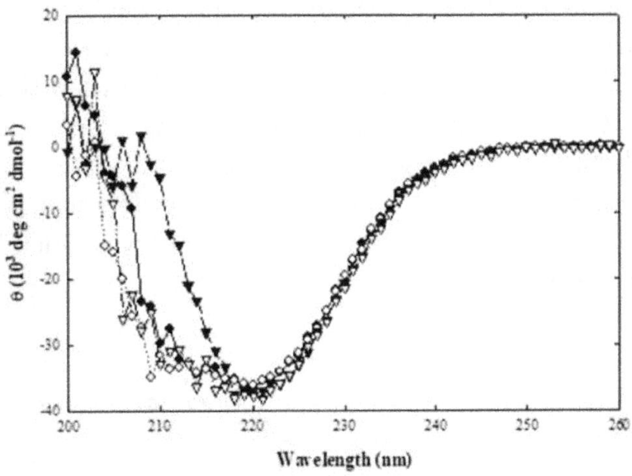

Figure 5. The far-UV CD spectroscopy of the HCV NS3 antigen recorded at 25°C. These recombinant NS3 antigens include fresh 37°C produced (•), fresh 25°C produced (o), long-term-stored 37°C produced (▼), and long-term-stored 25°C produced (∇) protein samples.

2.1.5. Thermal Stability of NS3 Proteins

By monitoring the CD signal at 222 nm in a temperature range of 4–96°C, the thermal denaturation of recombinant NS3 was measured. As shown in **Figure 6**, the CD values at 222 nm from 4 to 96°C revealed that the thermal unfolding of all NS3 proteins began at around 40°C and was completed above 65°C. When the unfolding fraction reached 50%, the observed temperatures were anticipated for the thermal denaturation (T_m) values at about 55–59°C (as shown in **Table 2**). For a more quantitative evaluation of the temperature effect, the unfolding curves were analyzed with both the two-state and the three-state transition models. Even though the unfolding curves exhibited a typical biphasic transition profile, when the unfolding curve was applied to the two-state unfolding model, it failed to obtain the corresponding thermodynamic parameters (data not shown). Moreover, the theoretical curves of the three-state transition model were best fitted to the experimental data, as shown in **Figure 6**; the corresponding thermodynamic factors are summarized in **Table 2**. One possible explanation is that two domains are independently unfolded, and two transition phases correspond to the unfolding of individual domains. The whole unfolding process was subjected to the second transition phase in which Tm was almost equal to a 50% fraction unfolded. In the second transition phase, the loss of stability of long-term-stored NS3 induced at 37°C was approximately 4.6 kcal/mol, a value that is explained by the disruption of several intra-molecular interactions. However, advanced experiments are required to clarify this phenomenon. In general, these CD values shown in **Figure 5** and **Figure 6** implied that all the recombinant NS3 possessed a similar structure and stability before the storage. After a six-month storage, however, the recombinant protein produced at a 37°C incubation was less stable than that of the protein produced at a 25°C incubation. There are three independent domains comprising this protein, including the helicase ATP-binding domain, the DEAD-like helicase C domain, and a domain containing a 7α-helix (see **Supplementary File**). Moreover, the TI (T_m Index) of the DEAD-like helicase C domain was much higher than that produced by other proteins.

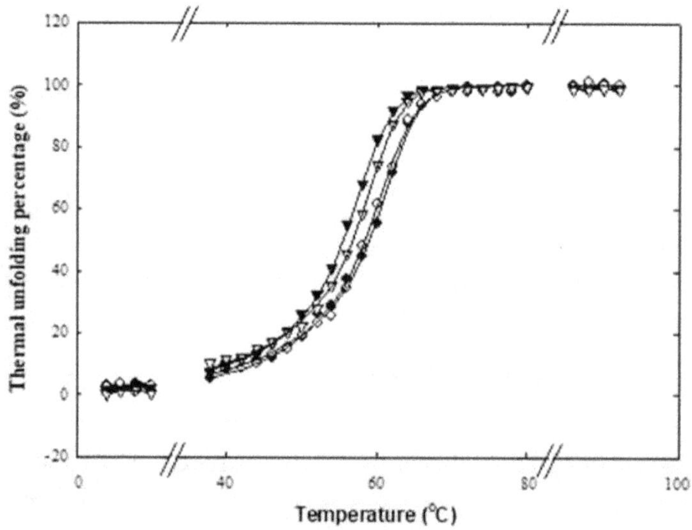

Figure 6. Thermal unfolding of HCV NS3 proteins recorded by far-UV CD spectroscopy. It was monitored by the change in ellipticity at a wavelength of 222 nm in the temperature range of 4–96°C. The thermal denaturation (T_m) of fresh 37°C produced (•), fresh 25 °C produced (o), long-term-stored 37°C produced (▼), and long-term-stored 25°C produced (∇) are obtained in order. Solid lines are the theoretical fitting curves with the thermodynamic parameters, also given in Table 2.

Table 2. Thermodynamic parameters for unfolding transition of NS3 Proteins produced at 37°C and 25 °C.

50% Unfolding		First Transition Phase						Second Transition Phase			
Sample [a]	T_m	T_m	ΔT_m	ΔH_m	ΔS_m	$\Delta\Delta G_m$	T_m	ΔT_m	ΔH_m	ΔS_m	$\Delta\Delta G_m$
	°C	°C	°C	kcal/mol	kcal/mol/K	kcal/mol	°C	°C	kcal/mol	kcal/mol/K	kcal/mol
F37	59	65.2	-	85	0.3	-	59.7	-	365	1.1	-
F25	59	64.0	-1.2	96	0.3	-0.3	59.1	-0.6	332	1.0	-0.7
L37	55	63.7	-1.5	81	0.	-0.4	55.5	-4.2	338	1.0	-4.6
L25	57	62.0	-3.2	89	0.3	-0.8	57.8	-1.9	387	1.2	-2.0

[a]: F37: Fresh 37 °C produced NS3; 25F: Fresh 25 °C produced NS3; 37L: Long-term-stored 37 °C produced NS3; 25L: Long-term-stored 25 °C produced NS3.

2.1.6. ELISA Specificity of NS3 Proteins

Two home-made HCV ELISA tests pre-coated by NS3 produced at 25°C or 37°C and two commercial kits, Abbott 2.0 and Ortho 3.0, were assessed by two BBI HCV panels (Boston Biomedica Inc., Easton, MA, USA), as shown in **Figure 7**A,B. In general, four ELISA tests demonstrated the same result: non-reactive specimens (S/Co < 1) contained negative samples and reactive specimens (S/Co > 1) contained positive and intermediate samples. It can be observed that the gap between the S/Co < 1 and the S/Co > 1 specimen presented on lane 4 was larger than the gap on lane 3. This indicated an improved specificity with 25°C incubation. However, the specificity of NS3 was not higher than the commercial kits. It was noted that these two commercial kits were multi-antigen ELISA and our lab-made microplates were single-antigen ELISA, with only NS3 as the antigen. The antigenicity of truncated recombinant NS3 protein produced at the lab was comparable with the commercial one, while its production strategies were found easy and effective via strategic low temperature expression. This method can be adopted in the near future to obtain potential products for HCV ELISA diagnosis.

The study of Elleuche et al. [27] and Lawyer et al. [28] respectively discussed the truncated-form protein's benefits for specific activities over the full-length protein. The main reason for the truncated form of the HCV NS3 used in this study was to reduce its self-hydrolytic activity, which raises NS3 purity and reduces false negatives for the detection of anti-NS3 in ELISA (data not shown).

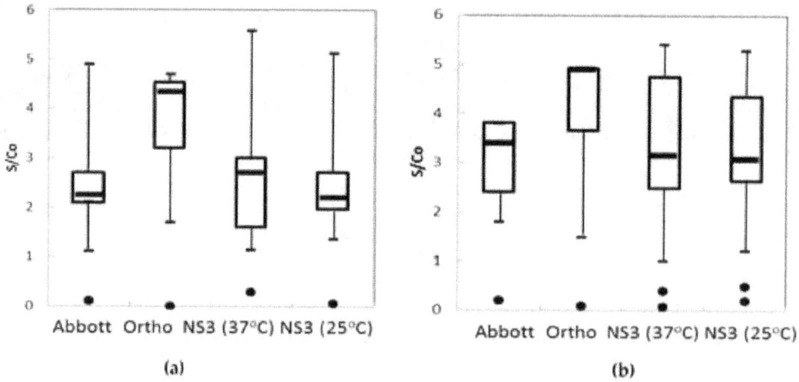

Figure 7. Box plots for ELISA specificity comparison of HCV NS3 antigens and two commercial kits. The experiments were performed using the Anti-HCV Low Titer Performance Panel PHV 105 (M) (a) and the Anti-HCV Mixed Titer Panel PHV 205 (b) as the standard specimens. The specimens of 7A include 7 positive samples, 1 negative sample (•), and 5 intermediate samples. The specimens of 7B include 20 positive samples, 2 negative sample (•), and 3 intermediate.

The expression at lower temperatures in an *E. coli* system helps to improve recombinant protein performance, which has been confirmed in the previous literature [29,30] and also confirmed in the results of our current study. Moreover, commercial biologic agents also mainly utilize lower temperatures for protein expression in an *E. coli* system, which illustrates this feature and advantage [31]. Additionally, the results of the current study are in line with the report of Sandomenico et al. [31], which confirms that the protein expression at a lower temperature (25°C) exhibits superior performance as compared to that of the higher temperature (37°C), as demonstrated by solubility, purity, antigenic efficacy, and stability.

3. MATERIALS AND METHODS

3.1. Materials

All reagents used in the study were of analytical or molecular biology grade and purchased from Sigma or Merck. *E. coli* BL21 (DE3) (BF$^-$ ompT *hsdS* $(r_B^- m_B^-)$ *dcm*$^+$ *gal*λ (DE3)) purchased from Stratagene was used for cloning and recombinant protein expression experiments. Abbott HCV EIA 2.0 and Ortho HCV EIA 3.0 were purchased from Abbott and Ortho-Clinical Diagnostics, respectively. Two commercial HCV seroconversion panels for ELISA evaluation were purchased from BBI Diagnostics (BBI Diagnostics Boston Biomedica, Inc., Easton, MA, USA). These two BBI panels contained 37 specimens. Thirteen specimens of these species were obtained from Anti-HCV Low Titer Performance Panel PHV 105 (M). There were 7 positive, 1 negative, and 5 intermediate samples. The other 24 specimens were obtained from Anti-HCV Mixed Titer Performance Panel PHV 205. There were 20 positive, 2 negative, and 3 intermediate samples.

3.2. Expression of the Recombinant NS3 Proteins

A plasmid (pET 21a) carrying a gene code for a truncated NS3 flanked by a T7 tag and a His-tag was a gift from General Biologicals Corp. (GBC: General Biologicals. Corp., Taiwan). The expressed DNA of truncated NS3 was located from 3525 to 4971 bps on the HCV genome (the HCV NS3 helicase domain, amino acid residues 1175–1657, total of 483 aa) (GenBank ID: P29846) [32] and it was ligated into a pET-21a plasmid double-digested by *BamHI* and *Hind*III. The resulting plasmid was pET21a-NS3, which was well-characterized for its recombinant DNA sequence and protein sequence. The plasmid was expressed in BL21 (DE3) and the molecular mass of recombinant protein was determined. A single colony of BL21 (DE3) harboring HCV NS3 plasmid was used to inoculate in 400 mL LB (Difco) broth containing 100 μg/mL ampicillin, incubated overnight at 37°C and then diluted to 40 L with LB-amp broth. After a 4 h incubation with shaking at 37°C, or after bacterial density reached about 0.8 at 600 nm by spectrophotometer, IPTG was added as an inducer to a final concentration of 0.5 mM. Comparative studies of recombinant protein activity/efficacy were performed by an additional incubation of 4 h at 37 °C and/or 18 h at 25°C.

3.3. Purification of the Recombinant NS3 Proteins

The harvested cells of BL21 (DE3) from the 40 L culture medium were suspended in 150 mL of Buffer A (50-mM Tris, 0.5 M-NaCl and 5-mM imidazole at pH 8.0) and lysed using a microfluidizer. The supernatant (crude extract) was separated by centrifuging (10,600× *g*, 60 min, 4°C) and applied to Talon affinity column (BD Bioscience, 50 mL), which was pre-equilibrated with a 2-fold volume of Buffer A. After washing the unbound proteins with a 10-fold volume of Buffer A by gravity at 4°C, the NS3 was eluted with 300 mL of the 100 mM imidazole (50 mM Tris, 0.5 M NaCl, and 100 mM imidazole at pH 8.0). Purified proteins were stored for further use in small aliquots (about 1 mL) at 4°C within one week (fresh NS3) or at -20°C over 6 months (long-term-stored NS3) for further experiments.

3.4. HPLC Performance

Identification was performed and purity of the recombinant proteins was analyzed using high-performance liquid chromatography (HPLC; 600E Multisolvent Delivery System, Waters Corporation, Milford, MA, USA) with an Ultra High-Resolution SEC column (BioSuite™ 125, 4 μm HR SEC, 7.8 mm × 300 mm, Waters Corporation, Milford, MA, USA) under 4 °C. About 100 μL of purified recombinant proteins was applied and eluted with 0.15 M phosphate buffer (pH 6.8) at a flow rate of 0.3 mL/min [33]. The absorbance data at 280 nm (996 Photodiode Array Detector, Waters

Corporation, Milford, MA, USA) were collected and processed using the Empower 2 Chromatography Data Software (Waters Corporation, Milford, MA, USA).

3.5. Western Blot Analysis and Protein Estimation

The purified recombinant NS3 was first analyzed by 12% SDS polyacrylamide gel electrophoresis (SDS-PAGE). Its specificity and stability were then observed by western blotting using anti-His as the primary antibody [34]. The protein concentration and secondary structure were assayed by Bradford method (Bio-Rad, Hercules, CA, USA) and Circular Dichroism spectra (details in the section below), respectively.

3.6. Electrospray Ionization Mass Spectrometric (ESI-MS) Analysis

Mass spectra were recorded with a quadrupole time-of-flight mass spectrometer (Micromass, Manchester, UK). This was used to scan at a ratio of mass to charge in the range of 100–2500 units (*m/z*), with a scan of 3 s/step and an interscan duration of 0.1 s/step. In all the ESI-MS experiments, the quadrupole scan mode was used under a capillary needle at 3 kV, a source block temperature of 80°C, and a desolvation temperature of 150 °C [35]. The desalted form of the NS3 proteins in 10% acetonitrile containing 0.1% formic acid used for the MS measurements was normally within the range of 5–10 μg. Data acquisition and processing were performed with the MassLynx software (Version 4.0).

3.7. CD Measurements

The stabilities of secondary structure were performed by Circular Dichroism (CD) spectrometry. It was recorded using an Aviv model 202 Circular Dichroism Spectrometer equipped with a 450-watt Xenon arc lamp. Far UV CD data were expressed in terms of the mean residue ellipticity (θ_{MRE}) in deg cm^2 $dmol^{-1}$ using the following:

$$\theta_{MRE} = (100\ \theta_{obs}\ Mw)/(nlc)(1)$$

where θ_{obs} is the observed ellipticity in degrees, MW is the protein molecular weight in g/mol, *n* is the number of amino acid residues, l is the path length of the cell in cm, c is the protein concentration in mg/mL and the factor of 100 originates from the conversion of the molar weight to mg/dmol unit. The concentration of the NS3 protein in CD measurements was about 0.1–0.3 mg/mL in the 1 cm cuvette.

3.8. Thermal Unfolding Experiments

Thermal unfolding curves of NS3 were experimentally obtained in a 20 mM phosphate buffer (pH 6.8) using an Aviv model 202 Circular Dichroism Spectrometer. The CD value at 222 nm was monitored while raising the temperature in the range of 4–96°C, with 2°C and 30 s intervals. The midpoint of thermal denaturation (T_m)

was extracted from the thermal unfolding, curved by normalization CD value, as in the following equation:

$$X = (1 - Y_T/Z) \times 100\% \quad (2)$$

where X is the percentage of thermal unfolding protein, Y_T is the CD value at a given temperature T, and Z is the average CD value between 70 and 96°C. The Tm was defined as the temperature when X = 50%.

The Tm and thermodynamic parameters of unfolding curves were analyzed by a three-state transition model. The enthalpy and entropy changes were calculated by the modified van 't Hoff equation [36]. The theoretical curve was fitted to the experimental data by non-linear least square fitting procedure to obtain the thermodynamic parameters for the individual transition phase. The difference in the free energy change of the unfolding between each NS3 and fresh 37°C incubated NS3 was estimated at the Tm of fresh 37°C incubated NS3 using the following formula:

$$\Delta G_m = \Delta T_m \Delta S_m \quad (3)$$

where ΔT_m is the difference in T_m between each NS3 and fresh NS3 incubated at 37 °C, and ΔS_m is the entropy change of fresh NS3 incubated at 37 °C at its T_m.

3.9. ELISA Analysis of the Recombinant NS3 Proteins

The purified recombinant NS3 diluted to 1.2 μg with the coating buffer (20 mM phosphate buffer, pH 6) was applied to microplates and the plates were then incubated at 4°C for at least 20 h. The microplates were washed, overcoated, and dried for 20 h. Specimens in two seroconversion panels (BBI) were diluted 20-fold with the Specimen Diluent C (3HC03-350 Lot C58C06SDP, GBC). A total of 100 μL of the diluted mixture was then added to the microplates for 1 h at 37°C as the primary antibody. After washing, the conjugated (100 μL, GBC) anti-human IgG-HRPO was then applied as the secondary antibody for 30 min at 37°C. The microplates were washed and developed by adding 100 μL of 3,3′,5,5′-tetramethylbenzidine in the dark for 30 min at room temperature. The peroxidase reaction was terminated by adding 100 μL of 2-N H_2SO_4, and the absorbance at 450–650 nm was measured using the ELISA reader (Molecular Devices). The absorbance was compiled statistically as S/Co value (S: sample value; Co: cutoff value). The Co value was calculated as the negative control OD value, plus the positive control OD value, divided by 4 (Co = NCx + PCx/4). Samples with an absorbance equal to or higher than the cut-off value, i.e., S/Co values greater than 1, were considered to be initially reactive in the assay.

4. CONCLUSIONS

This study confirms that the truncated form of the HCV NS3 protein as compared to the full-length protein exhibits better specific activity, purity, and application due to its reduced self-hydrolytic activity and false negative results in ELISA. Our results clearly show that the expression at lower temperatures in the *E. Coli* system improves recombinant protein performance. The incubation temperature shifting from 37°C to 25°C appeared to improve protein folding, solubility, the yield of the soluble fraction of HCV NS3 from 4.15 to 11.1 mgL^{-1}, and the storage shelf life of the cloned and over-expressed recombinant NS3. Such properties are probably due to a compact protein structure that increases the protein stability. When the specificity of NS3 expressed at 25°C was compared with that of another NS3 and two commercial kits by international standard panels, the truncated NS3 produced at 25°C had a better discriminating ability than the 37°C produced protein, and was competitive with the commercial kit (**Figure 7**). Hence, it may have the highest potential for HCV ELISA diagnosis. These characteristics and advantages can be seen in commercial biologics, which mostly use *E. coli* systems for protein expression. Moreover, the results of the current study are in line with the published reports, which confirms that the protein expression at lower temperatures (25°C) exhibits superior performance as compared to that at higher temperatures (37°C), as demonstrated by solubility, purity, antigenic efficacy, and stability. In conclusion, a lower temperature expression technology of proteins offers greater potential for the development of biological agents and in vitro diagnostics that are both more effective and commercially feasible.

5. CONCISE SUMMARY

The non-structural antigen protein 3 of the hepatitis C virus (HCV NS3), commonly-used for HCV ELISA diagnosis, possesses protease and helicase activities. To prevent auto-degradation, a truncated NS3 protein was designed by removing the protease domain. Firstly, it was over expressed in *E. coli* by IPTG induction under two different temperatures (25 and 37°C), and purified using affinity chromatography to attain homogeneity above 90%. The molecular mass of purified protein was determined to be approx. 55 kDa. While lowering the temperature from 37 to 25 °C, the yield of the soluble fraction of HCV NS3 was increased from 4.15 to 11.1 mgL^{-1} culture, which also improved the antigenic activity and specificity. The protein stability was investigated after long-term storage (for 6 months at -20°C) revealed no loss of activity, specificity, or antigenic efficacy. A thermal stability study on both freshly produced and stored HCV NS3 fractions at both temperatures showed that the unfolding curve profile properly obey the three-state unfolding mechanism. In the first transition phase, the midpoints of the thermal denaturation of fresh NS3 produced at 37°C and 25°C, and that produced after long-term storage at 37°C and

25°C, were 59.7°C, 59.1°C, 55.5°C, and 57.8°C, respectively. Microplates coated with the fresh NS3 produced at 25°C or at 37°C that were used for the HCV ELISA test and the diagnosis outcome were compared with two commercial kits—Abbott HCV EIA 2.0 and Ortho HCV EIA 3.0. Results indicated that the specificity of the HCV NS3 produced fresh at 25°C was higher than that of the fresh one at 37°C, hence showing potential for application in HCV ELISA diagnosis.

REFERENCES

1. Moriishi, K.; Mori, Y.; Matsuura, Y. Processing and pathogenicity of HCV core protein. *Uirusu* **2008**, *58*, 183–190. [Google Scholar] [CrossRef]
2. Dash,S.;Haque,S.;Joshi,V.;Prabhu,R.;Hazari,S.;Fermin,C.HCV-Hepatocellular carcinoma: New findings and hope for effective treatment. *Microsc. Res. Tech.* **2005**, *68*, 130–148. [Google Scholar] [CrossRef]
3. Muerhoff, A.S.; Jiang, L.; Shah, D.O.; Gutierrez, R.A.; Patel, J.; Garolis, C. Detection of HCV core antigen in human serum and plasma with an automated chemiluminescent immunoassay. *Transfusion* **2002**, *42*, 349–356. [Google Scholar] [CrossRef]
4. Lindsay, K. Therapy of chronic hepatitis C: Overview. *Hepatology* **1997**, *26*, 71S. [Google Scholar] [CrossRef]
5. Kolykhalov, A.A.; Agapov, E.V.; Blight, K.J.; Mihalik, K.; Feinstone, S.M.; Rice, C.M. Transmission of hepatitis C by intrahepatic inoculation with transcribed RNA. *Science* **1997**, *277*, 570. [Google Scholar] [CrossRef]
6. Encke, J.; Geissler, M.; Stremmel, W.; Wands, J.R. DNA-based immunization breaks tolerance in a hepatitis C virus transgenic mouse model. *Hum. Vaccin.* **2006**, *2*, 78–83. [Google Scholar] [CrossRef]
7. Rodrigue-Gervais, I.G.; Rigsby, H.; Jouan, L.; Sauvé, D.; Sékaly, R.P.; Willems, B. Dendritic cell inhibition is connected to exhaustion of CD8+ T cell polyfunctionality during chronic hepatitis C virus infection. *J. Immunol.* **2010**, *184*, 3134. [Google Scholar] [CrossRef][Green Version]
8. Lee, S.R.; Wood, C.L.; Lane, M.J.; Francis, B.; Gust, C.; Higgs, C.M. Increased detection of hepatitis C virus infection in commercial plasma donors by a third-generation screening assay. *Transfusion* **1995**, *35*, 845–849. [Google Scholar] [CrossRef]
9. Hidajat, R.; Nagano-Fujii, M.; Deng, L.; Tanaka, M.; Takigawa, Y.; Kitazawa, S. Hepatitis C virus NS3 protein interacts with ELKS-d and ELKS-a, members of a novel protein family involved in intracellular transport and secretory pathways. *J. Gen. Virol.* **2005**, *86*, 2197–2208. [Google Scholar] [CrossRef] [Green Version]

10. Tai, C.L.; Chi, W.K.; Chen, D.S.; Hwang, L.H. The helicase activity associated with hepatitis C virus nonstructural protein 3 (NS3). *J. Virol.* **1996**, *70*, 8477–8484. [Google Scholar] [CrossRef][Green Version]

11. Cao, C.; Shi, C.; Li, P.; Tong, Y.; Ma, Q. Diagnosis of hepatitis C virus (HCV) infection by antigen-capturing ELISA. *Clin. Diagn. Virol.* **1996**, *6*, 137. [Google Scholar] [CrossRef]

12. Ferrer, F.; Candela, M.J.; Garcia, C.; Martinez, L.; Rivera, J.; Vicente, V. A Comparative Study of Two Third-Generation Anti-Hepatitis C Virus ELISAs. *Haematologica* **1997**, *82*, 690–691. [Google Scholar]

13. Martin, P.; Fabrizi, F.; Dixit, V.; Quan, S.; Brezina, M.; Kaufman, E. Automated RIBA hepatitis C virus (HCV) strip immunoblot assay for reproducible HCV diagnosis. *J. Clin. Microbiol.* **1998**, *36*, 387–390. [Google Scholar] [CrossRef] [Green Version]

14. Poliakov, A.; Hubatsch, I.; Shuman, C.F.; Stenberg, G.; Danielson, U.H. Expression and purification of recombinant full-length NS3 protease-helicase from a new variant of Hepatitis C virus. *Protein Expr. Purif.* **2002**, *25*, 363. [Google Scholar] [CrossRef]

15. Poliakov, A.; Danielson, U.H. Refolding of the full-length non-structural protein 3 of hepatitis C virus. *Protein Expr. Purif.* **2005**, *41*, 298. [Google Scholar] [CrossRef]

16. Urbani, A.; Bazzo, R.; Nardi, M.C.; Cicero, D.O.; De Francesco, R.; Steinkühler, C. The metal binding site of the hepatitis C virus NS3 protease: A spectroscopic investigation. *J. Biol. Chem.* **1998**, *273*, 18760. [Google Scholar] [CrossRef] [Green Version]

17. Wardell, A.D.; Errington, W.; Ciaramella, G.; Merson, J.; McGarvey, M.J. Characterization and mutational analysis of the helicase and NTPase activities of hepatitis C virus full-length NS3 protein. *J. Gen. Virol.* **1999**, *80*, 701. [Google Scholar] [CrossRef]

18. Beran, R.K.; Pyle, A.M. Hepatitis C viral NS3-4A protease activity is enhanced by the NS3 helicase. *J. Biol. Chem.* **2008**, *283*, 29929–29937. [Google Scholar] [CrossRef][Green Version]

19. Greenfield, N.J. Applications of circular dichroism in protein and peptide analysis. *Trends Anal. Chem.* **1999**, *18*, 236–244. [Google Scholar] [CrossRef]

20. De Moraes, L.N.; Grotto, R.M.T.; Valente, G.T. A novel molecular mechanism to explain mutations of the HCV protease associated with resistance against covalently bound inhibitors. *Virus Gene.* **2019**, *274*, 197778. [Google Scholar]

21. Khu, Y.L.; Koh, E.; Lim, S.P.; Tan, Y.H.; Brenner, S.; Lim, S.G. Mutations that affect dimer formation and helicase activity of the hepatitis C virus helicase. *J. Virol.* **2001**, *75*, 205–214. [Google Scholar] [CrossRef] [PubMed][Green Version]

22. Marecki, J.C.; Aarattuthodiyil, S.; Byrd, A.K.; Penthala, N.R.; Crooks, P.A.; Paney, K.D. N-Naphthoyl-substituted indole thio-barbituric acid analogs inhibit the helicase activity of the hepatitis C virus NS3. *Bioorg. Med. Chem. Lett.* **2019**, *29*, 430–434. [Google Scholar] [CrossRef]

23. Hernández, S.; Díaz, A.; Loyola, A.; Villanueva, R.A. Recombinant HCV NS3 and NS5B enzymes exhibit multiple posttranslational modifications for potential regulation. *Virus Gene.* **2019**, *55*, 227–232. [Google Scholar] [CrossRef] [PubMed]

24. Gasteiger, E.; Hoogland, C.; Gattiker, A.; Duvaud, S.; Wilkins, M.R.; Appel, R.D.; Bairoch, A. Protein identification and analysis tools on the ExPASy server. In *The Proteomics Protocols Handbook*; Walker, J.M., Ed.; Humana Press: Totowa, NJ, USA, 2005. [Google Scholar]

25. Konermann, L.; Pan, J.; Wilson, D.J. Protein folding mechanisms studied by time-resolved electrospray mass spectrometry. *Biotechniques* **2006**, *40*, 135–141. [Google Scholar] [CrossRef]

26. Wilcox, W.C.; Long, D.; Sodora, D.L.; Eisenberg, R.J.; Cohen, G.H. The contribution of cysteine residues to antigenicity and extent of processing of herpes simplex virus type 1 glycoprotein D. *J. Virol.* **1988**, *62*, 1941–1947. [Google Scholar] [CrossRef] [PubMed][Green Version]

27. Elleuche, S.; Krull, A.; Lorenz, U.; Antranikian, G. Parallel N- and C-Terminal Truncations Facilitate Purification and Analysis of a 155-kDa Cold-Adapted Type-I Pullulanase. *Protein J.* **2017**, *36*, 56–63. [Google Scholar] [CrossRef]

28. Lawyer, F.C.; Stoffel, S.; Saiki, R.K.; Chang, S.H.; Landre, P.A.; Abrarnson, R.D.; Gelfand, D.H. High-level expression, purification, and enzymatic characterization of full-length thermus aquatlcus DNA polymerase and a truncated form deficient in 5’ to 3’ exonuclease activity. *Genome Res.* **1993**, *2*, 275–287. [Google Scholar] [CrossRef][Green Version]

29. Mühlmann, M.; Forsten, E.; Noack, S.; Büchs, J. Optimizing recombinant protein expression via automated induction profling in microtiter plates at diferent temperatures. *Microb. Cell Fact.* **2017**, *16*, 220. [Google Scholar] [CrossRef]

30. Chen, X.; Li, C.; Liu, H. Enhanced recombinant protein production under special environmental stress. *Front. Microbiol.* **2021**, *12*, 630814. [Google Scholar] [CrossRef]

31. Sandomenico, A.; Sivaccumar, J.P.; Ruvo, M. Evolution of Escherichia coli expression system in producing antibody recombinant fragments. *Int. J. Mol. Sci.* **2020**, *1*, 6324. [Google Scholar] [CrossRef]

32. Chen, P.J.; Lin, M.H.; Tai, K.F.; Liu, P.C.; Lin, C.J.; Chen, D.S. The Taiwanese hepatitis C virus genome: Sequence determination and mapping the 5' termini of viral genomic and antigenomic RNA. *Virology* **1992**, *188*, 102. [Google Scholar] [CrossRef]

33. Xue, H.; Wang, J.; Xie, J. Isolation, purification, and structure identification of antioxidant peptides from embryonated eggs. *Poult. Sci.* **2019**, *98*, 2360–2370. [Google Scholar] [CrossRef]

34. Kiely, P.R.; Eliades, L.A.; Kebede, M.; Stephenson, M.D.; Jardine, D.K. Anti-HCV confirmatory testing of voluntary blood donors: Comparison of the sensitivity of two immunoblot assays. *Transfusion* **2002**, *42*, 1053–1058. [Google Scholar] [CrossRef] [PubMed]

35. Anderson, M.D.; Breidinger, S.A.; Woolf, E.J. Effect of disease state on ionization during bioanalysis of MK-7009, a selective HCV NS3/NS4 protease inhibitor, in human plasma and human Tween-treated urine by high-performance liquid chromatography with tandem mass spectrometric detection. *J. Chromatogr. B Analyt. Technol. Biomed. Life Sci.* **2009**, *877*, 1047–1056. [Google Scholar] [CrossRef]

36. Honda, Y.; Fukamizo, T.; Okajima, T.; Goto, S.; Boucher, I.; Brzezinski, R. Thermal unfolding of chitosanase from *Streptomyces* sp. N174: Role of tryptophan residues in the protein structure stabilization. *Biochim. Biophys. Acta* **1999**, *1429*, 365. [Google Scholar] [CrossRef].

CHAPTER-9

EXPLORE THE POTENTIAL APPLICATIONS OF CYANIDE HYDRATASE DERIVED FROM *EXIDIA GLANDULOSA*: FREE CYANIDE REMOVAL FROM SIMULATED INDUSTRIAL EFFLUENTS

1. INTRODUCTION: AN OVERVIEW

Industrial activities, particularly those in mining, electroplating, and metal-processing sectors, often result in the release of cyanide-containing effluents, posing severe environmental and human health risks. Cyanide, a highly toxic compound, can persist in water bodies, leading to detrimental consequences for aquatic ecosystems and public health. The urgent need for efficient and sustainable methods to mitigate the impact of cyanide contamination has spurred research into innovative biotechnological solutions.

Cyanide is widely used in industry due to its chelating and electrolytic properties. For example, it is used to leach gold and silver from their ores. These processes generate significant amounts of cyanide waste [1,2]. Cyanide wastewaters also come from, e.g., electroplating, jewellery industry, (petro) chemical industry and coke making [1,3,4,5,6,7].

Cyanide occurs as free cyanide (fCN), i.e., CN^- and HCN, or as metal complexes. fCN is the most toxic; at pH ≤8.5, it occurs predominantly as volatile HCN, which is very hazardous [1]. The above industrial effluents also contain other chemicals such as sulfide, thiocyanate, ammonia, phenols or metals [4,5,6,7,8,9], which can complicate their treatment.

Organic pollutants have adverse effects on human health and the environment [10,11]. Cyanide is one of them, and is listed as toxic pollutant; according to the Environmental Protection Agency (EPA) effluent guidelines, the limit concentration of total cyanide is 1.30 mg/L for any one day in electroplating [12], while the limit for coke making is expressed in kg of total cyanide per ton of product (e.g., approximately 0.07 kg/t for any one day) [13].

The molecular mechanism of cyanide toxicity primarily consists in the formation of a complex with the iron cofactor in cytochrome c oxidase (EC 7.1.1.9), thereby affecting the respiratory chain. Consequently, multiple body systems (e.g., respiratory, cardiovascular, central nervous, sensory nervous, and endocrine) are severely affected by cyanide [14], with LD50 of 1–3 mg per kg of body weight (calculated for ingested HCN) [1]. In addition, aquatic ecosystems are extremely sensitive to low concentrations of fCN (≤0.01 mg/L and ≤0.15 mg/L for prolonged and acute exposure, respectively) [14].

Spontaneous processes based on volatilization, oxidation, precipitation, or microbial decomposition were traditionally used in the treatment of gold mine waste in “tailing ponds”. Later, the trend was toward more strictly controlled processes such as alkaline chlorination or oxidation with, e.g., SO_2 + air (Inco process), or SO_2 + air + H_2O_2 (CombinOx process) [1].

Nanomaterials are among the promising tools for eliminating or mitigating the harmful effects of environmental pollutants. For example, they were proposed for removing environmental contaminants such as industrial dyes [10,11,15], or nitrate ions [16]. Their uses for wastewater decyanation also emerged. These strategies used ZnO or Ga_2O_3/Pt nanoparticles acting as photocatalysts for the breakdown of fCN [17,18]. The current focus is on non-toxic nanoparticles produced through environmentally friendly methods, e.g., from biodegradable polymers [10,11].

Biological processes offer alternative ways to overcome the bottlenecks of the (physico)chemical approaches in terms of required space, time, energy or chemicals, or secondary waste. Microbial consortia (activated sludge) have already found full-scale applications in the treatment of coke making wastewaters [19].

In contrast, biocatalysts still have a very limited impact in this area, although cyanide hydratases (CynHs; EC 4.2.1.66) and cyanide dihydratases (CynDs; EC 3.5.5.1) seem to be promising tools for wastewater detoxification, as they catalyze the conversion of fCN to the less toxic formamide, and to a formic acid/ammonia mixture, respectively (**Figure 1**A). In particular, they are resistant to high fCN concentrations and do not require a cofactor. In addition, CynHs have the advantage of about 10-fold higher specific activity compared to CynDs [20]. CynHs also transform some nitriles such as 2-cyanopyridine (2CP), the products of which are the corresponding carboxylic acid and amide (**Figure 1**B).

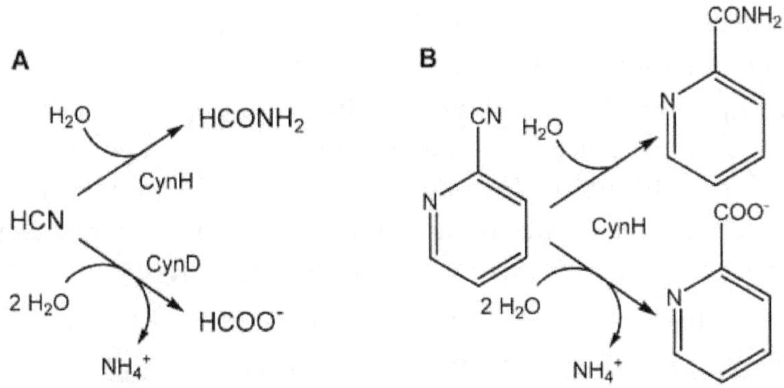

Figure 1. Transformation of (A) HCN by cyanide hydratase (CynH) or cyanide dihydratase (CynD), and (B) transformation of 2-cyanopyridine by CynH.

CynH has been considered a potential cyanide-degrading biocatalyst since it was first described in the fungus *Stemphylium loti* [21]. In the following years, other CynHs were obtained from *Gloeocercospora sorghi* [22], and genus *Fusarium* [23,24,25]. However, the CynH activities produced by the mycelia were mostly low: less than 0.05 U/mg dry weight in various fungal pathogens of plants [22] or 1.4 U/mg dry weight in *Fusarium solani* [24]. Only *Fusarium lateritium* showed a much higher activity of 102.5 U/mg dry weight [23]. Specific activities typically ranged from 4.6 to 85 U/mg protein in cell-free extract (CFE) and increased to 128–1109 U/mg protein by purification [22,23,24,25].

The presence of a CynH was also suspected in the bacterium *Serratia marcescens* [26] after a reaction product corresponding to formamide was found in HPLC. The putative CynH sequence had a shorter amino acid chain than typical CynHs (326 residues). Similar proteins are widely distributed in the genus *Serratia*, other Proteobacteria, or Actinobacteria according to BLAST searching in the NCBI database [27].

In contrast, there are only a few characterized CynDs. They originate from *Bacillus pumilus*, *Pseudomonas stutzeri* and *Alcaligenes xylosoxidans* [28] and their homologues seem to occur only in bacteria.

In the last two decades, several fungal CynHs have been obtained, whose genes are expressed in *Escherichia coli* [3,29,30,31,32,33]. The activities of the whole *E. coli* cells (153–600 U/mg dry weight) and CFE (736 U/mg protein) [33] were substantially higher in comparison with the endogenous enzymes, whereas the activities of the purified enzymes from the homologous and heterologous producers were similar

(100–1324 U/mg protein; [29,31,33]. Further enhancement of CynH action on fCN may be possible by the specific mutations recently proposed in silico [34].

However, the extent to which CynHs or CynDs can degrade fCN in different environments has scarcely been studied. Thus, fCN conversion by four recombinant CynHs was demonstrated for diluted Ag and Cu electroplating effluents containing 100 mM fCN [3]; it was almost complete in the presence of Ag but less than 70% in the presence of Cu. Another CynH was used with a spiked coke making effluent and degraded, e.g., 10 mM fCN to almost 100% [4]. In addition, an immobilized whole-cell CynD biocatalyst was used with diluted gold-mine wastewater and the maximum conversion was 98% starting from 17.6 mM fCN [2].

Recently, we have investigated Basidiomycota as a source of enzymes of the nitrilase superfamily, which include some putative CynHs [32,35]. The focus of this work is on a detailed study of one of these CynHs, which originates from the fungus *Exidia glandulosa*. The aim of this work was to determine its properties that are important for its future use: kinetic parameters, pH and temperature profiles, stability, and ability to remove cyanide under conditions relevant to wastewater treatment. Therefore, the degree of fCN conversion was studied at typical cyanide concentrations, alkaline pH (necessary to minimize the escape of highly toxic HCN), and the presence of contaminants accompanying cyanide in industrial effluents. Further steps toward the use of CynHs in wastewater treatment are proposed.

2. RESULTS

2.1. Enzyme Preparation and Properties

The sequence of enzyme NitEg was taken from databases (GenBank: KZV92691; UniProtKB: A0A165HZS1), where this protein was classified as "cyanide hydratase". The corresponding gene (*nitEg*) was synthesized, with codon usage bias optimized for *E. coli* .

2.1.1. Enzyme Overproduction

E. coli containing the optimized *nitEg* gene was grown to an optical density (determined at 600 nm; OD_{600}) of 12.0, which corresponds to approximately 3.6 mg dry weight/mL. Total whole-cell activity was typically 318 U/mL culture medium (88.4 U/mg dry weight), as determined by the rate of fCN consumption. Cells from 200 mL of culture were sonicated to obtain CFE containing 9.3 mg protein/mL (214 mg protein in total). Analysis of the CFE by SDS-PAGE revealed the major band to have an apparent molecular mass of approximately 43.85 kDa, which was essentially consistent with the theoretical molecular mass of NitEg carrying a His_6-tag (41,800 Da). The CFE had a specific activity of 280 U/mg as determined by the picric acid method.

2.1.2. Enzyme Purification and Activity

NitEg was purified in a single step. The amount of the purified enzyme was typically 12 mg from 200 mL of culture. The purification yield was 14.4% and the amount of NitEg accounted for approximately 40% of the total cell protein. Analysis of the purified NitEg by SDS-PAGE showed that the protein contained no significant impurities.

Formamide was the main product of fCN conversion by the purified enzyme (**Figure 2**). The concentration of formamide in the reaction mixture was about 22 mM after complete elimination of fCN (60 min). It was similar to the concentration of fCN in the control sample without enzyme after the same time. This suggests that the remainder of fCN (about 3 mM) was abiotically removed in both the reaction and the control.

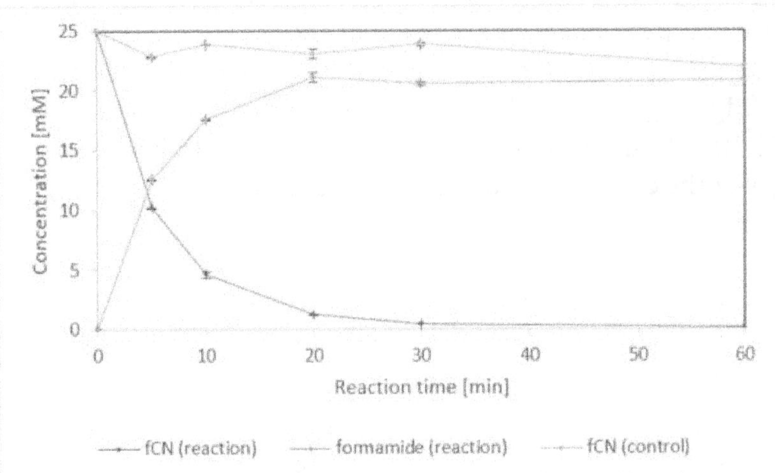

Figure 2. Transformation of 25 mM free cyanide (fCN) to formamide by enzyme NitEg (4.0 µg protein/mL) in 100 mM glycine/NaOH buffer, pH 9.0. Residual fCN concentration was determined by picric acid method. The enzyme was omitted from the control.

The specific activities calculated from the rates of fCN consumption and formamide production were approximately 697 U/mg and 784 U/mg protein, respectively. The kinetic parameters were calculated from the rates of formamide production (**Table 1**). The values of V_{max} and k_{cat} were relatively high, but K_M was also high. Nevertheless, the k_{cat}/K_M ratio indicates good catalytic efficiency of the enzyme.

Table 1. Specific activity and kinetic parameters of enzyme NitEg.

Activity Assay	Specific Activity [U/mg] 1	Vmax [U/mg]	KM [mM]	kcat [1/s]	kcat/KM [1/s/mM]
fCN consumption	697 ± 95	n.d.	n.d.	n.d.	n.d.
Formamide production	784 ± 32	1335 ± 81	22.2 ± 2.4	927 ± 80	42 ± 10

[1] at 25 mM free cyanide (fCN).

2.1.3. Effect of pH and Temperature

The pH and temperature profiles of NitEg were determined using 2CP as substrate. This was because 2CP is less sensitive to pH and temperature than fCN. 2CP was generally a good substrate for the CynHs we previously studied [32,33]. NitEg also showed acceptable activity for 2CP. The activity was expressed as the sum of the reaction products pyridine-2-carboxylic acid and pyridine-2-carboxamide (Figure 1B), and was determined to be approximately 6.6 U/mg. NitEg was most active at 40–45°C (Figure 3A) and exhibited a broad optimum at pH approximately 6–9 (Figure 3B). Its activity decreased significantly outside this pH range and at 50°C. The stability of the enzyme was acceptable in a similar pH range (Figure 3B).

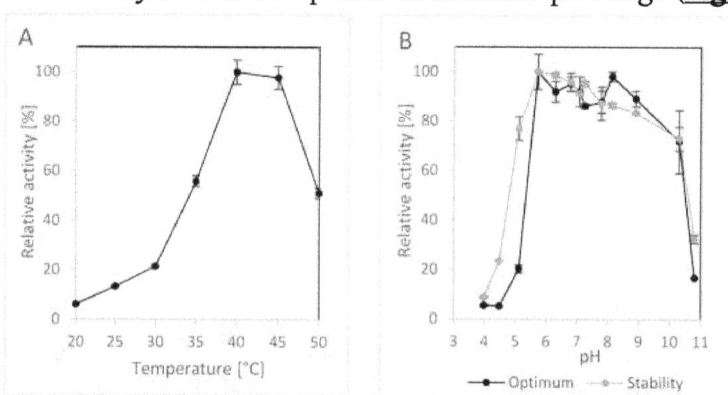

Figure 3. Effect of (A) temperature on the activity of enzyme NitEg and (B) pH on the activity and stability of NitEg. 2-Cyanopyridine (25 mM) was used as substrate.

2.1.4. Effect of Silver and Copper

The effect of metal ions (Cu^{2+}, Ag^{+}) prevalent in some electroplating baths [3] on the activity of NitEg was studied by measuring the rate of fCN consumption at pH 9.0 and 30°C (Table 2). Silver ions decreased the activity by about 1/3 at 0.1–1 mM and by 2/3 at 5 mM. No enzyme activity was detected at 10 mM concentration of silver ions. However, it should be noted that the concentration of fCN was also decreased in the controls (without enzyme) in the presence of silver (to about 15.7 mM and 5.7 mM at 5 mM and 10 mM Ag, respectively), apparently due to complex formation. A strong inhibition effect of Cu was observed at 5–10 mM.

Table 2. Effect of copper and silver ions on the activity of enzyme NitEg.

Metal Concentration (mM)	Relative Activity [%]	
	Ag+	Cu2+
-	100	100
0.1	66 ± 5	93 ± 7
1	61 ± 5	68 ± 8
5	31 ± 10 1	14 ± 3
10	n.d. 2	0

[1] 15.7 mM free cyanide (fCN) in control. [2] 5.7 mM fCN in control.

2.1.5. Temperature stability and shelf life

At temperatures selected according to the previous work (27–50°C) [3], the enzyme was pre-incubated for 24 h and the residual activities were determined using fCN as substrate. Under these conditions, the stability of the purified NitEg was acceptable at 27°C or 37°C, with about 60% and 40% of its initial specific activity, respectively, after 24 h. At 43°C, it retained approximately 40% of its initial activity after 3 h, but only 10% of it after 24 h, and it lost more than 90% of its activity at 50°C after 1 h (**Figure 4**).

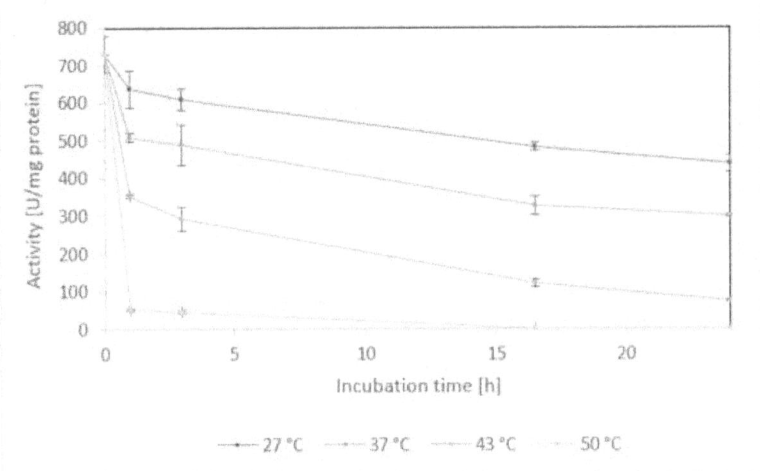

Figure 4. Temperature stability of enzyme NitEg. The purified enzyme (10.3 mg protein/mL) was incubated in 50 mM Tris/HCl buffer, pH 8.0, with 150 mM NaCl, at different temperatures without shaking. Specific activities were determined by the formamide assay.

At 4°C, the specific activity of purified NitEg (10.3 mg protein/mL) showed no statistically significant loss until day 69. Thereafter, the activity decreased, but was still almost 83% of the initial value after 98 days.

2.2. Enzyme Performance on Model Mixtures

The performance of purified NitEg was studied using model solutions that were prepared based on published wastewater analyses (see below). They mimicked the effluents of various industries in terms of concentrations of fCN and other contaminants, as well as pH.

2.2.1. Effect of Alkaline Media

First, the performance of NitEg was investigated with fCN at pH 9.0–10.5 (**Figure 5**). At pH 9.0, the removal of 25 mM fCN by 4 μg enzyme/mL was almost complete after 1 h. The abiotic removal accounted for approximately 10% within the same time. Higher pH resulted in less efficient fCN removal (approximately 89% and 54% at pH 9.5 and 10.0, respectively) under the same conditions. At pH 10.5, the enzyme was no more active in removing fCN (**Figure 5**A).

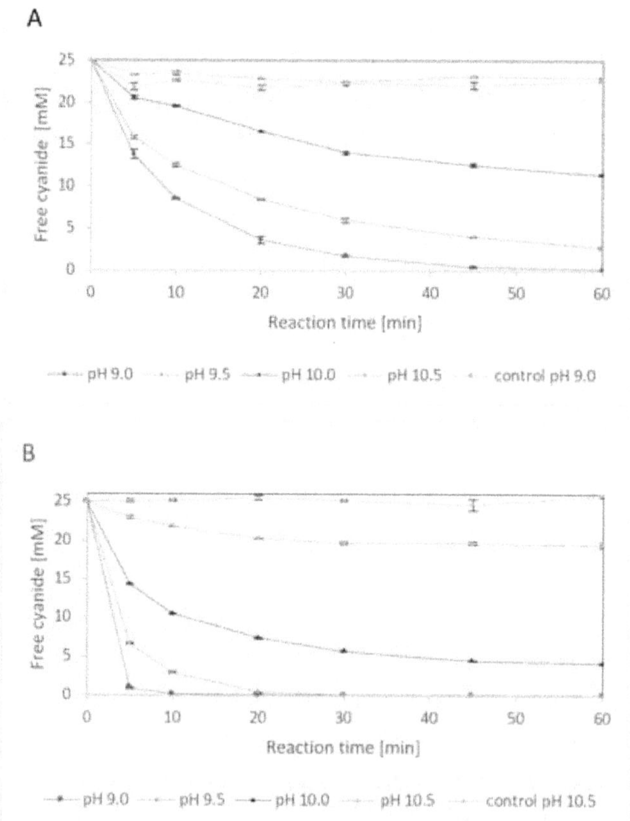

Figure 5. Effect of pH on the biocatalyzed degradation of 25 mM free cyanide (fCN) in 100 mM glycine/NaOH buffer by (A) 4.0 μg enzyme/mL and (B) 20 μg enzyme/mL. Controls were carried out at (A) pH 9.0 and (B) pH 10.5 without enzyme. Residual fCN concentration was determined by picric acid method.

Increasing the enzyme concentration to 20 µg/mL allowed removal of nearly 25 mM fCN at pH 9.0 or 9.5 after 30 min and over 80% fCN at pH 10 after 1 h. Some decrease in fCN concentration was observed even at pH 10.5 (**Figure 5**B). This suggested that the enzymatic reaction must proceed quite rapidly to achieve a high degree of fCN removal. This is probably due to the limited stability of the enzyme under the reaction conditions as indicated by the pH profile of enzyme stability (**Figure 3**B).

2.2.2. Enzyme Performance in the Presence of Phenol and Inorganic Salts

Coke making effluents typically contain ≤ 0.6 mM fCN, but occasionally up to 10 mM fCN. Their common constituents also include phenols, thiocyanate, and ammonia. A model mixture was prepared according to the reported content [**8**] of fCN, SCN^-, NH_4^+ and phenol in coke making effluents (**Figure 6**A). The upper limits of the concentrations were used, where the fCN concentration was 0.6 mM, and the pH was 9.1. Under these conditions, almost all of the fCN was degraded with 2.5 µg enzyme/mL after 90 min, with a negligible degree of abiotic elimination (**Figure 6**A).

A model petrochemical effluent was also prepared according to the literature [**5**]. Compared to the previous effluent, this contained almost eight times more fCN but less phenol or ammonia. It contained no SCN^-, but a significant concentration of S^{2-}; its pH was the same (9.1) (**Figure 6**B). The presence of S^{2-} (interfering compound [**36**]) did not allow the use of the picric acid method to determine fCN. This interference caused an attenuation of the signal at 0.029 mM (1 ppm) and 0.29 mM H_2S by 2.5% and 27.5%, respectively. In contrast, 2.9 mM and 29 mM H_2S increased the signal more than twofold and twentyfold, respectively [**36**]. Therefore, the fCN concentration in the model effluent containing more than 20 mM S^{2-} was monitored using the Spectroquant® kit, which is based on a different reagent (1,3-dimethylbarbituric acid). However, according to the manufacturer, this kit is not suitable for samples containing SCN^-.

An enzyme concentration of 5 µg/mL was sufficient to remove approximately 96% fCN after 45 min. No further decrease in fCN concentration could be detected, which may have been caused by the background signal generated by one or more components of the mixture (**Figure 6**B).

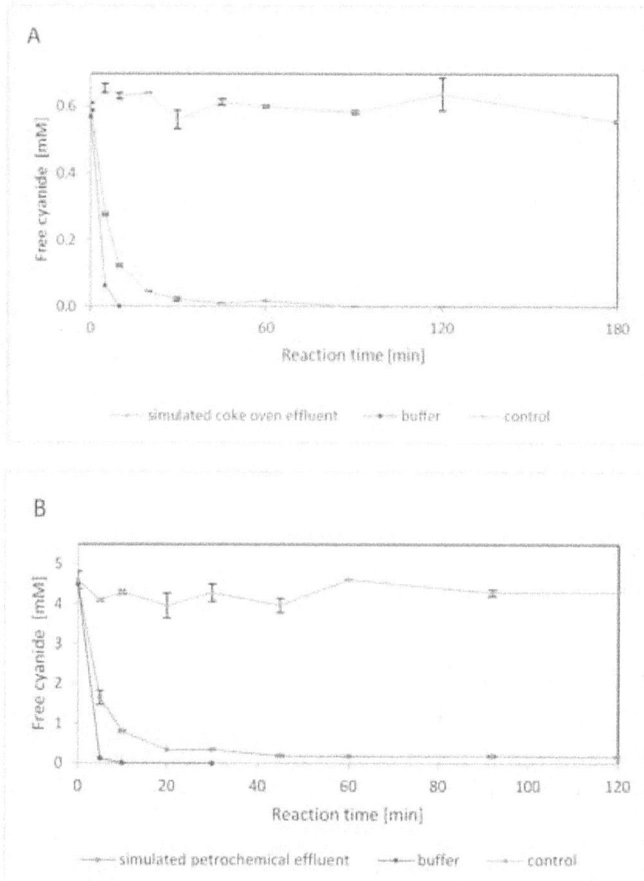

Figure 6. Biocatalyzed degradation of free cyanide (fCN) by enzyme NitEg in 100 mM glycine/NaOH, pH 9.1, containing (A) 0.6 mM fCN, 8.6 mM SCN^-, 10.7 mM NH_4^+ and 12.8 mM phenol (model coke plant effluent) [8], and (B) 4.6 mM fCN, 23.4 mM S^{2-}, 2.5 mM NH_4^+ and 0.64 mM phenol (model petrochemical effluent) [5]. Reactions were also performed in the same buffer with (A) 0.6 mM fCN or (B) 4.6 mM fCN without other additives. Enzyme concentration was (A) 2.5 µg/mL, and (B) 5.0 µg/mL. The enzyme was omitted from the controls. Residual fCN concentration was determined by (A) picric acid method or (B) Spectroquant® kit.

2.2.3. Enzyme Performance in the Presence of Heavy Metal Salts

The fCN-containing effluents from electroplating and precious metal mining contain heavy metals. The two electroplating wastes used in a previous work [3] contained undefined amounts of Cu or Ag. Another one contained about 30 mM Cu, 83 mM Ni and 0.9 mM Zn [37]. The fCN concentrations in these effluents can be extremely high (approximately 0.5–2 M) [2,3,37]. The CynD was not sufficiently effective in the original gold-mine effluent (528 mM fCN) and only removed 43% of the fCN [2]. Therefore, it was necessary to dilute this solution to 17.6 mM fCN [2] before applying the enzyme. Similarly, the electroplating wastewater was diluted to 100 mM fCN [3]. This can also cause the metal concentration to drop to a level tolerated by CynH. The above experiments have shown that NitEg does not work well at Ag or Cu concentrations of about 5 mM or more (**Section 2.1.4**).

First, the potential of NitEg to degrade 100 mM fCN was investigated (**Figure 7**A). Here, the pH of the model mixture was adjusted to 9.0, at which the stability of the fCN solutions was higher than at pH 8 used previously [3]. Different amounts of enzyme (14, 20, or 30 µg/mL) eliminated more than 97% from 100 mM fCN within 1 h. However, more than 2% of fCN remained in the reaction mixture within the monitored period (3 h) regardless of enzyme loading. The abiotic loss of fCN was about 17%.

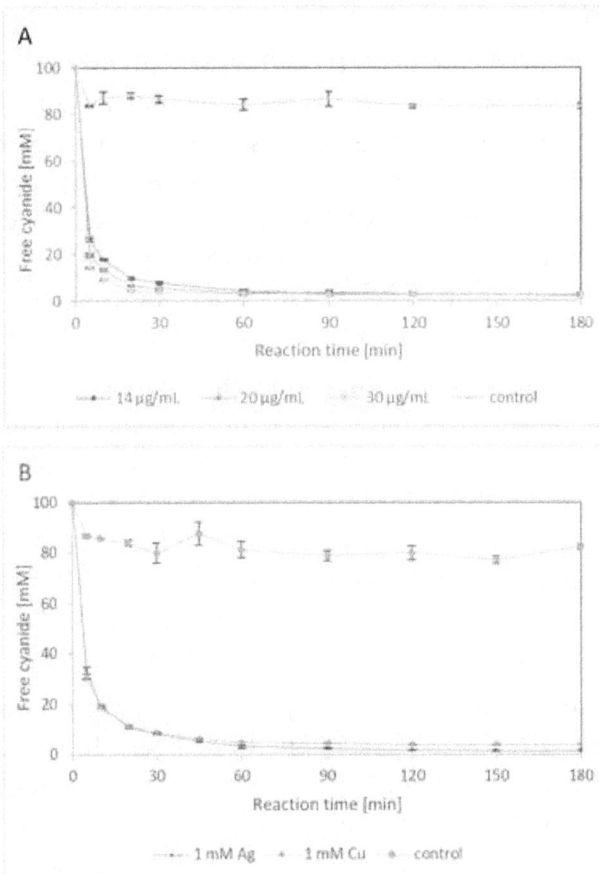

Figure 7. Biocatalyzed degradation of 100 mM free cyanide (fCN) in 100 mM glycine/NaOH, pH 9.0, by (A) 14, 20 or 30 µg enzyme/mL, and (B) 14 µg enzyme/mL in the presence of 1 mM $AgNO_3$ or 1 mM $CuSO_4$. Residual fCN concentration was determined by picric acid method. The enzyme and the metal salts were omitted from the control.

The simulated electroplating effluents contained 100 mM fCN and 1 mM $AgNO_3$ or 1 mM $CuSO_4$. The reactions were catalyzed by 14 μg enzyme/mL, which was sufficient to remove more than 97% fCN in the metal-free medium after 1 h (**Figure 7**A). The metal ions slightly decreased the initial reaction rates (**Figure 7**B), which was consistent with their effects on NitEg activity (**Section 2.1.4**). However, the percentage of fCN removal (96–98%) was similar to that in the absence of both of these metals. A copper concentration of 10 mM completely inhibited the catalytic process (not shown), which is also consistent with the results of the inhibition experiments above. The reaction in the presence of 10 mM silver was not studied because a metal cyanide complex was formed, which reduced the amount of fCN available to the enzyme (**Section 2.1.4**).

3. DISCUSSION

The presence of a large number of *cynH* genes in Ascomycota compared to Basidiomycota was previously demonstrated in our database searches [**31**,**32**]. The enzyme NitEg studied in this work is the first purified CynH from Basidiomycota. NitEg has an amino acid sequence identity of 86.29% (95% coverage) with CynH from *Neurospora crassa* (GenBank: XP_960160.2; UniProtKB: Q7RVT0) but NitEg is larger with its 366 amino acid residues (compared with 351 in *N. crassa*). The latter enzyme was previously purified [**3**]. In this work, we compare the properties of NitEg and the CynH from *N. crassa* as far as the published data allow.

NitEg was produced in *E. coli* following the general protocol we developed for the production of nitrilases (EC 3.5.5.1) and the related CynHs [**31**]. The yield of purification of NitEg was about 14.4%, which is similar to the yield of CynH from *Aspergillus niger* (17%) [**33**]. However, the purification of NitEg (a one-step protocol) was simpler due to the presence of the His_6-tag. We hypothesize that further optimization of the purification protocol can improve the yield of NitEg. Nevertheless, the amount of purified NitEg (approximately 12 mg) obtained from 200 mL of culture was sufficient for the aims of this work.

The CynH producing fungi *F. solani* and *Fusarium oxysporum* were cultivated at 10-L [**24**] or 45-L [**25**] scale, respectively. In contrast, heterologous production was performed at a small scale [**3**,**33**]. However, a related enzyme, nitrilase, was produced heterologously at 300-L scale. The growth medium was optimized and contained lactose as inducer. Under these conditions, the biomass yield was almost 13 g dry weight/L [**38**]. Therefore, we assume that a future scale-up of the production process of NitEg is very feasible using this method. It follows from the comparison of the specific activities of both enzymes (almost 500-fold higher in the CynH) that, after a suitable scale-up, the CynH production in the order of 10^6–10^7 U/L may be possible.

The use of whole cells of a recombinant organism for wastewater treatment could face regulatory obstacles, as it might be difficult to exclude cell escape. Enzyme purification would increase the cost of the catalyst but the use of CFE can be a suitable solution, as CFE has high specific activity and can be made free of residual whole cells.

Most CynHs are closely related with more than 60% sequence identity, but they differ in their specific activities. It is not entirely clear to what extent these results are influenced by different experimental conditions. Differences in pH profiles were also found between CynHs studied by the same method [3]. The operation of CynHs in alkaline media is of paramount importance, since similar conditions occur in industrial wastewaters. In this work, we investigated whether NitEg was able to fully remove fCN from alkaline media. It was found that this was possible for 25 mM fCN at pH 9–9.5, while more than 80% was still removed at pH 10.0. This was consistent with the pH activity profile of the enzyme with an optimum at pH approximately 6–9.

Alternative biocatalysts such as CynDs were investigated for the degradation of fCN in alkaline media [2,28,39,40]. However, even the improved CynDs still lagged behind CynHs such as NitEg, which had activity greater than 600 U/mg at pH 9.0. Moreover, at alkaline pH, NitEg remained active long enough to almost completely remove high concentrations of fCN. Since immobilization of CynD was shown to have a stabilizing effect under alkaline conditions [2], we hypothesize that it may have a positive effect on the operation of CynHs as well.

The studies of two CynDs and the CynH from *N. crassa* showed the importance of the C-terminus for the thermostability of both types of enzymes [28,40]. The CynH was more thermostable than CynDs with almost 90% activity after 4 h at 42°C [40], and about 50% activity after 1 day at 43°C [3]. In contrast, the CynHs from *G. sorghi* and *Aspergillus nidulans* retained very low residual activities under the same conditions [3]. The latter has a very similar sequence to the CynH from *A. niger* [33]. The CynH from *A. niger* was also investigated for stability, but using a different method (1 h preincubation), and was found to be unstable above 35°C. The stability of NitEg at 43°C was intermediate, with the enzyme retaining 10% of its initial activity after 1 day. However, significantly higher residual activities were observed after the same time at 27 or 37°C. The activity and stability of NitEg decreased sharply at 50°C. Therefore, this enzyme probably cannot be used under these conditions.

The kinetic parameters V_{max} and K_M were previously reported for the CynHs from *G. sorghi* (His_6-tagged) [41] and *A. niger* (without the tag) [33]. Both V_{max} and K_M were very high in these enzymes (V_{max} = 4400 U/mg and 6800 U/mg, K_M = 90 mM and 109 mM, respectively). In NitEg, both parameters were lower (V_{max} = 1335 U/mg, K_M

= 22 mM). Thus, the V_{max}/K_M ratio (U/mg/mM) was 60 in NitEg and similar in *A. niger* CynH (62) but lower in *G. sorghi* CynH (49). In contrast, K_M values were 2.6–7.3 mM in the CynDs, with V_{max} of 88–100 U/mg [41] and V_{max}/K_M ratio of 14–34 U/mg/mM.

The ability of NitEg to remove low concentrations of fCN did not appear to be significantly affected by the relatively high K_M. No residual fCN was found in the NitEg catalyzed degradation of 0.6 mM fCN. The reason for the incomplete (but still 96–98%) conversion of 100 mM fCN was probably the enzyme inactivation.

No entries for k_{cat} or k_{cat}/K_M were found for CynHs in the BRENDA database [42]. Here, they were calculated for the CynHs with known V_{max} (see above) and subunit molecular weights, which were similar in all these enzymes (approximately 40–42 kDa). Thus, the CynHs from *G. sorghi* and *A. niger*, and NitEg exhibit k_{cat} (1/s) of approximately 3000, 4500 and 927, respectively, and k_{cat}/K_M (1/s/mM) of approximately 33, 41 and 42, respectively. For the related nitrilases, the k_{cat}/K_M (1/s/mM) reported in the BRENDA database ranged from approximately 0.001 to 205 [42]. The highest value was found for nitrilase from *Pyrococcus* sp. for benzonitrile [43], while all other values were lower than for NitEg.

There are few examples demonstrating the use of CynHs or CynDs in real or simulated industrial effluents (**Table S3**). Four CynHs were used to degrade 100 mM fCN in the presence of Ag or Cu, or their absence [3]. The concentrations of the CynHs were 7–13 μg/mL in the reaction mixtures with Ag or the control (without metals), but they were increased 10-fold in the mixtures with Cu. The enzymes showed large differences in their reaction rates. However, complete degradation of fCN was only achieved in buffer, whereas in the presence of Ag and Cu approximately 90% and 70% of fCN was degraded, respectively. It is difficult to compare the performance of these enzymes with that of NitEg because the composition of the electroplating effluents used in the previous work is largely unknown (not even the metal content), and the pH of the reaction mixture was lower (about 8) than for NitEg (9.0).

Nevertheless, it can be stated with some caution that the rapid degradation of 100 mM fCN in the absence of metals was similar for CynH from *N. crassa* (best of four CynHs tested) [3] and NitEg (98–100% conversions after 1–2 h). In contrast, the other enzymes required 12 to 48 h for similar conversions [3]. In addition, we characterized NitEg with respect to properties not reported for the *N. crassa* CynH: kinetic parameters, shelf life, and fCN conversion in different environments. The properties found for NitEg were favorable and suggest that this enzyme may be superior in terms of its ability to catalyze fCN degradation.

The evaluation of NitEg suggests that the enzyme is unlikely to be significantly inhibited by the major constituents of industrial effluents (phenol, ammonia, sulfide or thiocyanate). However, cokemaking wastewater, e.g., contains a large number of minor organic constituents, which can be expressed collectively as chemical oxygen demand (COD) or biological oxygen demand (BOD) [44]. This is difficult to simulate in model mixtures. NitEg is also likely to be functional at certain concentrations of heavy metals (e.g., 1 mM of Ag or Cu). Although electroplating or gold mining wastewaters may contain higher concentrations of metals, they can be suitably diluted for enzyme action. These types of effluent are also more complex than the model mixtures, also containing other metals and metal cyanide complexes. Therefore, the aim of the future research will be the evaluation of NitEg or other cyanide-degrading biocatalysts in real wastewaters. Importantly, detailed analysis must be available for these wastewaters to ensure reproducibility, reliability and comparability of the data but this has rarely been the case in previous studies.

4. MATERIALS AND METHODS

4.1. Chemicals

Chemicals were obtained from standard suppliers and were of the highest purity: acetic acid (99%; Lach-ner, Neratovice, Czech Republic), acetonitrile (≥99.95%, HiPerSolv Chromanorm; VWR International, Radnor, PA, USA), ammonium chloride (≥99%; Lach-ner), boric acid (≥99.5%; Lach-ner), copper (II) sulphate (≥99%; Lach-ner), 2-cyanopyridine (99%; Sigma Aldrich, St. Louis, MO, USA), disodium hydrogen phosphate (≥99.6%; Lach-ner), ferric chloride (98%; Alfa Aesar, Haverhill, MA, USA), formamide (≥99%; VWR), glycine (≥99%; Reanal, Budapest, Hungary), hydrochloric acid (35%; Lach-ner), hydroxylamine hydrochloride (99%; Alfa Aesar), imidazole (research grade; Serva Electrophoresis GmbH, Heidelberg, Germany), isopropyl-β-D-thiogalactopyranoside (IPTG; research grade; Serva), methanol (100%; VWR), phenol (≥99%; VWR), phenylmethanesulfonyl fluoride (PMSF; Serva), picric acid (≥98%; Sigma Aldrich), phosphoric acid (≥85%; VWR), potassium cyanide (≥96%; Sigma Aldrich), potassium thiocyanate (≥99%; Lach-ner), 2-pyridinecarboxamide (98%; Sigma Aldrich), 2-pyridinecarboxylic acid (≥98%; Sigma Aldrich), silver nitrate (p.a.; Lach-ner), sodium carbonate (≥99%; VWR), sodium chloride (≥98%; Lach-ner), sodium dihydrogen phosphate (≥99%; Lach-ner), sodium dodecyl sulfate (SDS; ≥99%; Carl Roth GmbH + Co. KG, Karlsruhe, Germany), sodium hydroxide (≥98 %; Lach-ner), sodium sulfide (Alfa Aesar), and tris(hydroxymethyl) aminomethane (Tris; ≥99%; Merck KGaA, Darmstadt, Germany).

4.2. Strain

The gene encoding protein KZV92691.1 (NitEg) was optimized, synthesized and ligated into vector pET22b(+) by GeneArt (ThermoFisher Scientific, Waltham, MA, USA). The gene was fused to C-terminal His_6-tag coding sequence. The resulting construct was used to transform *E. coli* Origami B(DE3) cells (Merck).

4.3. Enzyme Induction and Purification

The transformed cells were grown in 2xYT medium (g/L: Tryptone (ThermoFisher Scientific) 16, yeast extract (ThermoFisher Scientific) 10, NaCl 5; pH 7.0) at 37 °C. After the OD_{600} reached 1.0 mM IPTG was added to a final concentration of 0.02 mM, and cultivation continued for 20 h at 20°C. Harvested cells were sonicated and NitEg was purified from the CFE by cobalt ion affinity chromatography on TALON® Metal Affinity Resin (Clontech Laboratories, Inc., Mountain View, CA, USA) as previously described for nitrilases [35], but with minor modifications: NitEg was eluted with 20 mM sodium phosphate buffer, pH 7.6, containing 300 mM NaCl, 0.1mM PMSF and 200 mM imidazole. The enzyme solution was concentrated using Amicon Ultra-30 K filter (Merck, Millipore, Burlington, MA, USA) and the buffer was replaced with 50 mM Tris/HCl buffer, pH 8.0, supplemented with 150 mM NaCl.

4.4. Enzyme Assays

The standard activity assay was performed using Eppendorf ThermoMixer Comfort (Eppendorf, Hamburg, Germany) at 30 °C and 850 rpm. Reaction mixtures (0.5 mL) in 1.5 mL Eppendorf tubes contained 100 mM glycine/NaOH buffer, pH 9.0, and an appropriate amount of whole cells, CFE, or purified enzyme. After 5 min of pre-incubation, the reaction was started by adding KCN at a final concentration of 25 mM. After 1–2 min of incubation, the reaction was stopped by adding 1 mL of methanol. Whole cells were removed by centrifugation (MiniSpin plus, Eppendorf; 14,500 rpm, 4 min). Centrifugation was omitted for samples containing CFE or purified enzyme. The residual fCN was determined spectrophotometrically by the picric acid method [36] with some modifications. Samples (0.01 mL) were mixed with 0.09 mL of 0.2 M Tris/HCl buffer, pH 8.0, and 0.2 mL of 0.5% picric acid in 0.25 M Na_2CO_3, and incubated at 100°C in a water bath for 5 min. The reaction was stopped by placing the test tube in an ice bath. After addition of 0.7 mL of distilled water, absorbance was determined at 520 nm against the control prepared in the same way but without fCN.

Formamide was determined spectrophotometrically [23] with some modifications. Samples (0.2 mL) were mixed with 0.4 mL of 2.3 M hydroxylamine/3.5 M NaOH (1/1) and incubated at 60 °C for 10 min. After the addition of 0.2 mL of 4 M HCl and 0.2 mL of 1.23 M $FeCl_3$, absorbance was determined at 540 nm against the control prepared in the same way but without formamide.

One unit of CynH activity was defined as the amount of enzyme that consumed 1 μmol of fCN/min or produced 1 μmol of formamide/min under the above conditions.

Enzyme kinetics was investigated using 2.5–25 mM KCN, and activities were determined using the formamide assay.

The thermostability of the purified enzyme was studied according to the literature [3] with some modifications. The enzyme was incubated at 27, 37, 43 or 50°C without shaking, and its specific activity was determined after 1, 3, 16.5 and 24 h using the formamide assay. The shelf life of the purified enzyme at 4°C was investigated within 98 days by periodic determination of its specific activity using the picric acid method.

The activity for 2-cyanopyridine (2CP) was determined analogously as that for fCN but using 50 mM Tris/HCl buffer, pH 8.0, with 150 mM NaCl, and the reaction was stopped after 10 min by adding 0.05 mL of 2M HCl per 0.5 mL of sample. The reaction mixtures were centrifuged and supernatants were used for HPLC analysis. The temperature optimum was determined at 20–50°C. The pH optimum was determined using Britton–Robinson ($CH_3COOH/H_3BO_3/H_3PO_4/NaOH$) buffers, pH 4.0–10.8, at 30°C. The pH stability was determined by pre-incubating the enzyme in the same buffers for 2 h at 30°C followed by determination of the residual activity at pH 8.0 and 30°C. The concentrations of substrate and reaction products (2-pyridinecarboxylic acid and 2-pyridinecarboxamide) were determined by reversed-phase HPLC (column ACE C8, 5 µm, 250 mm × 4 mm (Advanced Chromatography Technologies Ltd, Aberdeen, UK); mobile phase 10% acetonitrile in 5 mM sodium phosphate buffer, pH 7.2, flow rate 0.9 mL/min).

4.5. General Protocol for the Biocatalyzed Degradation of fCN

Reaction mixtures (1 mL) in 2-mL Eppendorf tubes were incubated in Eppendorf ThermoMixer Comfort (30°C, 850 rpm). Simulated effluents were prepared according to published data [5,8]. The fCN concentrations were 0.6, 4.6, 25 or 100 mM fCN. Optionally the reaction mixtures were supplemented with additives (Na_2S, KSCN, NH_4Cl, phenol, $CuSO_4$, $AgNO_3$; see **Figure 6** and **Figure 7** for details). The buffer was 100 mM glycine/NaOH (pH 9.0, 9.5, 10.0, or 10.5). Samples (0.1 mL) were withdrawn at various intervals (5–180 min) and the reaction was stopped with 0.2 mL methanol. The fCN concentrations were determined using the picric acid method (**Section 4.4**) or using the Spectroquant® kit (Merck).

5. CONCLUSIONS

In this work, proof-of-concept was provided for the performance of NitEg, a unique CynH from the Basidiomycota division, in fCN solutions of up to 100 mM concentrations at alkaline conditions. This CynH was studied as a potential biocatalyst using model mixtures prepared according to the literature. This ensures that the results obtained are reproducible and comparable with other biocatalysts used in the same way in the future. However, further investigation of NitEg or other CyHs in real wastewater will be necessary in the future, preferably with effluents that have been analyzed in detail. This work has shown that the enzyme is highly active, resistant to alkaline pH, tolerates high concentrations of fCN, and other industrial contaminants, and can be stored for extended periods under standard conditions. The enzyme is formed at high expression levels and its production is likely to be readily scalable. Immobilization can be useful to increase its robustness and enable its reuse. Other potential applications of NitEg or CynHs in general could be cyanide biosensors, some of which have been based on CynDs [45]. Importantly, NitEg is unique with respect to its origin, as CynHs are rare in Basidiomycota. Therefore, it may be of interest to study its natural role and potential impact on the interaction between *E. glandulosa* and its plant hosts.

6. CONCISE SUMMARY

Industries such as mining, cokemaking, (petro)chemical and electroplating produce effluents that contain free cyanide (fCN = HCN + CN^-). Currently, fCN is mainly removed by (physico)chemical methods or by biotreatment with activated sludge. Cyanide hydratases (CynHs) (EC 4.2.1.66), which convert fCN to the much less toxic formamide, have been considered for a mild approach to wastewater decyanation. However, few data are available to evaluate the application potential of CynHs. In this study, we used a new CynH from *Exidia glandulosa* (protein KZV92691.1 designated NitEg by us), which was overproduced in *Escherichia coli*. The purified NitEg was highly active for fCN with 784 U/mg protein, k_{cat} 927/s and k_{cat}/K_M 42/s/mM. It exhibited optimal activities at pH approximately 6–9 and 40–45°C. It was quite stable in this pH range, and retained approximately 40% activity at 37°C after 1 day. Silver and copper ions (1 mM) decreased its activity by 30–40%. The removal of 98–100% fCN was achieved for 0.6–100 mM fCN. Moreover, thiocyanate, sulfide, ammonia or phenol added in amounts typical of industrial effluents did not significantly reduce the fCN conversion, while electroplating effluents may need to be diluted due to high fCN and metal content. The ease of preparation of NitEg, its high specific activity, robustness and long shelf life make it a promising biocatalyst for the detoxification of fCN.

REFERENCES

1. Kuyucak, N.; Akcil, A. Cyanide and removal options from effluents in gold mining and metallurgical processes. *Miner. Eng.* **2013**, *50–51*, 13–29. [Google Scholar] [CrossRef]
2. Carmona-Orozco, M.L.; Panay, A.J. Immobilization of *E. coli* expressing *Bacillus pumilus* CynD in three organic polymer matrices. *Appl. Microbiol. Biotechnol.* **2019**, *103*, 5401–5410. [Google Scholar] [CrossRef]
3. Basile, L.J.; Willson, R.C.; Sewell, B.T.; Benedik, M.J. Genome mining of cyanide-degrading nitrilases from filamentous fungi. *Appl. Microbiol. Biotechnol.* **2008**, *80*, 427–435. [Google Scholar] [CrossRef]

4. Martínková, L.; Chmátal, M. The integration of cyanide hydratase and tyrosinase catalysts enables effective degradation of cyanide and phenol in coking wastewaters. *Water Res.* **2016**, *102*, 90–95. [Google Scholar] [CrossRef]
5. Jarrah, N.; Mu'azu, N.D. Simultaneous electro-oxidation of phenol, CN^-, S^{2-} and NH^{4+} in synthetic wastewater using boron doped diamond anode. *J. Environ. Chem. Eng.* **2016**, *4*, 2656–2664. [Google Scholar] [CrossRef]
6. Ibáñez, M.I.; Cabello, P.; Luque-Almagro, V.M.; Sáez, L.P.; Olaya, A.; Sánchez de Medina, V.; Luque de Castro, M.D.; Moreno-Vivián, C.; Roldán, M.D. Quantitative proteomic analysis of *Pseudomonas pseudoalcaligenes* CECT5344 in response to industrial cyanide-containing wastewaters using Liquid Chromatography-Mass Spectrometry/Mass Spectrometry (LC-MS/MS). *PLoS ONE* **2017**, *12*, e0172908. [Google Scholar] [CrossRef] [PubMed]
7. Yang, W.L.; Liu, G.S.; Chen, Y.H.; Miao, D.T.; Wei, Q.P.; Li, H.C.; Ma, L.; Zhou, K.C.; Liu, L.B.; Yu, Z.M. Persulfate enhanced electrochemical oxidation of highly toxic cyanide-containing organic wastewater using boron-doped diamond anode. *Chemosphere* **2020**, *252*, 126499. [Google Scholar] [CrossRef] [PubMed]
8. Papadimitriou, C.A.; Samaras, P.; Sakellaropoulos, G.P. Comparative study of phenol and cyanide containing wastewater in CSTR and SBR activated sludge reactors. *Bioresour. Technol.* **2009**, *100*, 31–37. [Google Scholar] [CrossRef]
9. Yu, X.B.; Xu, R.H.; Wei, C.H.; Wu, H.Z. Removal of cyanide compounds from coking wastewater by ferrous sulfate: Improvement of biodegradability. *J. Hazard. Mater.* **2016**, *302*, 468–474. [Google Scholar] [CrossRef]
10. Makhado, E.; Pandey, S.; Nomngongo, P.N.; Ramontja, J. Preparation and characterization of xanthan gum-cl-poly(acrylic acid)/o-MWCNTs hydrogel nanocomposite as highly effective reusable adsorbent for removal of methylene blue from aqueous solutions. *J. Colloid Interface Sci.* **2018**, *513*, 700–714. [Google Scholar] [CrossRef]
11. Pandey, S.; Do, J.Y.; Kim, J.; Kang, M. Fast and highly efficient catalytic degradation of dyes using κ-carrageenan stabilized silver nanoparticles nanocatalyst. *Carbohydr. Polym.* **2020**, *230*, 115597. [Google Scholar] [CrossRef]
12. Electroplating and metal finishing point source categories; Effluent limitations guidelines, pretreatment, standards and new source performance standards. *Fed. Regist.* **1983**, *48*, 32487. Available online: https://www.epa.gov/sites/default/files/2015-10/documents/electroplating-and-metal-finishing_proposed-rule_08-31-1982_47-fr-38462.pdf (accessed on 12 November 2021).
13. Code of Federal Regulations. 40, Subpart A-Cokemaking Subcategory. Available online: https://www.ecfr.gov/current/title-40/chapter-I/subchapter-N/part-420/subpart-A (accessed on 12 November 2021).

14. Cosmos, A.; Erdenekhuyag, B.O.; Yao, G.; Li, H.J.; Zhao, J.G.; Wang, L.J.; Lyu, X.J. Principles and methods of bio detoxification of cyanide contaminants. *J Mater. Cycles Waste Manag.* **2020**, *22*, 939–954. [Google Scholar] [CrossRef]

15. Rahdar, S.; Rahdar, A.; Sattari, M.; Hafshejani, L.D.; Tolkou, A.K.; Kyzas, G.Z. Barium/cobalt@polyethylene glycol nanocomposites for dye removal from aqueous solutions. *Polymers* **2021**, *13*, 1161. [Google Scholar] [CrossRef]

16. Rahdar, S.; Pal, K.; Mohammadi, L.; Rahdar, A.; Goharniya, Y.; Samani, S.; Kyzas, G.Z. Response surface methodology for the removal of nitrate ions by adsorption onto copper oxide nanoparticles. *J. Mol. Struct.* **2021**, *1231*, 129686. [Google Scholar] [CrossRef]

17. Baeissa, E.S.; Mohamed, R.M. Enhancement of photocatalytic properties of Ga_2O_3-SiO_2 nanoparticles by Pt deposition. *Chin. J. Catal.* **2013**, *34*, 1167–1172. [Google Scholar] [CrossRef]

18. Bagabas, A.; Alshammari, A.; Aboud, M.F.A.; Kosslick, H. Room-temperature synthesis of zinc oxide nanoparticles in different media and their application in cyanide photodegradation. *Nanoscale Res. Lett.* **2013**, *8*, 516. [Google Scholar] [CrossRef][Green Version]

19. Fan, L.; Yao, H.; Deng, S.; Jia, F.; Cai, W.; Hu, Z.; Guo, J.; Li, H. Performance and microbial community dynamics relationship within a step-feed anoxic/oxic/anoxic/oxic process (SF-A/O/A/O) for coking wastewater treatment. *Sci. Total Environ.* **2021**, *792*, 148263. [Google Scholar] [CrossRef]

20. Martínková, L.; Veselá, A.B.; Rinágelová, A.; Chmátal, M. Cyanide hydratases and cyanide dihydratases: Emerging tools in the biodegradation and biodetection of cyanide. *Appl. Microbiol. Biotechnol.* **2015**, *99*, 8875–8882. [Google Scholar] [CrossRef]

21. Fry, W.E.; Millar, R.L. Cyanide degradation by an enzyme from *Stemphylium loti*. *Arch. Biochem. Biophys.* **1972**, *151*, 468–474. [Google Scholar] [CrossRef]

22. Wang, P.; VanEtten, H.D. Cloning and properties of a cyanide hydratase gene from the phytopathogenic fungus *Gloeocercospora sorghi*. *Biochem. Biophys. Res. Commun.* **1992**, *187*, 1048–1054. [Google Scholar] [CrossRef]

23. Cluness, M.J.; Turner, P.D.; Clements, E.; Brown, D.T.; O'Reilly, C. Purification and properties of cyanide hydratase from *Fusarium lateritium* and analysis of the corresponding *chy1* gene. *J. Gen. Microbiol.* **1993**, *139*, 1807–1815. [Google Scholar] [CrossRef][Green Version]

24. Barclay, M.; Tett, V.A.; Knowles, C.J. Metabolism and enzymology of cyanide/metallocyanide biodegradation by *Fusarium solani* under neutral and acidic conditions. *Enzym. Microb. Technol.* **1998**, *23*, 321–330. [Google Scholar] [CrossRef]

25. Yanase, H.; Sakamoto, A.; Okamoto, K.; Kita, K.; Sato, Y. Degradation of the metal-cyano complex tetracyanonickelate (II) by *Fusarium oxysporum* N10-. *Appl. Microbiol. Biotechnol.* **2000**, *53*, 328–334. [Google Scholar] [CrossRef]

26. Kushwaha, M.; Kumar, V.; Mahajan, R.; Bhalla, T.C.; Chatterjee, S.; Akhter, Y. Molecular insights into the activity and mechanism of cyanide hydratase enzyme associated with cyanide biodegradation by *Serratia marcescens*. *Arch. Microbiol.* **2018**, *200*, 971–977. [Google Scholar] [CrossRef]

27. Basic Local Alignment Search Tool. Available online: **https://blast.ncbi.nlm.nih.gov** (accessed on 12 November 2021).

28. Park, J.M.; Sewell, B.T.; Benedik, M.J. Cyanide bioremediation: The potential of engineered nitrilases. *Appl. Microbiol. Biotechnol.* **2017**, *101*, 3029–3042. [Google Scholar] [CrossRef]

29. Nolan, L.M.; Harnedy, P.A.; Turner, P.; Hearne, A.B.; O'Reilly, C. The cyanide hydratase enzyme of *Fusarium lateritium* also has nitrilase activity. *FEMS Microbiol. Lett.* **2003**, *221*, 161–165. [Google Scholar] [CrossRef][Green Version]

30. Kaplan, O.; Veselá, A.B.; Petříčková, A.; Pasquarelli, F.; Pičmanová, M.; Rinágelová, A.; Bhalla, T.C.; Pátek, M.; Martínková, L. A comparative study of nitrilases identified by genome mining. *Mol. Biotechnol.* **2013**, *54*, 996–1003. [Google Scholar] [CrossRef] [PubMed]

31. Veselá, A.B.; Rucká, L.; Kaplan, O.; Pelantová, H.; Nešvera, J.; Pátek, M.; Martínková, L. Bringing nitrilase sequences from databases to life: The search for novel substrate specificities with a focus on dinitriles. *Appl. Microbiol. Biotechnol.* **2016**, *100*, 2193–2202. [Google Scholar] [CrossRef]

32. Rucká, L.; Chmátal, M.; Kulik, N.; Petrásková, L.; Pelantová, H.; Novotný, P.; Příhodová, R.; Pátek, M.; Martínková, L. Genetic and functional diversity of nitrilases in Agaricomycotina. *Int. J. Mol. Sci.* **2019**, *20*, 5990. [Google Scholar] [CrossRef][Green Version]

33. Rinágelová, A.; Kaplan, O.; Veselá, A.B.; Chmátal, M.; Křenková, A.; Plíhal, O.; Pasquarelli, F.; Cantarella, M.; Martínková, L. Cyanide hydratase from *Aspergillus niger* K10: Overproduction in *Escherichia coli*, purification, characterization and use in continuous cyanide degradation. *Proc. Biochem.* **2014**, *49*, 445–450. [Google Scholar] [CrossRef]

34. Malmir, N.; Fard, N.A.; Mgwatyu, Y.; Mekuto, L. Cyanide hydratase modification using computational design and docking analysis for improved binding affinity in cyanide detoxification. *Molecules* **2021**, *26*, 1799. [Google Scholar] [CrossRef]

35. Rucká, L.; Kulik, N.; Novotný, P.; Sedova, A.; Petrásková, L.; Příhodová, R.; Křístková, B.; Halada, P.; Pátek, M.; Martínková, L. Plant nitrilase homologues in fungi: Phylogenetic and functional analysis with focus on nitrilases in *Trametes versicolor* and *Agaricus bisporus*. *Molecules* **2020**, *25*, 3861. [Google Scholar] [CrossRef]

36. Fisher, F.B.; Brown, J.S. Colorimetric determination of cyanide in stack gas and waste water. *Anal. Chem.* **1952**, *24*, 1440–1444. [Google Scholar] [CrossRef]

37. Pérez-Cid, B.; Calvar, S.; Belén Moldes, A.; Cruz, J.M. Effective removal of cyanide and heavy metals from an industrial electroplating stream using calcium alginate hydrogels. *Molecules* **2020**, *25*, 5183. [Google Scholar] [CrossRef]

38. Xue, Y.P.; Wang, Y.P.; Xu, Z.; Liu, Z.Q.; Shu, X.R.; Jia, D.X.; Zheng, Y.G.; Shen, Y.C. Chemoenzymatic synthesis of gabapentin by combining nitrilase-mediated hydrolysis with hydrogenation over Raney-nickel. *Catal. Commun.* **2015**, *66*, 121–125. [Google Scholar] [CrossRef]

39. Wang, L.; Watermeyer, J.M.; Mulelu, A.E.; Sewell, B.T.; Benedik, M.J. Engineering pH-tolerant mutants of a cyanide dihydratase. *Appl. Microbiol. Biotechnol.* **2012**, *94*, 131–140. [Google Scholar] [CrossRef]

40. Crum, M.A.; Park, J.M.; Mulelu, A.E.; Sewell, B.T.; Benedik, M.J. Probing C-terminal interactions of the *Pseudomonas stutzeri* cyanide-degrading CynD protein. *Appl. Microbiol. Biotechnol.* **2015**, *99*, 3093–3102. [Google Scholar] [CrossRef]

41. Jandhyala, D.M.; Willson, R.C.; Sewell, B.T.; Benedik, M.J. Comparison of cyanide-degrading nitrilases. *Appl. Microbiol. Biotechnol.* **2005**, *68*, 327–335. [Google Scholar] [CrossRef] [PubMed]

42. BRENDA. The Comprehensive Enzyme Information System. Available online: **https://brenda-enzymes.org** (accessed on 2 November 2021).

43. Dennett, G.V.; Blamey, J.M. A new thermophilic nitrilase from an antarctic hyperthermophilic microorganism. *Front. Bioeng. Biotechnol.* **2016**, *4*, 5. [Google Scholar] [CrossRef][Green Version]

44. Mondal, A.; Sarkar, S.; Nair, U.G. Comparative characterization of cyanide-containing steel industrial wastewater. *Water Sci. Technol.* **2021**, *83*, 322–330. [Google Scholar] [CrossRef] [PubMed]

45. Ketterer, L.; Keusgen, M. Amperometric sensor for cyanide utilizing cyanidase and formate dehydrogenase. *Anal. Chim. Acta* **2010**, *673*, 54–59. [Google Scholar] [CrossRef] [PubMed]

CHAPTER-10

CLAVISPORA LUSITANIAE ABS7 HARBORS AN ALKALOTHERMOPHILIC AMYLOPULLULANASE: PURIFICATION, CHARACTERIZATION AND POTENTIAL APPLICATION IN LAUNDRY DETERGENT

1. INTRODUCTION: AN OVERVIEW

In the realm of enzymatic exploration, the identification and characterization of novel enzymes hold significant promise for various industrial applications. Among these, the alkalothermophilic amylopullulanase derived from the yeast Clavispora lusitaniae ABS7 has emerged as a subject of keen interest. This enzyme exhibits unique properties, making it well-suited for purification and detailed characterization. As we delve into the purification process and unravel the distinctive features of this enzyme, a particular focus lies on its potential application in laundry detergent—an area where enzymatic catalysts play a pivotal role. This investigation not only sheds light on the enzymatic prowess of Clavispora lusitaniae ABS7 but also opens avenues for its practical utilization in enhancing the efficiency of laundry detergents.

According to the catalytic reactions, the International Enzymes Commission has categorized seven classes of enzymes: EC1, oxidoreductases; EC2, transferases; EC3, hydrolases; EC4, lyases; EC5, isomerases; EC6, ligases; and EC7, translocases [1]. Among hydrolases, amylases produced by fungi are the most widely used commercial enzymes to meet the ever-increasing demands of the global enzyme market. They are widely used in industry and have been of a great interest in the food, detergent, pharmacy, textile, paper, and bioethanol industries [2,3,4]. The global industrial enzymes market should increase to $7.0 billion by 2023 compared to $5.5 billion in 2018 at a compound annual growth rate of 4.9% for the period 2018–2023 [5].

In automatic dishwasher and laundry, detergent formulations are fortified with alkaline amylases (higher than 8.0) [6,7] to improve the ability of the detergents to remove tough stains, making the detergent environmentally friendly [8]. Detergent enzymes represent one of the largest and most successful applications of modern industrial biotechnology since they account for about 40% of the total worldwide enzyme production [9].

In the formulation of the enzymatic detergent, amylases are the second enzyme used after proteases, and about 90% of liquid detergents contain these enzymes [10]. Amylases are usually produced using bacteria such as *Bacillus licheniformis* and molds such as *Aspergillus oryzae* and *Aspergillus niger*. However, few studies have been done on the production of these in yeasts.

On an industrial scale, submerged fermentation (SMF) and solid-state fermentation (SSF) are frequently used for the production of microbial amylases. SMF is used to produce bio products from a broth medium such as molasses or a liquid medium. This method provides remarkable humidity, which is crucial for the growth of microorganisms; it allows easy sterilization, controllable temperature, and pH, etc. [11].

SSF is used for the production of amylases from easily recycled solid wastes (as cheap substrate) such as wheat bran, potato peels, citrus waste, and paper. This method provides low humidity, and it requires simple equipment. However, SSF is slower than SMF in the utilization of substrates by microorganisms [11]. In our work, we used liquid fermentation because the waste used as the basic medium for the production of amylopullulanase is whey.

Purification of glycosyl hydrolases such as amylases has been reported in some fungi. For the purification of amylases, their concentration using precipitation with ammonium sulfate followed by dialysis or by solvents such as acetone is recommended before chromatography. The enzymes were purified using gel filtration chromatography and/or ion exchange chromatography and then characterized [12,13,14,15].

Since the application of amylase is increasing in various industrial areas, the demand for novel amylases, mainly thermostable amylolytic enzymes, is increasing worldwide in industry [16]. The advantages of using thermostable amylases in industrial processes include the low risk of contamination, the cost of external cooling, and high diffusion rate [17].

In Algeria, in order to improve the economy by reducing the cost of enzymes, it would be interesting to produce thermostable amylases. In this light, this study describes the purification, characterization, and potential application in detergent of the amylopullulanase (α-amylase and pullulanase activities) from *Clavispora lusitaniae* ABS7. To the best of our knowledge, this study is the first research on the purification and characterization of amylopullulanase from a yeast strain.

2. RESULTS AND DISCUSSION

The yeast strain *Clavispora lusitaniae* ABS7 has presented a clear lysis zone for both enzymes. As the inductor, starch was used for the α-amylase activity (YPSA medium) and pullulan for pullulanase (YPPA medium). In their presence, the amylolytic yeast secretes α-amylase and pullulanase will diffuse and hydrolyze starch (YPSA medium) (**Figure 1**A) and pullulan (YPPA medium) (**Figure 1**B) to form a clear halo around the colonies [**18**].

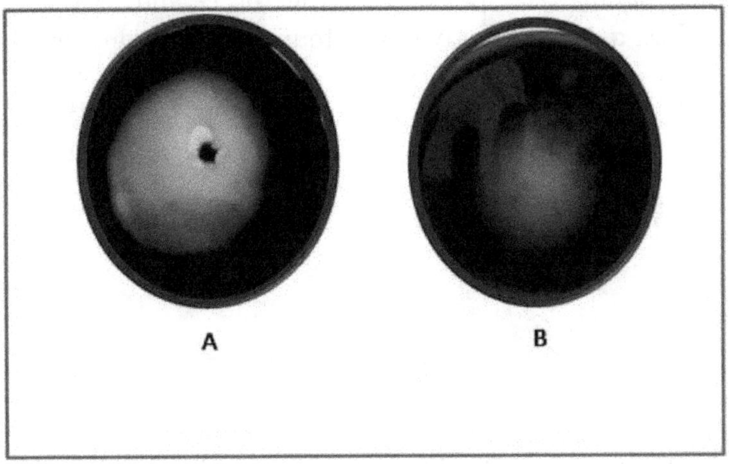

Figure 1. Lysis zones of starch (A) and pullulan (B) in the presence of amylolytic enzymes produced by *Clavispora lusitaniae* ABS7.

According to Ramachandran et al. [**19**], out of 150 starch assimilating yeasts in nature, only a few strains were able to hydrolyze both α-1,4 and α-1,6 linkages carbohydrate polymers such as *Lipomyces kononenkoae* and *Cryptococcus* sp. S-2 [**20**].

2.1. Purification of Enzymes

Clavispora lusitaniae ABS7 produced α-amylase and pullulanase activities of 346,340 IU and 325,900 IU, respectively, in the enzymatic extract after lyophilization. The use of acetone precipitation led to the concentration of protein with an increase in the specific activity of the α-amylase from 79.97 to 352.41 U/mg and of the pullulanase from 75.25 to 291.33 U/mg with recoveries of 45.79% and 40.23%, respectively (**Table 1**).

Table 1. A summary of the purification of the amylolytic enzyme produced by *C. lusitaniae* ABS7.

Purification Step	Total Protein (mg/mL)	Total Activity (IU)		Specific Activity IU/mg	Purification (Fold)	Yield (%)
Lyophilized Extract	4330.6	α-amylase	346,340	79,975	1	100
		Pullulanase	325,900	75.25	1	100
Acetone precipitation	450	α-amylase	158,587	352.41	4.40	45.79
		Pullulanase	131,102.5	291.33	3.87	40.23
Sephacryl S 200	75.22	α-amylase	106,122	1410.82	17.64	30.64
		Pullulanase	85,386.1	1135.15	15.08	26.2

Sephacryl S-200 gel filtration chromatography allowed an increase of the α-amylase and pullulanase specific activities of 17 and 15 times, respectively. However, the best α-amylase activity was found in the second peak (**Figure 2**).

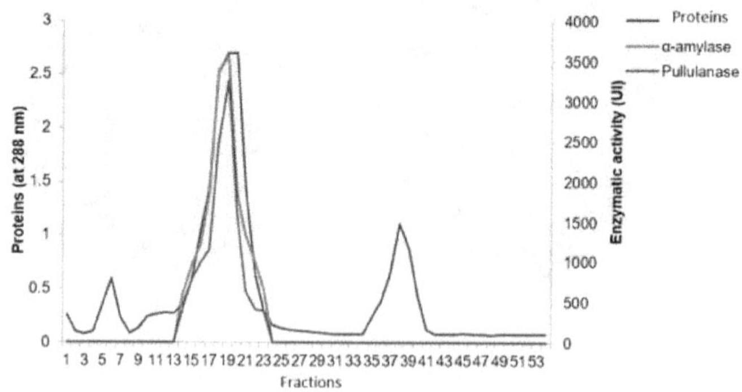

Figure 2. Chromatographic profile of α-amylase and pullulanase on Sephacryl S-200.

The elution in the DEAE cellulose column (**Figure 3**) showed the existence of three protein fractions containing the α-amylase activity.

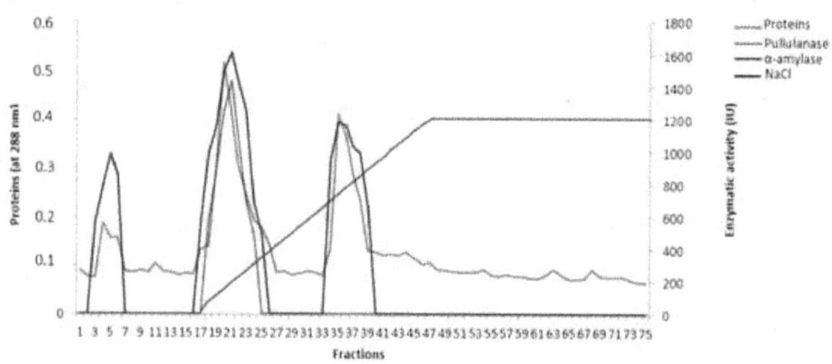

Figure 3. Chromatographic profile of α-amylase and pullulanase DEAE cellulose.

In addition to the α-amylase activity, the second peak also contains a pullulanase activity (**Figure 3**). **Table 1** shows that the specific activity of α-amylase and pullulanase increased 50 and 44 times with recoveries of 23.9% and 21%, respectively. Consequently, the second peak was the most interesting with a total amylolytic activity that hydrolyses o α-1,4 and α-1,6 bonds.

2.2. SDS-PAGE Analysis

SDS-PAGE analysis of the purified enzyme showed a single protein band with an apparent molecular weight of 75 kDa (**Figure 4**a,b). This indicates that the purified enzyme is a monomer. Vishnu et al. [21] described a monomeric enzyme of *L amylophilus* GV6 of 90 kDa corresponding to a pullulanase Type I.

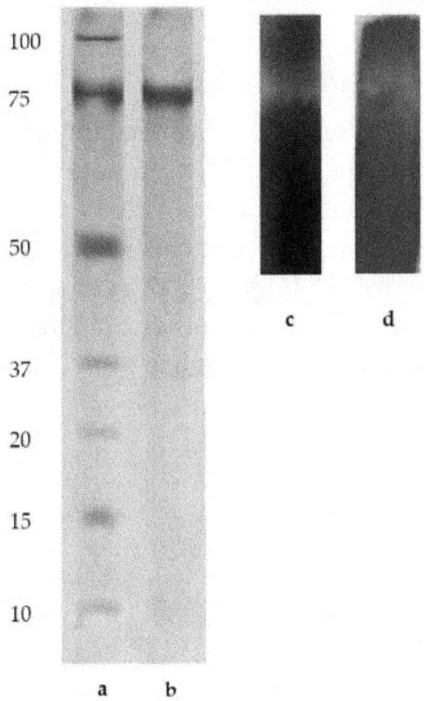

Figure 4. SDS-PAGE electrophoretic profile of the purified enzyme. Revelation of protein bands by Coomassie blue. (a) Marker proteins, (b) amylopullulanase of *Clavirospora lusitaniae* ABS 7, (c) revelation with the lugol, and (d) revelation with the Congo red.

According to the bibliographies, the molecular mass of the purified amylopullulanase varies between 74 and 450 kDa [22], and that of pullulanases varies between 54 and 134 kDa [23]. Kar et al. [24] revealed that the purified amylopullulanase produced from *Streptomyces erumpens* MTCC 7317 had a molecular mass 45.0 kDa.

The revelation with lugol and Congo red showed that this single band possesses both *α*-amylase and pullulanasic activities (**Figure 4**c,d). The amylolytic enzyme from our yeast strain *Clavispora lusitaniae* ABS7 appears to be a bi-functional amylopullulanase enzyme with two active sites for α-amylase and pullulanase.

It was found that on SDS-PAGE, the alkaline amylopullulanase from *Bacillus* sp. KSM-1378,showedasingleproteinwithtwodifferentactivesites,onetohydrolyze-1,4 bonds and the other to cut α-1,6 bonds [25]. The same results were obtained with the same enzyme of *Clostridium thermohydrosulfuricum*, *Bacillus circulans* F-2, *L. amylophilus* GV6, and Thermoanaerobacter ethanolicus [21,26,27,28,29], while the study of an alkaline amylopullulanase from alkalophilic *Bacillus* sp. KSM-1378 has shown that the two catalytic activities of the enzyme involve two different active sites [30].

2.3. Thin Layer Chromatography (TLC)

In order to confirm the presence of the two amylolytic enzymes, α-amylase and pullulanase, as well as their hydrolysis products (derived from starch and pullulan), a TLC was carried out with the purified enzyme. After incubation with pullulan, the purified enzyme produced exclusively a mixture of glucose, maltose, and maltotriose identified by TLC (**Figure 5**). The enzyme attacked both glycoside α-1,6 and α-1,4 linkages of pullulan and other branched polysaccharides such as starch to produce glucose and maltose (**Figure 5**).

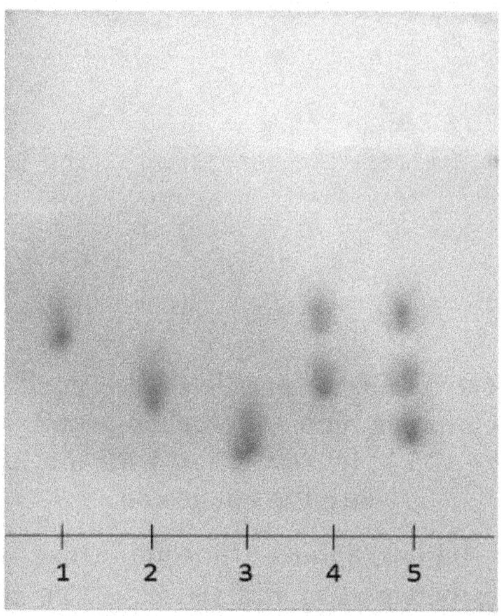

Figure 5. TLC chromatography and analysis of hydrolyzed products from starch and pullulan. 1—glucose, 2—maltose, 3—maltotriose, 4—the enzyme and starch, and 5—the enzyme and pullulan.

It can be suggested that the enzymatic mechanism of glucose and maltose formation from pullulan occurs after the hydrolysis of α-1,6 bonds, which gives maltotriose. The latter will be hydrolyzed to glucose and maltose. The cleavage of α-1,6 bonds of starch amylopectin gave chains of maltodextrins much longer than those of maltotriose. It is likely that the maltotriose chains are hydrolyzed faster to produce exclusively glucose and maltose [22]. On the other hand, the main product from starch is obviously maltose [4].

The electrophoretic profile by SDS-PAGE showed a single band with both α-amylase and pullulanase activities; this result was also confirmed by TLC. The studied enzyme was revealed as an endo-type enzyme and thus it is a pullulanase of Type II or amylopullulanase [31,32]. It was found that pullulanase of *Bacillus cereus* H1.5 cannot attack dextran, which contains α-1,6 bonds; however, it hydrolyzes pullulan into maltotriose (the main product) and other polysaccharides such as starch into maltose and glucose. Therefore, it can be classified as a Type II pullulanase or amylopullulanase [31]. Leveque et al. [33] showed that all thermostable pullulanases from thermophilic archaea are of Type II, whereas no amylopullulanase Type I has been characterized in these thermophilic microorganisms.

2.4. Physicochemical Parameters of the Studied Enzymes

2.4.1. Effect of Temperature on the Amylopullulanase Activity

The influence of temperature on the activity of alkaline amylopullulanase with pullulan and starch as substrates was determined by measuring the activity at different temperatures from 40 to 100°C at pH 9.

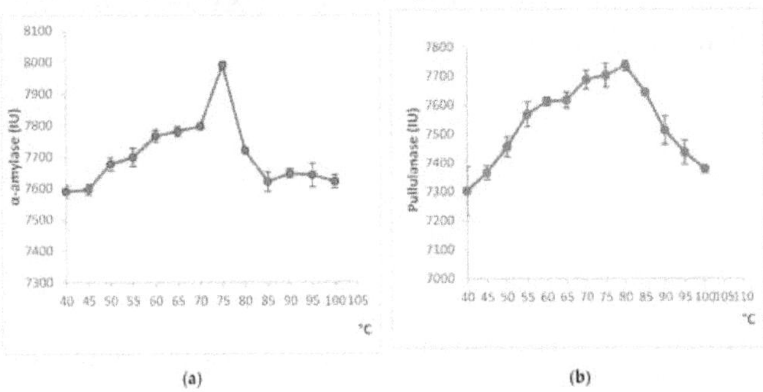

Figure 6. Effect of temperature on the (a) α-amylase and (b) pullulanase activities.

The activity of the enzymes highly depends on the temperature (highly significant difference with $F = 6$ ($p < 0.001$) for the amylase and $F = 10.3$ ($p = 0.000$) for the pullulanase). The pullulanase and α-amylase activities of the studied yeast strain *C. lusitaniae* ABS7 exhibited optimum temperatures of 80 and 75°C, respectively (Figure 6). De Souza and Magalhães [4] found that, among the ten species of yeasts and molds studied, none showed an optimum enzymatic activity up to 75°C.

Moreover, the study of Nakamura et al. [34] revealed that *C. lusitaniae* produced a phytase with an optimum temperature of 70 °C.

The temperature for optimal α-amylase activity was 70 °C for *Lipomyces kononenkoae* [35], 40°C for *Schwanniomyces alluvius* [36], 50 °C for *Cryptococcus flavus* [37], and 55°C for *Sporidiobolus pararoseus* PH-Gra1 [38]. The optimum temperatures of the purified α-amylase from *Bacillus substilis*, *B. licheniformis* AI20, and *Haloarcula* sp. were found to be 45, 60–80, and 35–40°C, respectively [39,40].

A previous work [41] reported the optimal temperature and pH of amylopullulanase from *Streptococcus infantarius* ssp. as 37°C and 6.8, respectively.

The temperature optima for both pullulanase and amylase activities from *Thermoanaerobacter* strain B6A was 75°C [42]. The Pullulanase Type II from *Bacillus cereus* H1.5 [31] and from *Thermococcus hydrothermalis* [43] showed optimal activity at 55 and 105°C, respectively. The optimum temperatures for the pullulanase action from *Aerobacter aerogenes* [44] and from alkalophilic *Bacillus* sp. S-1 [45] were 50 and 60°C, respectively.

2.4.2. Effect of pH on α-Amylase and Pullulanase Activities

Analysis of the experimental results of the two activities by the (ANOVA) method reveals that the pH considerably affects the α-amylase activity (F = 63.2 (p = 0.000)) and the pullulanase activity (F = 60 (p = 0.001)). The alpha amylase and the pullulanase of the yeast *Clavispora lusitaniae* ABS7 presented a wide range of activity from pH 5 to 12, with an optimum pH of 9 (**Figure 7**). Beyond this pH value, a decline in activities is observed.

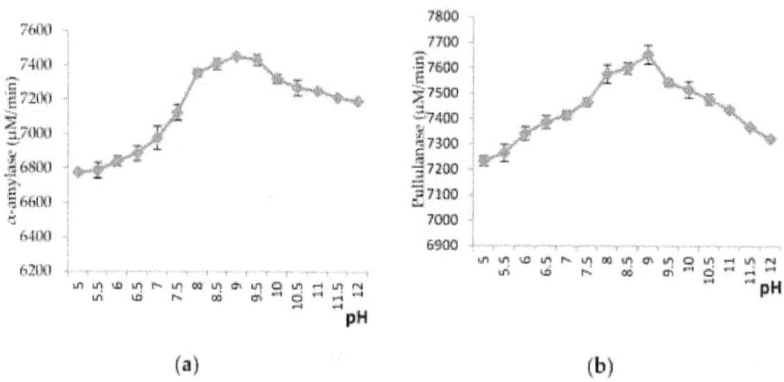

Figure 7. Effect of pH on the α-amylase activity (a) and pullulanase (b).

Different studies found the optimum pH of 5 to 6; 5 is very common for bacterial pullulanase such as *Exiguobacterium acetylicum*, *Thermoanaerobacter* strain B6A, *Pyrococcus furiosus*, *Pyrococcus woesei*, and *Thermococcus* strain TY [46,47].

Furthermore, Kim et al. [45], described a Type I pullulanase of a mesophilic and alkalophilic bacteria *Bacillus* sp. S-1 with an optimum pH range from 8 to 10. Additionally, Asha et al. [48] showed that the optimum pH of the purified alkaline pullulanase isolated from *Bacillus halodurans* was found to be 10.

Microbial thermostable amylases have the optimum pH ranging from 5 to 10.5 [49]. The extracellular amylase from the yeast *Schwanniomyces alluvius* had an optimum pH of 6.3 [36]. The amylase from *Lipomyces kononenkoae* was monomeric, with an optimum pH of 4.5 to 5.0 [35]. The amylase activity of *Cryptococcus flavus* was optimal at pH 5.5 [37] and that of *Sporidiobolus pararoseus* PH-Gra1 at pH 6.5 [38]. The optimal pH of the purified amylases from *Bacillus substilis* [39], *B. licheniformis* AI20 [40], and *Streptomyces* sp. Al-Dhabi-46 [15] was found to be 6.0, 6–7.5, and 8, respectively.

The *Talaromyces pinophilus* α-amylase (TpAA) was most active at pH 4.0–5.0 [50] and the maximum activity of α-amylase from *Trichoderma harzianum* against soluble starch was determined at pH 4.5 and 40°C [51].

2.4.3. Study of Thermal Stability of α-Amylase and Pullulanase Activities

The thermal stability of the enzyme was studied after incubation at different temperatures (75 and 100°C for amylase and 80 and 100 °C for pullulanase) for 30–180 min at pH 9 for both activities.

The results show that the α-amylase maintained 51.76% of its initial activity after 120 min of incubation at 100°C and 88% after 3 h at 75°C. However, a decrease of about 61.4% of its activity after 180 min of incubation at 100 °C was registered (Figure 8).

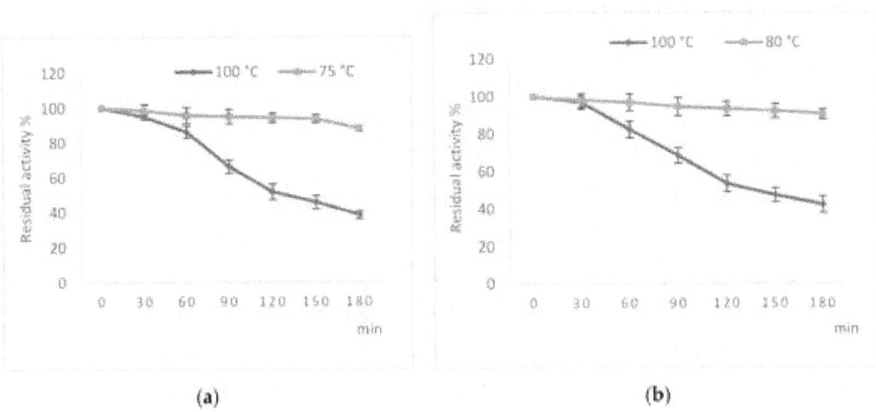

Figure 8. Stability of the α-amylase (a) and pullulanase (b) activities of the purified enzyme.

For pullulanase activity, the enzyme kept 53.4% and 42% of its initial activity after incubation at 100°C for 120 and 180 min, respectively. In addition, the enzyme maintained 91% of its activity after 180 min of incubation at 80°C.

From the above analysis, it can be revealed that the α-amylase and pullulanase activities retain more than 50% of their initial activities after heat-treatments at 75 and 85°C for 2 h, respectively.

In fact, the amylase resistance to thermal denaturation can be explained by the presence of calcium in the medium, which stabilizes the enzyme and increases its activity [48,52]. The substrate also has a stabilizing effect on the enzyme [53]. The thermostability of the studied α-amylase and pullulanase is also due to the presence of certain groups of amino acids and their sequence (Cyst, Tyr, Ser, Glu, Asp, Arg, Lys, and Leu) [53]. Declerck et al. [54] revealed that the amylases tend to adjust their conformational flexibility to achieve optimal catalytic efficiency in the temperature range that are supposedly functional. The thermophilic proteins are generally more rigid than psychrophilic proteins and more flexible than their mesophilic homologous [55].

Protein molecules do not have a fixed structure; nevertheless, they exhibit a dynamic character with a conformational flexibility [56]. Several studies [57] performed on mesophilic and thermophilic proteins suggested that sufficient molecular flexibility (via atomic movements) exists to facilitate the conformational changes, necessary for enzymatic activity (for example, fixing and releasing the substrate, etc.).

2.5. Effect of Different Salts and Chemical Reagents on the α-Amylase and Pullulanase Activities of C. lusitaniae ABS7

2.5.1. Effect of Salts

The effect of various metal ions on the α-amylase and pullulanase activities is shown in **Figure 9**. ANOVA of the experimental results shows that the α-amylase and the pullulanase are affected by salts ($F = 288.62$ ($p = 0.000$) and $F = 408.89$ ($p = 0.000$), respectively). The pullulanase activity of *C. lusitaniae* ABS7 was decreased by Zn^{2+}, Mn^{2+}, and Na^{+}, while that of α-amylase was increased by 25.62%, 20.28%, and 39.73%, respectively. The results indicated that Fe^{2+} inhibited the activity of both enzymes with a decrease of 20%, 19% for α-amylase and 20.46% for pullulanase. On the other hand, the Cu^{2+} ions stimulated the α-amylase activity and slightly inhibited the pullulanase activity (3.62%). In addition, the α-amylase activity was slightly decreased by Mg^{2+} (0.6%) and Ca^{2+} and Hg^{2+} (11.17%), while these ions stimulated the pullulanase activity.

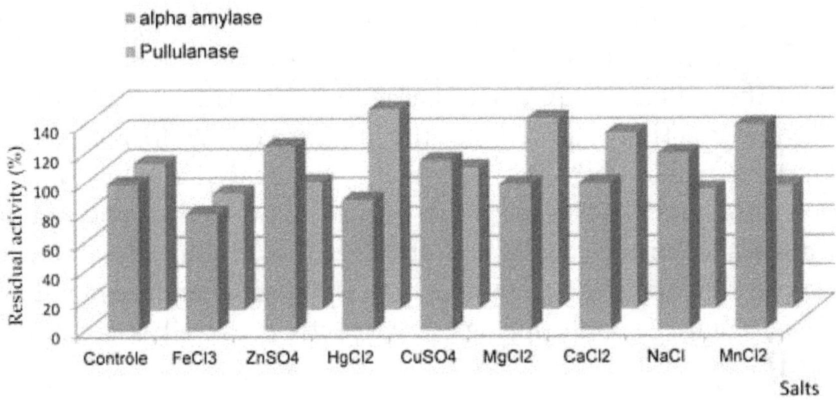

Figure 9. Effect of salts on α-amylase and pullulanase activities.

Asha and al. [48] showed that Cu^{2+}, Zn^{2+}, Mn^{2+} caused a decrease in alkaline pullulanase activity of *Bacillus halodurans*, while Ca^{2+} had a stimulating effect on the enzyme. Thus, Ca^{2+} might be required for stabilization and maintenance of the enzyme conformation. Mrudula et al. [32] have shown that 5 mM of Mg^{2+}, Ca^{2+}, Cu^{2+}, Fe^{3+}, Zn^{2+}, Hg^{2+}, Cd^{2+}, and Li^{2+} ions inhibit both the α-amylase and pullulanase activities of *Clostridium thermosulfurogenes* SVM17. The results of Qiao et al. [46] indicate that the activity of the *Exiguobacterium acetylicum* pullulanase increases in the presence of Fe^{2+} and Mn^{2+} and decreases in the presence of Cu^{2+}.

It has been revealed that calcium ions increase the amylase activity [31,33,58] and maintain the initial activity of α-amylase of the yeast *Cryptococcus flavus* [59] and *Cryptococcus* sp. S-2 [20]. The thermal stability of *Streptomyces avermitilis* α-amylase [60] and *Bacillus cereus* H1.5 pullulanase [31] is increased by calcium ions because they render the protein molecule more rigid. In addition, the calcium-stabilizing effect on the thermostability of the enzyme can be explained by the release of hydrophobic residues in the protein [58].

Vishnu et al. [21] showed a 2.5 mM $CaCl_2$ increase the α-amylase and pullulanase activities of *Lactobacillus amylophilus* GV6, whereas at 5 mM, it becomes an inhibitor of both enzymes.

The moderate inhibition of α-amylase and pullulanase by these ions was observed in other studies on α-amylase from several bacteria species such as *Bacillus* sp. LI711 [61], as well as the pullulanase of *B. stearothermophilus* KP1064 [62].

Lin et al. [17] explain that the inhibitory effect caused by Hg^{2+} and Cu^{2+} ions on the *Bacillus* sp. TS23 α-amylase activity may be due to the competition between the exogenous cations and the cations associated with the proteins, leading to a decrease in activity. The differential behavior of the activities of α-amylase and pullulanase regarding certain metal ions (such as Ca^{2+}, Hg^{2+}, and Mg^{2+}) may be due to the presence of two different active sites, one for the α-amylase and the other for the pullulanase [63]. Compared to the control, the effect of divalent ions on amylolytic activity showed that the presence of Mg^{2+} increased amylolytic activity by 146%, while Mn^{2+}, Fe^{3+}, Ca^{2+}, and Na^{+} increased amylolytic activity to 141%, 116%, 112%, and 111%, respectively [41]. It is known that most amylases are considered metalloenzymes, which are enzymes that require metal ions (usually Ca^{2+}) to maintain their stable native structure. However, there are metal-activated amylases that require Ca^{2+} only during catalytic activity as well as Ca^{2+}-independent amylases [64].

Some extracellular amylases are not activated by Ca^{2+} [65]; others are activated and stabilized by other divalent metal ions such as *B. licheniformis* 2618 amylase, which requires Mg^{2+} [66]. With the exception of Hg^{2+}, which partially inhibited the *L. amylophilus* NRRL B-4437 and *L. amylovorus* ATCC 33,620 α-amylase, various metal ions, such as 1 mM Ca^{2+}, Cu^{2+}, and Ba^{2+}, stimulated the *L. amylophilus* amylase activity while they inhibited the L. amylovorus α-amylase activity [67]. The activity of α-amylase from *Thermomyces lanuginosus* F1 increased in the presence of Mn^{2+}, Co^{2+}, Ca^{2+}, Zn^{2+}, and Fe^{2+} [68].

2.5.2. Influence of Different Chemical Reagents

The influence of different chemical reagents on the α-amylase and pullulanase activities was also studied (**Figure 10**). The effect of those chemical reagents on the two activities is very significant, F = 1089.63 (p = 0.000) for the amylase and F = 259.98 (p = 0.0001) for the pullulanase. It was noticed that the inhibitory effect of the urea is more important on the α-amylase than on pullulanase activities. In contrast to that of tween 80, it is less strong on α-amylase than on pullulanase. The absence of inhibition of α-amylase by EDTA, a strong chelating agent of metal, suggested that the isolated enzyme could not be a metalloenzyme. The presence of $CaCl_2$ has no effect on the α-amylase activity (**Figure 9**). It allows concluding that Ca^{2+} ions are not necessary for the activity of this α-amylase and probably important for its stability and the maintenance of its conformation.

In contrast, EDTA inhibits pullulanase activity, with a 35% loss of its residual activity. This result indicates that the pullulanase is a metalloenzyme and the activity is $CaCl_2$ dependent. This is due to the presence of calcium ions, which increases the activity (**Figure 9**). Iefuji et al. [20] reported the null effect of EDTA on the yeast α-amylase in the yeast *Cryptococcus* sp. S-2. It appears that SDS and mercaptoethanol inhibit both amylase and pulllanasic activities from *Clavispora lusitaniae* ABS7.

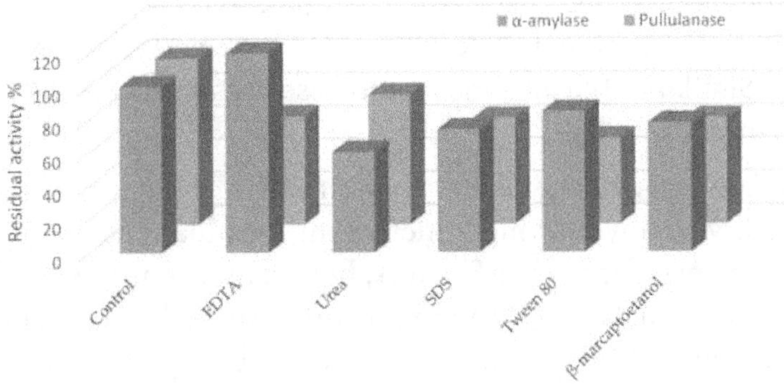

Figure 10. Effect of different chemical reagents on α-amylase and pullulanase activities.

Arabaci and Arikan [69] found that EDTA has no effect on the amylopullulanase of *Geobacillus thermoleovorans* NPI, while 5% marcaptoethanol inhibits it. The study of Ara et al. [25] showed that SDS inhibits the amylase activity of *Bacillus* sp. KSM-1378. The inhibitory effect of urea, guanidine-HCl, and disodium EDTA on α-amylase was also revealed in *Thermomyces lanuginosus* F1 [68].

2.6. Compatibility Test with Various Commercial Laundry Detergents

The α-amylase and pullulanase activities of *C. lusitaniae* ABS7 show significant compatibility with all detergents in commercial detergents (Figure 11). A highly significant difference with $F = 277.10$ ($p = 0.000$) for the α-amylase and $F= 258.41$ ($p = 0.000$) for the pullulanase was noted.

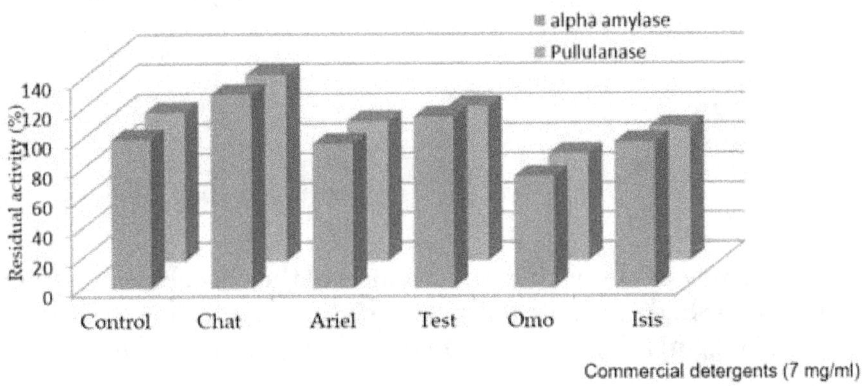

Figure 11. Stability and compatibility of α-amylase and pullulanase activities of *Clavispora lusitaniae* ABS7.

The amylases are used in the detergents to degrade the residues of foods such as potatoes, chocolate, etc., to dextrins and other smaller oligosaccharides. The suitability of any hydrolytic enzymes for inclusion in detergent formulation depends on its stability and compatibility with detergent ingredients. In the presence of commercial detergents such as Chat, Ariel, Test, Omo, and Isis, the alkaline α-amylase maintained 130%, 97%, 115%, 75%, and 98%, respectively, of its initial activity. Similarly, the alkaline pullulanase preserved 125%, 94%, 104%, 72%, and 90% of its initial activity.

The α-amylase activity was at a maximum with the laundry detergent chat at 45°C. The stability of any enzyme in detergent formulations mainly depends on different components, such as surfactants, bleaching agents, and stabilizers used in the detergent formulations [70]. Consequently, partial loss of α-amylase activity in certain detergents can be attributed to the inhibitory effect of one or several components of these detergents.

On the other hand, some components of the detergent may have a stimulatory effect on α-amylase and pullulanase of *C. lusitaniae* ABS7 [70] (increase in enzymatic activity in the presence of detergent compared to that of the control without detergent). The observation for other hydrolytic enzymes in the presence of detergent components has already been reported [70].

2.7. Wash Performance Analysis

Stain removal ability to the purified alkaline amylopullulanase was evaluated by using chocolate and jam stained on cotton tissues. **Figure 12** shows that treatment of the chocolate–jam stains by detergent (Chat) supplemented with the purified alkaline amylopullulanase resulted in a perfect elimination of stains from cotton fabrics compared to stain removal by detergent or enzyme alone.

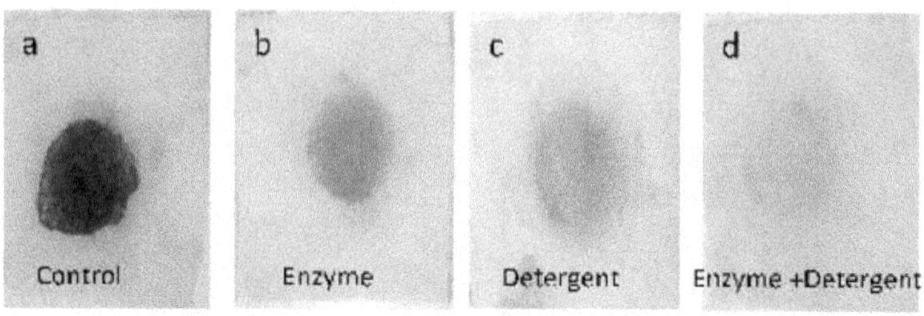

Figure 12. Test analysis of washing performance of the chocolate–jam stain on the pieces of tissue washed; (a) control (tap water), (b) enzyme (500 IU), (c) detergent (7 mg/mL); (d) enzyme (500 IU) + detergent (7 mg/mL).

Finally, the results of this study show that the studied amylopullulanase presented an excellent stability and a significant compatibility with the commercial laundry detergents at 45 °C (usually used for washing). This is favourable for its inclusion in the formulations for automatic dishwashers and laundries. In this light, the *Clavispora lusitaniae* ABS7 can be exploited industrially as a microbial cell factory for high-level alkalothermostable amylopullulanase, which could possibly represent a potential alternative to the use of other detergent enzymes that are not able to work properly in an alkaline environment.

3. MATERIALS AND METHODS

3.1. Yeast

Yeast strain ABS7 was isolated at the Microbiological Engineering and Applications Laboratory, Mentouri University, Constantine, Algeria, from wheat grains (*Triticum turgidum var. Durum*) cultivated and stored in an arid region (Biskra, Algerian Sahara). The yeast was identified as *Clavispora lusitaniae* ABS7 by Microbiology and Molecular Genetics Laboratory at INRA–CNRS, Thiverval-Grignon, France. The yeast was maintained on YPGA agar slants comprised of yeast extract 1%, glucose 2%, peptone 1%, and agar 1.5%. Cultures were maintained at 30°C for 24 h and then stored at 4 or -80°C in Cryo-Beads.

3.1.1. Ability of Yeasts to Produce Amylolytic Enzymes

The ability of *Clavispora lusitaniae* ABS7 to produce alpha-amylase and pullulanase was obtained by the agar diffusion method (plate–test–agar). This method is a semi-quantitative method, which is described as follows: In Petri dishes containing 40 mL of YPSA (for amylase) and YPPA (for pullulanase) medium containing 0.05% chloramphenicol (to avoid any contamination), a well 6 mm in diameter is hollowed out using the inverted end of the Pasteur pipette, and 60 μL of the yeast suspension is placed therein. After incubation at 40°C for 48 h, the hydrolysis zones are revealed after the addition of 10 mL of Lugol, which gives a transparent zone compared to the blue zones containing unhydrolyzed starch. For the pullulanase activity, 10 mL of Congo red 1% is added. After 15 min, successive rinses with NaCl (1N) are carried out to remove the excess dye, until a transparent zone appears; areas containing unhydrolyzed pullulan are shown in red (18). The composition of YPPA is as follows: yeast extract 1%, pullulan 2%, peptone 1%, and agar 1.5% and that of YPSA is yeast extract 1%, starch 2%, peptone 1%, and agar 1.5%

3.1.2. Study of the Inoculum

For preparation of the inoculum, 40 mL of YPGA medium was poured into 250 mL Erlens Meyers and inoculated with a pure strain of *Clavispora lusitaniae* ABS7. After incubation for 48 h at 40°C, 50 mL of sterile distilled water was added, and the cell suspension was obtained by manual stirring.

The cell count was performed by direct counting using a Thoma cell (0.1 mm, 1/400 mm^2). An inoculum of 2.5×10^6 cells/mL was used.

3.2. Culture Media

3.2.1. Base Medium

The culture medium used is based on whey at pH 4.46, obtained from the manufacture of cheeses. Whey contains the soluble components of milk: lactose, proteins, mineral salts, and traces of fat [53].

The one that is used in our work is supplied by the dairy of the Rkima brothers, industrial zone Palma, Constantine. It is stored at 4°C for a short period of 24–48 h or at -20°C for a longer period.

3.2.2. Whey Processing

The whey was filtered through the gauze to remove impurities. Before use, the whey underwent a thermocoagulation treatment under the combined action of pH (adjusted to 4.6) and temperature at 100°C for 30 min in order to precipitate the caseins, which will was then removed by centrifugation at 4000× *g* for 15 min (or by filtration). The supernatant or the filtrate (100% without dilution) constituted the base medium and was used to prepare the culture medium [71].

3.3. Production of α-Amylase and Pullulanase in Fermenter

In a Fermenter 2 L (Sartorius, Dourdan, France), *C. lusitaniae* ABS7 was cultured in an optimized whey-based medium. Different substances were added to the basal medium, such as starch: 3.34 g/L, yeast extract: 0.429 g/L, salt solution (KH_2PO_4: 850 mg/L, K_2HPO_4: 150 mg/L, $MgSO_4$, $7H_2O$: 500 mg/L and $CaCl_2$ $6H_2O$: 100 mg/L): 9.5 mL/L, and trace elements solution ($CuSO_4$, $5H_2O$: 40 µg/L, KI: 100 µg/L, $FeCl_3$, $6H_2O$: 200 µg/L and $MnSO_4$, $4H_2O$: 400 µg/L): 4.65 mL/L. The incubation was carried out at 54 °C for 40 h with stirring at 135 rpm [72].

3.4. Enzyme Activity and Protein Concentration Assays

The extracellular α-amylase and pullulanase activities were measured by incubating 0.5 mL of an appropriately diluted enzyme sample with 0.5 mL of 1% (*w*/*v*) starch solution or pullulan solution in Tris HCl buffer pH 7, 8 at 40°C for 30 min, respectively. The reaction was stopped using 3,5-dinitrosalicylic acid. One unit of α-amylase or pullulanase activity was defined as the amount of enzyme that produced reducing sugar equivalent to 1 µmoles of maltose/min [73]. Protein concentration was measured using the method of Lowry [74] using the bovine serum albumin as standard.

3.5. Purification of the Enzyme

After 28 h of incubation, the cells were removed by centrifugation at 8000× *g* for 30 min at 4°C. The supernatant was lyophilized. The lyophilizate was re-dissolved in 0.2 M Tris HCl pH 8 buffer and was used as the enzyme source.

3.5.1. Protein Precipitation with Acetone

The cell-free extract from fermentation broth was partially purified by the acetone precipitation method [48]. Four times, the volume of chilled acetone was added to the extract, and it was left to precipitate overnight at -20°C. A pellet was obtained by centrifugation at 10.000 rpm for 10 min. The pellet was dissolved in a minimum quantity of 0.2 M Tris HCl buffer (pH 8). Acetone precipitated sample was redissolved in the same buffer.

3.5.2. Sephacryl S200 Chromatography

A total of 2.7 mL of the enzyme preparation was applied to the Sephacryl S200 (Fisher, Illkirch-Graffenstaden, France) column (1 m × 1.6 cm), equilibrated with 0.2 M Tris HCl buffer (pH 8). The elution was carried out with the same buffer, at a flow rate of 0.5 mL/min. Fractions of 2 mL were collected. On each collected fraction, the optical density at 280 nm and the α-amylase and pullulanase activities were measured. The active fractions were pooled and concentrated.

3.5.3. Ion Exchange Chromatography

One mL of the concentrated enzyme was applied directly to DEAE cellulose (Sigma-Aldrich, St. Quentin Fallavier, France), a column (10 × 1 cm) previously equilibrated with Tris HCl buffer 0.2 M (pH 8). After washing through all unbound protein, the enzyme was eluted using the same buffer containing 1.5 M NaCl at a flow rate of 0.5 mL/min. Elution is performed with a gradient of in 0.2 M Tris HCl buffer at pH 8. Fractions of 2 mL were collected at a flow rate of 0.5 mL/min. The active fractions were pooled and concentrated with a 10 kDa membrane cut-off.

3.6. Electrophoresis SDS PAGE

The molecular weight of the pure protein was estimated by SDS-PAGE on 10% homogeneous polyacrylamide gel [75].

3.7. Thin Layer Chromatography (TLC)

The purified enzyme was incubated at 40 °C with 1% pullulan or starch. Samples were withdrawn after 6 h and, subjected to thin-layer chromatography (TLC aluminum sheets silica gel 60 F254) (Merck, Darmstadt, Germany). Each sample was analyzed using butanol/acetic acid/water (3:1:1, *v*/*v*/*v*) as the solvent system and methanol/sulfuric acid (1:1, *v*/*v*) as developing reagents. Glucose, maltose, and maltotriose were used as standards.

3.8. Enzyme Characterization

3.8.1. Effect of Temperature on Amylase Activity

The optimal temperature for the activity of the enzyme was determined at temperatures of 20 to 100 °C with an increment of 5 °C [76].

3.8.2. Effects of pH on Amylase Activity

The effect of pH on the enzyme activity was determined by incubating the purified enzyme between pH 5 and 12 using the standard assay condition. The buffers used were 0.5 M citrate-Na_2HPO_4 buffer (pH 5), 0.02 M phosphate buffer (pH 6–8), and 0.1 M glycine-NaOH buffer (pH 8.5–12) [31].

3.8.3. Thermostability

Temperature stability of the α-amylase activity was tested by pre-incubating the enzyme at 75 and 100°C and that of the pullulanase at 80 and 100°C at various times ranging from 0 to 180 min.

The enzyme solution was distributed using the same volume in separate tubes, which were heated together in a water bath at a carefully controlled temperature. The different samples were removed one after the other at predetermined times and instantly cooled in an ice bath. After each heat treatment, the **α**-amylase and pullulanase activities were measured [76].

3.9. Effect of Metals Ions and Chelating Agent

The effect of metal ions on the α-amylase and pullulanase activities was determined by adding 5 mM of different ions to the standard assay. The used metals were $FeCl_3$, the $ZnSO_4$, $HgCl_2$, $CuSO_4$, $MgCl_2$, $CaCl_2$, NaCl, and $MnCl_2$. Each metal ion was separately incubated with alkaline amylopullulase at 60 °C for 30 min in 0.2 M Tris HCl buffer (pH 8) and then the α-amylase and pullulanase activities were measured. The activity of the enzyme alone in the same buffer and pH was taken to be 100%.

Moreover, other chemical substances are tested such as EDTA at 2 mM, urea (2 M), SDS at 1% (*w/v*), β-marcaptoéthanol (1%), and Tween 80 (1%) on both enzymatic activities. The enzymatic activities were determined by pre-incubating the enzyme in the presence of each reagent at 40°C at pH 8.

3.10. Compatibility Test with Various Commercial Laundry Detergents

To confirm the potential of alkaline amylopullulanase from *C. lusitaniae* ABS7 as a laundry detergent additive, we tested its compatibility and stability towards some commercial laundry detergents available in the local market, such as Ariel, Cat, Test, Omo, and Isis. Before the enzyme stability test, the detergent solutions (7 mg/mL) were preheated to 100°C for 60–90 min to destroy the endogenous enzyme activity [77]. Then, the detergent and enzyme were mixed in a ratio of 1:1 (*v/v*) and incubated at 45°C for 1 h, and the residual activity was determined. The enzyme activity of a control (without detergent), incubated under the similar conditions, was taken as 100% [77].

3.11. Analysis of the Wash Performance

To determine the effectiveness of purified alkaline amylopullulanase for its use as a bio-detergent additive, wash performance was evaluated by determining the ability of the enzyme to remove the chocolate stain on cotton fabrics. The chocolate was heated to 70°C and was used with jam as an application on clean cotton fabrics (7 × 7 cm) dried overnight in a hot air oven [77].

To test the wash performance, every piece of the dirty clothes was dipped in Erlenmeyer flasks containing:

- 25 mL of tap water (control).
- 20 mL of tap water and 5.0 mL of the purified alkaline amylopullulanase (500 U/mL).
- 20 mL of tap water and 5.0 mL of heated detergent (7 mg/mL).
- 20 mL of tap water and 5.0 mL of heated detergent (7 mg/mL), containing 500 U/mL of the purified alkaline enzyme.

All flasks were incubated at 37°C for 60 min stirring 200 rpm. After incubation, the tissue pieces are removed, rinsed with water, and dried [77,78].

4. CONCLUSIONS

This study allowed us to isolate an amylolytic yeast *Clavispora lusitaniae* ABS, which possesses both α-amylase and pullulanase extracellular activities. This property thus provides it with the ability to hydrolyze the α-1,4 and α-1,6 glycosidic bonds of polysaccharides. These two activities are probably localized in two distinct active sites of a Type II amylopullulanase with saccharifying power. Pullulanase is a calcium-dependent metallo-enzyme. The activity of α-amylase is independent of calcium, although it is essential for its stability and for the maintenance of the structure of the enzyme.

Clavispora lusitaniae ABS7, isolated from wheat grains from an arid Saharan zone, is thermophilic and alkalophilic and produces enzymes that are thermostable and active in an alkaline environment. The properties of the amylopullulanase of *Clavispora lusitaniae* ABS7 designate it for industrial application, more particularly in the field of the "starch" and detergent industries. Studies of its compatibility with various commercial laundry detergents have shown that it offers excellent stability and compatibility with commercial detergents. The amylopullulanase from *Clavispora lusitaniae* ABS was better-suited to different industrial processes such as starch and laundry detergent industries.

5. CONCISE SUMMARY

In the present study, α-amylase and pullulanase from *Clavispora lusitaniae* ABS7 isolated from wheat seeds were studied. The gel filtration and ion-exchange chromatography revealed the presence of α-amylase and pullulanase activities in the same fraction with yields of 23.88% and 21.11%, respectively. SDS-PAGE showed a single band (75 kDa), which had both α-amylase (independent of Ca^{2+}) and pullulanase (a calcium metallo-enzyme) activities. The products of the enzymatic reaction on pullulan were glucose, maltose, and maltotriose, whereas the conversion of starch produced glucose and maltose. The α-amylase and pullulanase had pH optima at 9 and temperature optima at 75 and 80°C, respectively. After heat treatment at 100°C for 180 min, the pullulanase retained 42% of its initial activity, while α-amylase maintained only 38.6%. The cations Zn^{2+}, Cu^{2+}, Na^{+}, and Mn^{2+} increased the α-amylase activity. Other cations Hg^{2+}, Mg^{2+}, and Ca^{2+} were stimulators of pullulanase. Urea and Tween 80 inhibited both enzymes, whereas EDTA only inhibited pullulanase. In addition, the amylopullulanase retained its activity in the presence of various commercial laundry detergents. The performance of the alcalothermostable enzyme of *Clavispora lusitaniae* ABS7 qualified it for the industrial use, particularly in detergents, since it had demonstrated an excellent stability and compatibility with the commercial laundry detergents.

REFERENCES

1. Tao, Z.; Dong, B.; Teng, Z.; Zhao, Y. The Classification of Enzymes by Deep Learning. *IEEE Access* **2020**, *8*, 89802–89811. [**Google Scholar**] [**CrossRef**]
2. Balakrishnan, M.; Jeevarathinam, G.; Kumar, S.K.S.; Muniraj, I.; Uthandi, S. Optimization and scale-up of α-amylase production by *Aspergillus oryzae* using solid-state fermentation of edible oil cakes. *BMC Biotechnol.* **2021**, *21*, 33. [**Google Scholar**] [**CrossRef**] [**PubMed**]
3. Pandey, G.; Munguambe, D.M.; Tharmavaram, M.; Rawtani, D.; Agrawal, Y.K. Halloysite nanotubes—An efficient 'nano-support' for the immobilization of α-amylase. *Appl. Clay Sci.* **2017**, *136*, 184–191. [**Google Scholar**] [**CrossRef**]
4. De Souza, P.M.; Magalhães, P.O. Application of microbial α-amylase in industry—A review. *Braz. J. Microbiol.* **2010**, *41*, 850–861. [**Google Scholar**] [**CrossRef**] [**PubMed**]
5. BBC Research, Global Markets for Enzymes in Industrial Applications. Available online: **https://www.bccresearch.com/market-research/biotechnology/global-markets-for-enzymes-in-industrial-applications-bio030k.html** (accessed on 1 October 2018).
6. Kim, T.U.; Gu, B.G.; Jeong, J.Y.; Byun, S.M.; Shin, Y.C. Purification and characterization of maltotetraose forming alkaline *Bacillus* strain GM 8901. *Appl. Environ. Microbiol.* **1995**, *61*, 3105–3112. [**Google Scholar**] [**CrossRef**] [**Green Version**]
7. Far, B.E.; Ahmadi, Y.; Khosroushahi, A.Y.; Dilmaghani, A. Microbial Alpha-Amylase Production: Progress, Challenges and Perspectives. *Adv. Pharm. Bull.* **2020**, *10*, 350–358. [**Google Scholar**] [**CrossRef**]
8. Hmidet, N.; Jemil, N.; Nasri, M. Simultaneous production of alkaline amylase and biosurfactant by *Bacillus methylotrophicus* DCS1: Application as detergent additive. *Biodegradation* **2018**, *30*, 247–258. [**Google Scholar**] [**CrossRef**]

9. Ito, S.; Kobayashi, T.; Ozaki, K. Development of Detergent Enzymes. *J. Appl. Glycosci.* **2000**, *47*, 243–251. [**Google Scholar**] [**CrossRef**][**Green Version**]

10. Mitidieri, S.; Souza Martinelli, A.H.; Schrank, A.; Vainstein, M.H. Enzymatic detergent formulation containing amylase from *Aspergillus niger*: A comparative study with commercial detergent formulations. *Bioresour. Technol.* **2006**, *97*, 1217–1224. [**Google Scholar**] [**CrossRef**]

11. Raul, D.; Biswas, T.; Mukhopadhyay, S.; Das, S.K.; Gupta, S. Production and partial purification of alpha Amylase from *Bacillus subtilis* (MTCC 121) using solid state fermentation. *Biochem. Res. Int.* **2014**, *2014*, 568141. [**Google Scholar**] [**CrossRef**] [**PubMed**][**Green Version**]

12. Djekrif, D.S.; Gheribi, A.Z.; Meraihi, Z.; Bennamoun, L. Application of a statistical design to the optimization of culture medium for α-amylase production by *Aspergillus niger* ATCC 16404 grown on orange waste powder. *J. Food Eng.* **2006**, *73*, 190–197. [**Google Scholar**] [**CrossRef**]

13. Siroosi, M.; Borjian, B.F.; Amoozegar, M.A.; Babavalian, H.; Hassanshahian, M. Halophilic Amylase Production and Purification from *Haloarcula* sp. Strain D61. *Biointerface Res. Appl. Chem.* **2021**, *11*, 7382–7392. [**Google Scholar**]

14. Naidu, K.; Maseko, S.; Kruger, G.; Lin, J. Purification and characterization of α-amylase from *Paenibacillus* sp. D9 and *Escherichia coli* recombinants. *Biocatal. Biotransf.* **2020**, *38*, 124–134. [**Google Scholar**] [**CrossRef**]

15. Al-Dhabi, N.A.; Esmail, G.A.; Ghilan, A.K.M.; Arasu, M.V.; Duraipandiyan, V.; Ponmurugan, K. Isolation and purification of starch hydrolysing amylase from *Streptomyces* sp. Al-Dhabi46- obtained from the Jazan region of Saudi Arabia with industrial applications. *J. King Saud Univ. Sci.* **2020**, *32*, 1226–1232. [**Google Scholar**] [**CrossRef**]

16. Simair, A.A.; Qureshi, A.S.; Khushk, I.; Ali, C.H.; Lashari, S.; Bhutto, M.A.; Mangrio, G.S.; Lu, C. Production and Partial Characterization of α-Amylase Enzyme from *Bacillus* sp. BCC 50-01 and Potential Applications. *BioMed Res. Int.* **2017**, *2017*, 9173040. [**Google Scholar**] [**CrossRef**][**Green Version**]

17. Lin, L.L.; Chyau, C.C.; Hsu, W.H. Production and properties of a raw starch degrading amylase from the thermophilic and alkaliphilic *Bacillus* sp. TS23. *Biotechnol. Appl. Biochem.* **1998**, *28*, 61–68. [**Google Scholar**]

18. Moubasher, H.; Wahsh, S.S.; El-Kassem, N.A. Purification of pullulanase from Aureobasidium pullulans. *Microbiology* **2010**, *79*, 759–766. [**Google Scholar**] [**CrossRef**]

19. Ramachandran, N. Amylolytic enzymes from the yeast *Lipomyces kononenkoae*. *Biologia* **2005**, *60*, 103–110. [**Google Scholar**]

20. Iefuji, H.; Chino, M.; Kato, M.; Iimura, Y. Raw-starch-digesting and thermostable α-amylase from the yeast *Cryptococcus* sp. S2-: Purification, characterization, cloning and sequencing. *Biochem. J.* **1996**, *318*, 989–996. [Google Scholar] [CrossRef][Green Version]

21. Vishnu, C.; Naveena, B.J.; Altaf, M.D.; Venkateshwar, M.; Reddy, G. Amylopullulanase—A novel enzyme of *L. amylophilus* GV6 in direct fermentation of starch to L(+) lactic acid. *Enzyme Microb. Technol.* **2006**, *38*, 545–550. [Google Scholar] [CrossRef]

22. Zareian, S.; Khajeh, K.; Ranjbar, B.; Dabirmanesh, B.; Ghollasi, M.; Mollania, N. Purification and characterization of a novel amylopullulanase that converts pullulan to glucose, maltose, and maltotriose and starch to glucose and maltose. *Enzyme Microb. Technol.* **2010**, *46*, 57–63. [Google Scholar] [CrossRef]

23. Wasko, A.; Polak-Berecka, M.; Targonski, Z. Purification and characterization of Pullulanase from *lactococcus lactis*. *Prep. Biochem. Biotechnol.* **2011**, *41*, 252–261. [Google Scholar] [CrossRef] [PubMed]

24. Kar, S.; Ray, R.C.; Mohapatra, U.B. Purification, characterization and application of thermostable amylopullulanase from Streptomyces erumpens MTCC 7317 under submerged fermentation. *Ann. Microbiol.* **2012**, *62*, 931–937. [Google Scholar] [CrossRef]

25. Ara, K.; Saeki, K.; Igarashi, K.; Takaiwa, M.; Uemura, T.; Hagihara, H.; Kawai, S.; Ito, S. Purification and characterization of an alkaline amylopullulanase with both α-1,4 and α-1,6 hydrolytic activiy from alkalophilic *Bacillus* sp. KSM1378-. *Biochim. Biophys. Acta (BBA) Gen. Subj.* **1995**, *1243*, 315–324. [Google Scholar] [CrossRef]

26. Kim, C.H.; Kim, Y.S. Substrate specificity and detailed characterization of a bifunctional amylase-pullulanase enzyme from Bacillus circulans F2 having two different active sites on one polypeptide. *Eur. J. Biochem.* **1995**, *227*, 687–693. [Google Scholar] [CrossRef]

27. Melasniemi, H. Characterization of a-amylase and pullulanase activities of *Clostridium thermohydrosulfuricum*. *Biochem. J.* **1987**, *246*, 193–197. [Google Scholar] [CrossRef] [PubMed][Green Version]

28. Kim, C.-H.; Kim, D.-S.; Taniguchi, H.; Maruyama, Y. Purification of a amylase-pullulanase bifunctional enzyme by high-performance size-exclusion and hydrophobic-interaction chromatography. *J. Chromatogr. A* **1990**, *512*, 131–137. [Google Scholar] [CrossRef]

29. Mathupala, S.P.; Lowe, S.E.; Podkovyrov, S.M.; Zeikus, J.G. Sequencing of the amylopullulanase (apu) gene of *Thermoanaerobacter ethanolicus* 39E, and identification of the active site by site-directed mutagenesis. *J. Biol. Chem.* **1993**, *268*, 16332–16344. [Google Scholar] [CrossRef]

30. Ara, K.; Igarashi, K.; Saeki, K.; Ito, S. An Alkaline Amylopullulanase from Alkalophilic *Bacillus* sp. KSM1378-; Kinetic Evidence for Two Independent Active Sites for the α1,4- and α1,6- Hydrolytic Reactions. *Biosci. Biotechnol. Biochem.* **1995**, *59*, 662–666. [Google Scholar] [CrossRef]

31. Hii, L.S.; Ling, T.C.; Mohamad, R.; Ariff, A.B. Characterization of Pullulanase Type II from *Bacillus cereus* H1.5. *Am. J. Biochem. Biotechnol.* **2009**, *5*, 170–179. [Google Scholar] [CrossRef]

32. Mrudula, S.; Gopal, R.; Seenayya, G. Purification and characterization of highly thermostable amylopullulanase from a thermophilic, anaerobic bacterium *Clostridium thermosulfurogenes* SVM17. *Malays. J. Microbiol.* **2011**, *7*, 97–106. [Google Scholar] [CrossRef]

33. Lévêque, E.; Janeček, Š.; Haye, B.; Belarbi, A. Thermophilic archaeal amylolytic enzymes. *Enzyme Microb. Technol.* **2000**, *26*, 3–14. [Google Scholar] [CrossRef]

34. Nakamura, Y.; Fukuhara, H.; Sano, K. Secreted Phytase Activities of Yeasts. *Biosci. Biotechnol. Biochem.* **2000**, *64*, 841–844. [Google Scholar] [CrossRef]

35. Prieto, J.A.; Bort, B.R.; Martínez, J.; Randez-Gil, F.; Buesa, C.; Sanz, P. Purification and characterization of a new alpha-amylase of intermediate thermal stability from the yeast *Lipomyces kononenkoae*. *Biochem. Cell Biol.* **1995**, *73*, 41–49. [Google Scholar] [CrossRef]

36. Simões-Mendes, B. Purification and characterization of the extracellular amylases of the yeast *Schwanniomyces alluvius*. *Can. J. Microbiol.* **1984**, *30*, 1163–1170. [Google Scholar] [CrossRef]

37. Wanderley, K.J.; Torres, F.A.G.; Moraes, L.M.P.; Ulhoa, C.J. Biochemical characterization of Î±-amylase from the yeast *Cryptococcus flavus*. *FEMS Microbiol. Lett.* **2004**, *231*, 165–169. [Google Scholar] [CrossRef][Green Version]

38. Kwon, Y.M.; Choi, H.S.; Lim, J.Y.; Jang, H.S.; Chung, D. Characterization of Amylolytic Activity by a Marine-Derived Yeast *Sporidiobolus pararoseus* PH-Gra1. *Mycobiology* **2020**, *48*, 195–203. [Google Scholar] [CrossRef]

39. Demirkan, E. Production, purification, and characterization of α-amylase by *Bacillus subtilis* and its mutant derivates. *Turk. J. Biol.* **2011**, *35*, 705–712. [Google Scholar]

40. Abdel-Fattah, Y.R.; Soliman, N.A.; El-Toukhy, N.M.; El-Gendi, H.; Ahmed, R.S. Production, Purification, and Characterization of Thermostable α-Amylase Produced by *Bacillus licheniformis* Isolate AI20. *J. Chem.* **2013**, *2013*, 673173. [Google Scholar] [CrossRef][Green Version]

41. Rodríguez-Saavedra, C.; Rodríguez-Sanoja, R.; Guillén, D.; Wacher, C.; Díaz-Ruiz, G. *Streptococcus infantarius* 25124 isolated from pozol produces a high molecular weight amylopullulanase, a key enzyme for niche colonization. *Amylase* **2021**, 5, 1–12. [Google Scholar] [CrossRef]

42. Saha, B.C.; Lamed, R.; Lee, C.Y.; Mathupala, S.P.; Zeikus, J.G. Characterization of an *endo*-Acting Amylopullulanase from *Thermoanaerobacter* Strain B6A. *Appl. Environ. Microbiol.* **1990**, *56*, 881–886. [Google Scholar] [CrossRef][Green Version]

43. Erra-Pujada, M.; Chang-Pi-Hin, F.; Debeire, P.; Duchiron, F.; O'Donohue, J. Purification and properties of the catalytic domain of the thermostable pullulanase Type II from *Thermococcus hydrothermalis*. *Biotechnol. Lett.* **2001**, *23*, 1273–1277. [Google Scholar] [CrossRef]

44. Ueda, S.; Ohba, R. Purification, Crystallization and Some Properties o-Extracellular Pullulanase from *Aerobacter aerogenes*. *Agric. Biol. Chem.* **1972**, *36*, 2381–2391. [Google Scholar] [CrossRef][Green Version]

45. Kim, C.H.; Choi, H.I.; Lee, D.S. Purification and Biochemical Properties of an Alkaline Pullulanase from Alkalophilic *Bacillus* sp. S-l. *Biosci. Biotechnol. Biochem.* **1993**, *57*, 1632–1637. [Google Scholar] [CrossRef]

46. Qiao, Y.; Peng, Q.; Yan, J.; Wang, H.; Ding, H.; Shi, B. Gene cloning and enzymatic characterization of alkali-tolerant Type I pullulanase from *Exiguobacterium acetylicum*. *Lett. Appl. Microbiol.* **2015**, *60*, 52–59. [Google Scholar] [CrossRef] [PubMed]

47. Hii, S.L.; Tan, J.S.; Ling, T.C.; Ariff, A.B. Pullulanase: Role in Starch Hydrolysis and Potential Industrial Applications. *Enzyme Res.* **2012**, *2012*, 921362. [Google Scholar] [CrossRef][Green Version]

48. Asha, R.; Niyonzima, F.N.; Sunil, S.M. Purification and properties of pullulanase from *Bacillus halodurans*. *Int. Res. J. Biol. Sci.* **2013**, *2*, 35–43. [Google Scholar]

49. Lim, S.J.; Hazwani-Oslan, S.N.; Oslan, S.N. Purification and characterisation of thermostable α-amylases from microbial sources. *BioResources* **2019**, *15*, 2005–2029. [Google Scholar] [CrossRef]

50. Xian, L.; Wang, F.; Luo, X.; Feng, Y.L.; Feng, J.X. Purification and Characterization of a Highly Efficient Calcium-Independent α-Amylase from *Talaromyces pinophilus* 95-1. *PLoS ONE* **2015**, *10*, e0121531. [Google Scholar] [CrossRef] [Green Version]

51. Mohamed, S.A.; Azhar, E.I.; Ba-Akdah, M.M.; Tashk, N.R.; Kumosani, T.A. Production, purification and characterization of α-amylase from *Trichoderma harzianum* grown on mandarin peel. *Afr. J. Microbiol. Res.* **2011**, *5*, 930–940. [Google Scholar] [CrossRef][Green Version]

52. Saini, R.; Saini, H.S.; Dahiya, A. Amylases: Characteristics and industrial applications. *J. Pharmacogn. Phytochem.* **2017**, *6*, 1865–1871. **[Google Scholar]**

53. Larpent, J.P.; Larpent-Gourgaud, M. *Mémento Technique de Microbiologie*, 3rd ed.; Lavoisier-Tec & Doc: Paris, France, 1997; Volume 8, pp. 217–240. **[Google Scholar]**

54. Declerck, N.; Machius, M.; Joyet, P.; Wiegand, G.; Huber, R.; Gaillardin, C. Hyperthermostabilization of *Bacillus licheniformis* α-amylase and modulation of its stability over a °50C temperature range. *Protein Eng. Des. Sel.* **2003**, *16*, 287–293. **[Google Scholar] [CrossRef][Green Version]**

55. Lonhienne, T.; Zoidakis, J.; Vorgias, C.E.; Feller, G.; Gerday, C.; Bouriotis, V. Modular structure, local flexibility and cold-activity of a novel chitobiase from a psychrophilic antarctic bacterium. *J. Mol. Biol.* **2001**, *310*, 291–297. **[Google Scholar] [CrossRef]**

56. Unsworth, L.D.; van der Oost, J.; Koutsopoulos, S. Hyperthermophilic enzymes—Stability, activity and implementation strategies for high temperature applications: Properties and applications of hyperthermozymes. *FEBS J.* **2007**, *274*, 4044–4056. **[Google Scholar] [CrossRef]**

57. Koutsopoulos, S.; van der Oost, J.; Norde, W. Temperature-dependent structural and functional features of a hyperthermostable enzyme using elastic neutron scattering: Dynamics of a Hyperthermostable Enzyme. *Proteins* **2005**, *61*, 377–384. **[Google Scholar] [CrossRef]**

58. Al-Quadan, F.; Akel, H.; Natshi, R. Characteristics of a novel, highly acid- and thermo-stable amylase from thermophilic *Bacillus* strain HUTBS62 under different environmental conditions. *Ann. Microbiol.* **2011**, *61*, 887–892. **[Google Scholar] [CrossRef]**

59. Galdino, A.S.; Silva, R.N.; Lottermann, M.T.; Alvares, A.C.M.; Moraes, L.M.P.D.; Torres, F.A.G.; Freitas, S.M.D.; Ulhoa, C.J. Biochemical and Structural Characterization of Amy1: An Alpha-Amylase from *Cryptococcus flavus* Expressed in *Saccharomyces cerevisiae*. *Enzyme Res.* **2011**, *2011*, 157294. **[Google Scholar] [CrossRef][Green Version]**

60. Hwang, S.Y.; Nakashima, K.; Okai, N.; Okazaki, F.; Miyake, M.; Harazono, K.; Ogino, C.; Kondo, A. Thermal Stability and Starch Degradation Profile of α-Amylase from *Streptomyces avermitilis*. *Biosci. Biotechnol. Biochem.* **2013**, *77*, 2449–2453. **[Google Scholar] [CrossRef] [PubMed][Green Version]**

61. Bernhardsdotter, E.C.M.J.; Ng, J.D.; Garriott, O.K.; Pusey, M.L. Enzymic properties of an alkaline chelator-resistant α-amylase from an alkaliphilic *Bacillus* sp. isolate L1711. *Process Biochem.* **2005**, *40*, 2401–2408. **[Google Scholar] [CrossRef]**

62. Suzuki, Y.; Imai, T. *Bacillus stearothermophilus* KP 1064 pullulan hydrolase. *Appl. Microbiol. Biotechnol.* **1985**, *21*, 20–26. [Google Scholar] [CrossRef]

63. Rüdiger, A.; Jorgensen, P.L.; Antranikian, G. Isolation and characterization of a heat-stable pullulanase from the hyperthermophilic archaeon *Pyrococcus woesei* after cloning and expression of its gene in *Escherichia coli*. *Appl. Environ. Microbiol.* **1995**, *61*, 567–575. [Google Scholar] [CrossRef] [PubMed][Green Version]

64. Zohra, R.R.; Qader, S.A.; Pervez, S.; Aman, A. Influence of different metals on the activation and inhibition of α-amylase from thermophilic *Bacillus firmus* KIBGE-IB28. *Pak. J. Pharm. Sci.* **2016**, *29*, 1275–1278. [Google Scholar] [PubMed]

65. Freer, S.N. Purification and characterization of the extracellular alpha-amylase from *Streptococcus bovis* JB1. *Appl. Environ. Microbiol.* **1993**, *59*, 1398–1402. [Google Scholar] [CrossRef][Green Version]

66. Divakaran, D.; Chandran, A.; Pratap Chandran, R. Comparative study on production of a-Amylase from *Bacillus licheniformis* strains. *Braz. J. Microbiol.* **2011**, *42*, 1397–1404. [Google Scholar] [CrossRef][Green Version]

67. Pompeyo, C.; Gomez, M.; Gasparian, S.; Morlon-Guyot, J. Comparison of amylolytic properties of *Lactobacillus amylovorus* and of *Lactobacillus amylophilus*. *Appl. Microbiol. Biotechnol.* **1993**, *40*, 266–269. [Google Scholar] [CrossRef]

68. Odibo, F.J.C.; Ulbrich-Hofmann, R. Thermostable α-Amylase and Glucoamylase from *Thermomyces lanuginosus* F1. *Acta Biotechnol.* **2001**, *21*, 141–153. [Google Scholar] [CrossRef]

69. Arabacı, N.; Arıkan, B. Isolation and characterization of a cold-active, alkaline, detergent stable α-amylase from a novel bacterium *Bacillus subtilis* N8. *Prep. Biochem. Biotechnol.* **2018**, *48*, 419–426. [Google Scholar] [CrossRef] [PubMed]

70. Joo, H.S.; Chang, C.S. Production of an oxidant and SDS-stable alkaline protease from an alkaophilic *Bacillus clausii* I52- by submerged fermentation: Feasibility as a laundry detergent additive. *Enzyme Microb. Technol.* **2006**, *38*, 176–183. [Google Scholar] [CrossRef]

71. Lorient, D.; Closs, B.; Courthaudon, J.L. Connaissances nouvelles sur les propriétés fonctionnelles des protéines du lait et des dérivés. *Lait* **1991**, *71*, 141–171. [Google Scholar] [CrossRef]

72. DJekrif, D.S.; Gillmann, L.; Cochet, N.; Bennamoun, L.; Ait-Kaki, A.; Labbani, K.; Nouadri, T.; Meraihi, Z. Optimization of thermophilic pullulanase and α-amylase production by amylolytic yeast. *Int. J. Micro. Biol. Res.* **2014**, *6*, 559–569. [Google Scholar]

73. Bernfeld, P. Amylase α and β. In *Methods in Enzymology*; Colowick, S.P., Kaplan, O.N., Eds.; Academic Press: New York, NY, USA, 1955; pp. 140–146. [Google Scholar]

74. Lowry, O.H.; Rosebrough, N.; Farr, A.; Randall, R.J. Protein measurement with the folin phenol reagent. *J. Biol. Chem.* **1951**, *193*, 265–275. [Google Scholar] [CrossRef]

75. Laemmli, U.K. Cleavage of Structural Proteins during the Assembly of the Head of Bacteriophage T4. *Nature* **1970**, *227*, 680–685. [Google Scholar] [CrossRef] [PubMed]

76. Arikan, B. Highly thermostable, thermophilic, alkaline, SDS and chelator resistant amylase from a thermophilic *Bacillus* sp. isolate A15-3. *Bioresour. Technol.* **2008**, *99*, 3071–3076. [Google Scholar] [CrossRef]

77. Hmidet, N.; Ali, N.E.-H.; Haddar, A.; Kanoun, S.; Alya, S.K.; Nasri, M. Alkaline proteases and thermostable α-amylase co-produced by Bacillus licheniformis NH1: Characterization and potential application as detergent additive. *Biochem. Eng. J.* **2009**, *47*, 71–79. [Google Scholar] [CrossRef]

78. Rameshkumar, A.; Sivasudha, T. Optimization of Nutritional Constitute for Enhanced Alpha amylase Production Using by Solid State Fermentation Technology. *Int. J. Microbiol. Res.* **2011**, *2*, 143–148. [Google Scholar]

CHAPTER-11

INFLUENCE OF SOIL SURFACE CHEMISTRY ON THE ADSORPTION AND FUNCTIONALITY OF UREASE DERIVED FROM A CRUDE PROTEIN EXTRACT: IMPLICATIONS FOR BIOCEMENTATION APPLICATIONS

1. INTRODUCTION: AN OVERVIEW

The interaction between soil surface chemistry and enzymes plays a crucial role in various biogeochemical processes. In this study, we delve into the impact of soil surface chemistry on both the adsorption and functionality of urease, a key enzyme derived from a crude protein extract. Understanding these interactions is of paramount importance as they have significant implications, particularly in the context of bio-cementation applications. Bio-cementation, a promising field at the intersection of biotechnology and geotechnical engineering, relies on the enzymatic processes to enhance soil stability. By exploring the intricate relationship between soil surface chemistry and urease activity, this research aims to contribute valuable insights that can inform and optimize bio-cementation strategies for sustainable soil improvement.

Biocementation is a sustainable engineering technique that has received significant attention as an environmentally friendly alternative to chemical stabilization methods for soils [1,2,3,4,5]. Chemical stabilization poses significant harm to the environment, mainly because it utilizes materials that are major contributors to greenhouse gas emissions, such as cement, fly ash, lime, and the leaching of toxic contaminants into the environment [5,6,7,8]. Instead, biocementation is based on the generation of calcium carbonate precipitates from urea hydrolysis, a chemical reaction that is only feasible when catalyzed by the urease enzyme [9]. Biocementation methods consist of an injection of a mixture containing urea, calcium, and the catalyst urease into the soil. Urease can be added in the form of whole bacterial cells, crude extracts, or purified proteins. The generated calcium carbonate deposits between the soil grains bridge them like a cement, which then leads to a stronger soil matrix [10].

Biocementation via the enzyme-induced calcite precipitation (EICP) technique uses purified urease enzymes, mostly from the jack bean plant, and it has proven to be an effective technique to increase soil strength [11,12,13,14]. However, enzyme extraction from plants is a resource-intensive and time-consuming process; protein purification can further increase costs significantly. To overcome this limitation, we work herein with a new technique called bacterial enzyme-induced calcite precipitation (BEICP). In BEICP, free microbial urease is used as part of a total bacterial protein extract obtained from *Sporsarcina pasteurii*, a commonly found soil bacterium that produces urease in high quantities [15]. Biocementation via BEICP has proven to be effective in cementing coarse and fine-grained containing soils [16,17] to varying degrees. However, to be successfully implemented, a deeper fundamental understanding of the mechanisms leading to biocementation is necessary.

In the BEICP technique, the spatial distribution of adsorbed, enzymatically active urease dictates the locations where calcium carbonate precipitation, and therefore cementation, will take place. In our previous work, we reported that urease adsorbed to Ottawa silica sand and silt retained some enzymatic activity, even when present in a complex protein mixture [10]. However, an overall loss of urease activity after adsorption was observed (i.e., the sum of urease activity in the soil and supernatant was lower compared to the original activity of the total protein suspension), with greater losses observed for samples with higher silt contents. These observations raised the question of how activity loss relates to the overall amount of protein adsorbed and soil surface chemistry. Understanding this relationship is especially important for BEICP applications, because soils have highly heterogeneous surfaces often coated with metal oxides and hydrophobic compounds, with varying sizes of positively and negatively charged patches [18,19].

In particular, protein adsorption is highly dependent on the surface chemistries of the protein and the sorbent surface, as well as the presence of other proteins. For example, proteins commonly exhibit high affinity towards hydrophobic surfaces, even if they are hydrophilic in nature [20,21,22,23,24,25,26], but that often leads to protein denaturation and sharp decreases in enzymatic activity [27]. Although soils generally display low levels of hydrophobicity [28,29], hydrophobic soils, defined as having contact angles ≥ 90°, have been documented worldwide and are especially common in surface soils that often dry out or have been exposed to fires [28,30,31,32,33,34,35,36]. It is also noteworthy that hydrophobic soils are associated with increased soil erosion [37,38,39,40], for which soil stabilization via biocementation may be a suitable mitigation method [41,42,43,44]. In addition, low amounts of adsorbed proteins and the subsequent retention of their enzymatic activity have been observed in soils containing aluminium and iron oxides, even in single-protein adsorption studies [19,22,45].

Therefore, the overall goal of this paper was to investigate the behaviour of *S. pasteurii* total bacterial proteins, specifically urease, upon adsorption onto soil samples with varying degrees of hydrophobicity and electrical charges, and the resulting enzymatic activities. We hypothesized that higher amounts of adsorbed proteins including urease would result from favourable protein–soil surface interactions, such as those resulting from opposing electrostatic charges or hydrophobicity. This increased amount of adsorbed proteins, in turn, was expected to increase overall urease enzymatic activity in the soil to different degrees, depending on the configuration of the adsorbed urease. To test this hypothesis, batch adsorption experiments were conducted in sand mixtures where electrostatic and hydrophobic driving forces were expected to be dominant. An amount of 100% organosilane breathable-soil waterproofing agent and iron oxide was used to produce hydrophobic and positively-charged sands, respectively. The ranges of hydrophobicity or the positive charges in soils were varied by changing the ratios of mixing between coated and uncoated sands. The total amount of adsorbed proteins, and specifically urease, were measured though spectrophotometric and proteomic analyses, and enzymatic activity assays were performed in soil and supernatant portions following adsorption.

2. RESULTS AND DISCUSSION

2.1. Surface Coverage

In this study, soil mixtures containing different levels of hydrophobicity and positive charge patches were considered. Adsorption is a surface phenomenon and is largely controlled by the surface chemistry of the adsorbent and adsorbate [46]. Sorbent surfaces with greater numbers of available adsorption sites have higher adsorption capacities, but the likelihood of protein adsorption and surface coverage is dictated by the affinity of the proteins to the solid surface. At high surface coverage levels, adsorbed proteins are tightly packed; adsorption is more likely to occur in a multilayer fashion and might result in blockage of the active site [23,45]. At low surface coverage, intermolecular interactions are limited, but protein denaturing caused by conformational changes in the molecular structure is more likely [47,48,49,50,51].

It is worth noting that although the total surface areas among the samples were markedly different, the total amount of protein adsorbed onto each soil mixture was similar. This suggests that differences in surface area between the different samples tested did not impact total protein adsorption. The hydrophobic-containing soil samples (denoted as "HF") displayed the highest surface coverages; while increasing the hydrophobicity level from 10% HF to 20% HF did not significantly affect the surface coverage of these samples (p-value > 0.05), and the values for 100% HF were approximately 25% higher. In addition, surface coverages in all hydrophobic-containing samples were considerably higher than in the sand samples (p-value < 0.05). The surface coverages for sand and iron soil (denoted as "iron") mixtures were not statistically different from each other, even though the total surface area of the 100% iron sample was 32 times larger than the sand-only surface area. Despite this, there appears to be a trend of lower surface coverage for samples with higher iron contents. Similar observations linking soil iron content with lower adsorbed protein masses have been reported in natural soils [52,53,54].

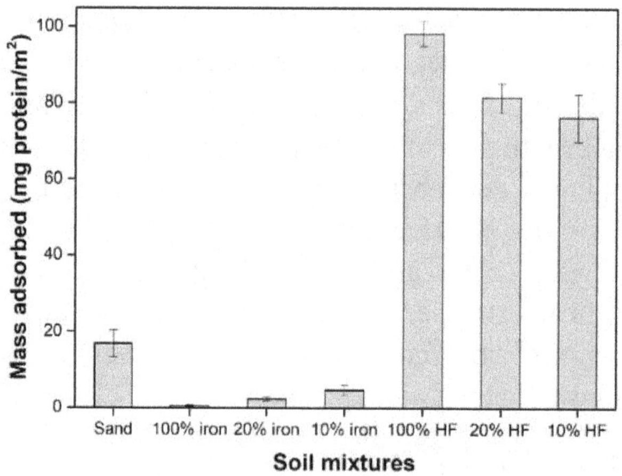

Figure 1. The surface coverage for samples containing sand, mixtures of sand and iron-coated sand ("iron"), and mixtures of sand and hydrophobic-coated sand ("HF"). The initial protein concentration used was 4 mg/mL, with a ratio of 2 mg protein per gram of soil. Surface coverage was calculated by normalizing the amount of total protein adsorbed by the total surface area of each sample. Values shown are averages of three replicates and error bars represent the standard error.

2.2. Protein Adsorption

Similar amounts of total proteins were adsorbed in all soil samples tested (Figure 2, p-values > 0.05) even though the specific surface area of the 100% iron sample was approximately 32 and 150 times larger than sand and 100% HF, respectively. Similar findings were reported for protein adsorption onto soil mineral surfaces for various proteins including urease, where the amount of adsorbed protein did not positively correlate to the surface area available [18,19,22,45]. The differences in surface coverages combined with the equal amounts of adsorbed proteins indicate that most proteins in the *S. pasteurii* total protein extract have a higher affinity towards hydrophobic surfaces. In fact, several studies have shown that most proteins have a high affinity towards hydrophobic surfaces, including hydrophilic proteins such as BSA, with increased amounts of adsorbed proteins displayed on surfaces with higher hydrophobicity levels [20,21,22,23,24,25,26,55]. Furthermore, though hydrophobic amino acid residues are mostly ingrained inside the protein's molecular structure, a hydrophobic protein core can become exposed due to structural unfolding upon initial binding via electrostatic interactions. This unfolding then gives rise to numerous hydrophobic binding sites in the protein, that will in turn enhance subsequent adsorption [56,57,58,59].

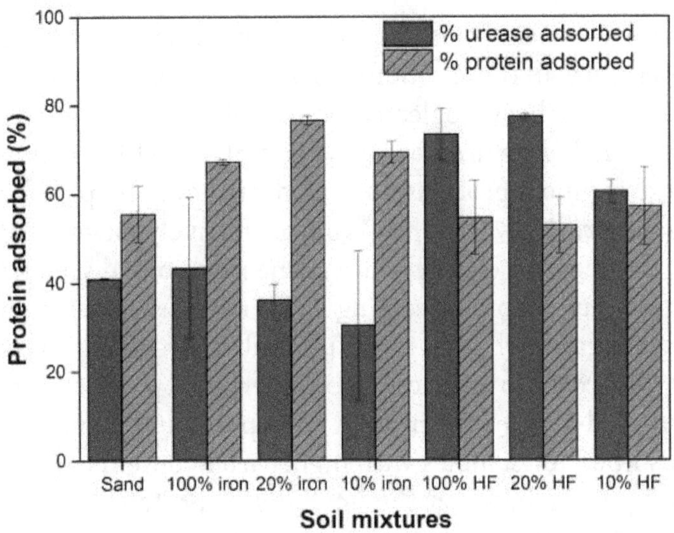

ss

Figure 2. The percentage of overall protein adsorbed, and urease adsorbed, from the initial total protein extract. The total amount of adsorbed proteins was obtained through Nanodrop measurements at 280 nm wavelengths before and after adsorption. Urease concentrations were obtained from targeted PRM LC/MS analysis. Results shown are the averages of three and two independent experiments for the % adsorbed urease and % total proteins adsorbed, respectively. Error bars represent the standard error.

While the total mass of adsorbed proteins was similar across all samples, the percentage of urease adsorbed onto hydrophobic-containing mixtures was significantly higher than in sand samples (Figure 2, p-value < 0.05). These results suggest urease was preferentially adsorbed onto 20% HF and 100% HF samples compared to other proteins. This could be a result of surface chemistry modifications arising from multilayers of adsorbed proteins in the tightly packed surfaces of 20%

and 100% HF samples, which might have resulted in a sorbent surface that was more favourable for urease adsorption, as well as an increased affinity of urease towards hydrophobic surfaces. Hence, the presence of other proteins may have facilitated urease adsorption onto the 20% and 100% HF mixtures.

No significant differences were found in the percentage of adsorbed urease among iron-containing mixtures, even when compared to sand (p-value > 0.05). In addition, our results indicate that urease adsorption onto sand- and iron-containing soil mixtures was less favourable in comparison to the total proteins in the mixture. It is unclear if increased iron contents had a more detrimental effect on urease adsorption, but previous studies have found low levels of urease adsorption onto clay minerals coated with iron [22]. Studies have found that the presence of urease is negatively related to the amount of iron present, either in natural soils or in experimental settings [60].

Furthermore, the adsorption experiments herein were conducted at pH 8 and ambient temperature. In these conditions, about 87% of all proteins present in the initial protein mixture have isoelectric points (pI) lower than 8 based on our label-free LC/MS data, including urease which has a pI of 4.6 [61], pointing to an overall negative charge for most proteins during adsorption. The pIs of Ottawa sand and iron-coated sand are 2 [62] and 9.3 [46], respectively, suggesting that the sand surface had a net negative charge whereas the iron-coated sand had a net positive surface charge. Results from streaming current measurements confirmed a shift towards less net negative charges in samples with increased iron contents (data not shown). Thus, electrostatic interactions between most proteins and the iron-coated sand were expected to enhance protein adsorption and result in higher amounts of proteins adsorbed for samples with higher iron contents. However, our results showed no differences between the amounts of protein adsorbed among those samples (**Figure 2**). The surface of iron and sand soil mixtures is mostly composed of silica, iron, and hydroxyls groups. The complete absence of hydrophobic groups in these surfaces suggest that electrostatic interactions were dominant. Possibly, the negatively charged proteins that adsorbed first onto the positively charged patches on the iron-coated surface created a repulsive environment that hindered the subsequent electrostatic attachment of negatively charged proteins [58]. These findings are in agreement with previous studies that have also found low amounts of proteins adsorbed onto aluminum- and iron oxide-coated surfaces, especially in soils and clay minerals [19,22,45].

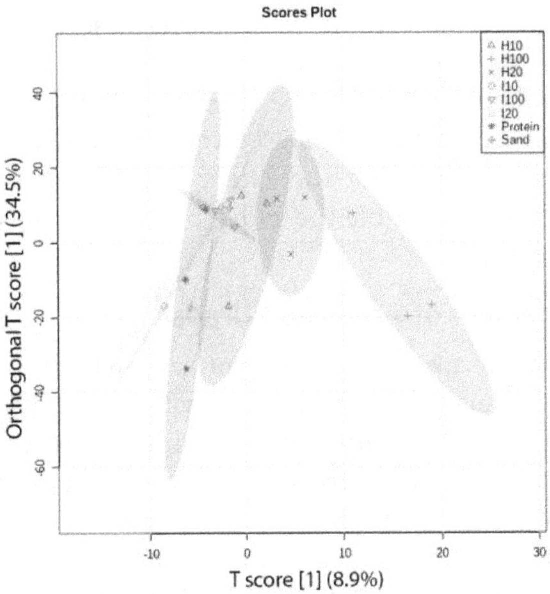

Figure 3. A score plot for all proteins present in the initial total protein extract and in supernatant portions of samples following adsorption. The score plot employs the orthogonal projections to latent structures discriminant analysis (OPLS-DA) modeling method and was developed through the MetaboAnalyst online tool [63]. Displayed regions reflect a 95% confidence interval. Prior to analysis, the data was log transformed and auto-scaled (mean centered and divided by the standard deviation of each variable). Sand = uncoated sand; Protein = initial protein sample; I10, I20, I100 = 10, 20, and 100% iron-coated sand, respectively; H10, H20, H100 = 10, 20, and 100% hydrophobic sand, respectively.

Label-free LC/MS proteomic analysis was conducted on the initial total protein extract and in supernatant portions of each sample following adsorption. These data revealed that after adsorption, supernatant samples from hydrophobic soil mixtures were markedly different from the initial protein extract, especially 20% and 100% HF samples (**Figure 3**). On the other hand, iron-containing samples were more closely related to sand and the original protein extract, indicating that the protein profile in these supernatant samples upon adsorption did not differ as much from the initial protein extract. It is also worth noting that hydrophobic samples were more distinctive from each other compared to the iron-containing soil mixtures, which were closely clustered. The significantly higher surface coverage displayed in hydrophobic samples (**Figure 1**) suggests that protein adsorption happened in a multilayer fashion. Thus, the underlying soil surface was replaced as the adsorbent surface by a layer of adsorbed proteins as the adsorption sites were occupied, which in turn gave rise to a new sorbent surface with a different surface chemistry than the soil grain surfaces. Therefore, these observations suggest that the change in surface chemistry due to multilayer adsorption was able to attract a more diverse pool of proteins present in the crude protein extract, which would explain why hydrophobic samples were markedly distinct from each other and from all other samples (**Figure 3**). This diverse pool of proteins is potentially able to attach to any surface, but the extent of protein adsorption achieved is still dependent on how compatible the sorbent surface chemistry would be to the proteins in the mixture.

2.3. Enzymatic Activity of Adsorbed Urease

The enzymatic activity of adsorbed urease, normalized to the free urease activity of the initial total protein extract, is shown as activity yield in **Figure 4**. Hydrophobic samples showed the highest activity among all samples, and the retained enzymatic activity increased for samples with higher hydrophobic contents. In addition, urease activity values for hydrophobic soil mixtures were significantly higher than the values for sand- and iron-containing soil mixtures (p-value < 0.1). Low urease activity was retained in soil mixtures containing iron. The 10% and 20% iron samples were found to not be statistically different from each other, or from sand, in terms of retained urease activity.

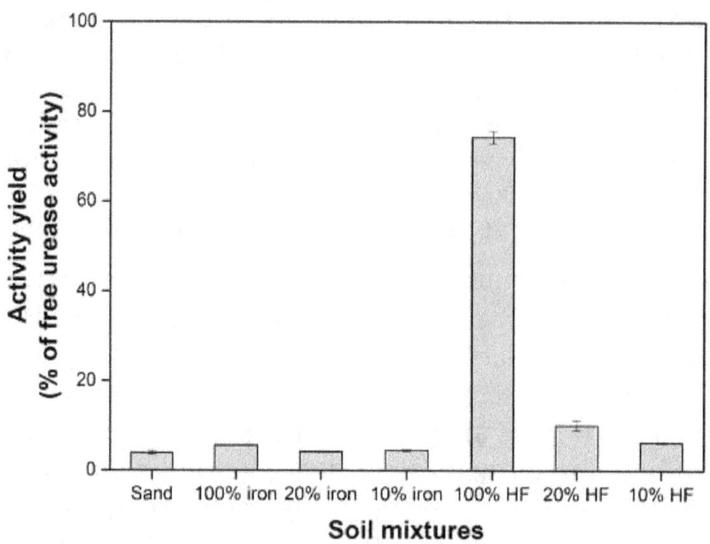

Figure 4. Urease activity yield upon adsorption. Activity yield was calculated as the % of activity of adsorbed urease in each soil mixture normalized to the activity urease in the initial total protein extract. Data shown represent the average of three independent experiments, with error bars indicating one standard error.

The 100% iron sample displayed higher urease activity than sand ($p = 0.07$). Although the presence of iron has been correlated with a lower amount of adsorbed urease or the presence of urease in natural soils, several studies found an increased activity of several enzymes, including urease, in the presence of iron as shown [52,53,54]. Overall, urease activity decreased in the order of 100% HF > 20% HF > 10% HF > 100% iron > sand, 10% iron/20% iron. These findings agree with the values obtained for the catalytic efficiency of adsorbed urease, defined as the ratio of $Vmax/K_M$ from Michaelis–Menten experiments, which shows the decrease in catalytic efficiency and follows the order 100% HF > 100% iron > sand.

Total urease activity, defined as the sum of the activity retained in the soil and supernatant portions of each sample after adsorption, showed that the decrease in urease activity observed in the soil portions of all soil mixtures was not solely a result of adsorption intricacies. Except for 100% HF, all samples displayed high urease activity in their supernatant portions. This suggests that urease adsorption was distinctively more favourable in the 100% HF samples, where more urease was absorbed into the soil. Additional experiments with 50% HF confirmed similar trends to 100% HF. For all other samples, some urease stayed in the supernatant after adsorption, as indicated by residual urease activity in the supernatant. Nevertheless, all samples displayed an overall loss of total urease activity; the highest activity loss was observed in samples containing 10% and 20% iron (up to 58%), whereas the 100% HF samples displayed the lowest loss of activity (less than 25%).

Protein denaturation upon adsorption is highly dependent on surface coverage, especially on whether or not proteins have room to spread over the surface [23,64]. Although hydrophobic interactions have been shown to be more detrimental to the molecular structure of adsorbed proteins [27], some studies found that electrostatic interactions played a major role in protein denaturation upon adsorption to solid surfaces compared to hydrophobic interactions [65,66]. Therefore, aside from surface coverage, it is not the nature of the adsorption-driving force that dictates the likelihood of protein unfolding, but rather the extent of the protein's affinity towards the sorbent surface, regardless of the driving force. We can hypothesize that the high packing density observed in hydrophobic soil mixtures of 100% HF samples (**Figure 1**) had a strong effect on hindering the spreading of adsorbed proteins and their denaturation, which explains why a smaller decrease in overall urease activity was found in those samples. However, although the 20% HF and 10% HF samples had much higher surface coverage than sand, 100% iron, and 50% iron samples (**Figure 1**), the overall losses of activity in those samples were similar. It has been shown that the dissociation of jack bean urease into its subunits has little effect on urease activity, and subunits are often functionally independent [56,67,68]. As such, even though samples such as sand and the iron-containing soil mixtures might have experienced a higher degree of protein unfolding due to their low surface coverage of proteins, the structural changes might have happened in a way that did not affect activity as much, and the overall level of activity loss ended up being similar to those found in 10% HF and 20% HF samples. Moreover, surface topography has been highlighted as an important variable that can constrain the mobility and agglomeration of adsorbed proteins [69,70], which in turn would also affect the degree of protein unfolding. In this study, surface topography was not controlled for and likely varied widely across tested soil samples, based on SEM pictures of the coated and uncoated soil grains.

3. MATERIALS AND METHODS

3.1. Materials Used

Commercial-grade Ottawa silica sand 20/30 (Gilson Company Inc., Lewis Center, OH, USA), composed of 98.7% silica and specific gravity of 2.65, was used as the primary sand material in this study. Pure culture *S. pasteurii* (ATCC 11859) cells were used as the source of crude total bacterial proteins. All chemicals were obtained from Fisher Scientific (Walthan, MA, USA) unless otherwise noted.

3.2. Total Protein Extract

Growth media and cell harvesting of *S. pasteurii* followed previously described procedures [10] with the following modifications. Upon extraction of the crude bacterial total protein by sonication, the protein extracts were not dialyzed prior to storing at -20°C. Aqueous protein concentrations were quantified spectrophotometrically using Nanodrop measurements at 280 nm.

3.3. Soil Pretreatments

Commercial-grade Ottawa silica sand 20/30 (Gilson Company Inc., Lewis Center, OH, USA) was used as uncoated sand, and hereafter defined as "sand". Hydrophobic-coated sand was obtained by mixing 100% organosilane-based breathable-soil waterproofing agent from TerraSil (Zydex Industries, Gotri, Vadodara, India) in a 10% ratio by weight with the Ottawa sand. The mixture was left at ambient temperature overnight and then dried at 110°C until completely dry. The dried coated sand was sieved through a U.S Sieve No. 20 and No.30 (0.85 and 0.6 mm openings, respectively); the portion retained between the two sieves is herein defined as "100% HF".

Iron-coated sand was obtained following the IOCS-1 procedure [71] designed to yield sand particles with a 9.3 isoelectric point. In brief, approximately 80 g of Ottawa silica sand was mixed with 200 mL of 2.5 M $FeCl_3$ solution and heated to 110°C for 3 h. The temperature was then raised to 550°C and heating continued for an additional 3 h. Sand was cooled to room temperature in open air, after which it was rinsed with DI water and air-dried. Next, sand from the previous step was mixed in a 1:2 ratio with a 2.1 M solution of $Fe(NO_3)_3$ plus 1.5% *w/w* ratio of a 10 M NaOH solution. This mixture was heated to 110°C overnight or until completely dried. The resulting coated sand aggregates were then mechanically broken and sieved through a U.S Sieve No. 20 and No. 30 (0.85 and 0.6 mm openings, respectively); the portion retained between the two sieves is herein defined as "100% iron". Uncoated sand was mixed with the 100% HF or 100% iron in various ratios.

3.4. Soil Characterization

Contact angle: Contact angle measurements were conducted on 100% HF samples using the sessile drop method, using nanopure water as the probe liquid [72]. In brief, a water droplet was placed on top of the sand surface and a handheld camera was used to visually capture the drop shape on the surface. The tangent to drop profile was aligned with the base using an online protractor tool (**https://www.ginifab.com/feeds/angle_measurement/**, accessed on 28 October 2021), and yielded a value of 106°, which confirmed that the organosilane-coated sand was indeed hydrophobic.

Streaming current: To confirm iron-containing soil mixtures were indeed more positively charged in samples with higher iron contents, streaming current was measured in all samples using the Laboratory Charge Analyzer (Chemtrac, Atlanta, GA, USA). All measurements were conducted in 50 mM HEPES buffer with 4% EDTA at pH 8 and ambient temperature.

Surface area: The specific surface area of each soil mixture was measured by N_2 physisorption analysis using a Micromeritics 3Flex surface characterization analyzer (Micromeritics, Atlanta, GA, USA) (**Table S1**). Approximately 3.5 ± 0.5 g of samples was used for BET analysis from 0.05 to 0.3 P/P_0 using nitrogen as the carrier gas at 77 K. Prior to analysis, all samples were degassed at 200°C for 120 min to remove any adsorbates.

3.5. Adsorption Experiments

Batch adsorption experiments were conducted using 2 g of each soil mixture. Soil mixtures containing 10% and 20% (*w/w*) ratios of iron coated sand or hydrophobic coated sand to uncoated sand along with samples containing 100% iron, 100% HF, and 100% sand were considered in this study. Batch adsorption experiments were conducted at room temperature (22–25°C) and end-over rotation for 16 h, using a total protein extract with 4 mg/mL protein concentration. Following adsorption, protein soil mixtures were separated via centrifugation at 8000× *g* for 5 min at 20°C to obtain the soil and supernatant fractions from each sample. Supernatants were filtered through 0.2 μm pore size syringe filters, and soil samples were washed with buffer (50 mM HEPES buffer with 4% EDTA at pH 8) to remove loosely bound proteins. These two fractions are herein defined as "supernatant" and "soil". Samples containing only soil in buffer and the crude protein extract without soil were also tested as controls. All adsorption experiments were conducted in 50 mM HEPES buffer with 4% EDTA (pH 8). Experiments were conducted in triplicates using different total protein extracts.

3.6. Urease Activity Measurements

Urease activity was determined through measurements of urea consumption over time according to method A in Rahmatullah and Boyde [73]. Soil samples were suspended in 8 mL of 50 mM HEPES buffer with 4% EDTA at pH 8, and 2 mL of supernatant samples were used. Sub-samples were taken following the addition of 2.08 mM urea approximately every 30 s for 3 min for soil fractions, and every 15 s for 90 min for supernatants. Urease activity was determined based on the linear slope of each reaction and is expressed in U, where 1 U is equal to 1 μmol of urea consumed per minute.

3.7. Proteomics Analysis

Supernatant samples derived from adsorption experiments were analyzed using label-free liquid chromatography/mass spectrometry (LC/MS) relative quantification and targeted parallel reaction monitoring (PRM) LC/MS absolute quantification techniques at the Protein Facility at Iowa State University. In brief, crude protein extracts were reduced with DTT for the label-free analysis. The Cys were modified with iodoacetamide and then digested overnight with trypsin/Lys-C. Samples were desalted using C18 MicroSpin Columns (Nest Group SEM SS18V) prior to drying in a SpeedVac. PRTC standard (Pierce part #88320) was spiked into each sample to serve as an internal control. The peptides were then separated by LC and analyzed through MS/MS by fragmenting each peptide with an Exactive Hybrid Quadrupole-Orbitrap Mass Spectrometer with an HCD fragmentation cell (Thermo Fisher Scientific, Waltham, MA, USA). Data was normalized using the PRTC peak areas. The Proteome Discoverer program version 2.2 (Thermo Fisher Scientific, Waltham, MA, USA) was used to create a peak list. Only proteins with peptide spectrum matches (PSMs) higher than three were used for data analysis [74]. The resulting intact and fragmentation pattern was compared to a theoretical fragmentation pattern using the UniProt database for *S. pasteurii* [75] in order to identify proteins based on the peptides present. The retention time and masses of urease subunits α, β, and γ obtained from the label-free untargeted quantification analysis were used to produce an inclusion list of peptides for the PRM runs. A total number of 11, 4, and 3 peptides present in the urease α, β, and γ subunit samples, respectively, displaying a high number of PMS were selected for the PRM inclusion list. In the PRM analysis, myoglobin was spiked in each sample as an internal standard at 0.5 μg/μL concentration. An MS1 scan was used to find the peptides of interest from the generated inclusion list and MS/MS scan was used to confirm the peptide's identity for quantification. The mass of each urease subunit was obtained based on the ratio of the peak areas of myoglobin standard, the protein of interest, and the known amount of myoglobin injected.

3.8. Statistical Analysis

Statistical analysis was performed using the R Stats Package version 4.1.1. Compared samples were subjected to an f-test for equality of variances and followed by unpaired or paired t-tests based on the results of the f-test. A p-value less than 0.05 was considered as statistical significance.

4. CONCLUSIONS

Understanding how protein adsorption onto soils results in enzymatic activity is vital for enzyme-based sustainable engineering applications such as biocementation. The findings from this study showed that similar amounts of adsorbed proteins from the original crude protein extract adsorbed onto all soil surfaces regardless of surface chemistry. However, when considering the adsorbed urease, our findings suggests that urease was preferentially adsorbed onto 20% HF and 100% HF samples. Higher levels of urease activity were retained in hydrophobic-containing samples, and urease activity decreased in the order of 100% HF > 20% HF > 10% HF > 100% iron > sand, 10% iron/20% iron. Similar levels of enzymatic activity were retained in iron-containing soil mixtures and sand, indicating that although favourable electrostatic interactions were expected to enhance adsorption and urease activity in iron-containing mixtures, our results showed otherwise. Finally, although higher amounts of urease and retained activity were found in hydrophobic-containing soils, the overall loss of activity in 10% HF and 20% HF samples were similar to values obtained for sand and 100% iron (ranging from 40–47%). The 10% iron and 20% iron samples displayed the highest activity loss, with up to 55 and 58%, respectively, whereas 100% HF exhibited the lowest loss of activity of less than 25%.

The findings from this study show that preferential adsorption of targeted proteins in a complex protein mixture is possible, even when competition for adsorption sites is high, as noted in the 100% HF soils. The fact that low amounts of proteins adsorbed onto iron-containing soils, even though that interaction should be electrostatically favourable, highlights the complexity of protein adsorption and the importance of soil characterization prior to the implementation of biocementation to assess the likelihood of achieving successful cementation. Furthermore, our results suggest that surface coverage can play a major positive role in mitigating the loss of enzymatic activity upon adsorption. Thus, surface manipulations to yield high surface coverage, such as TerraSil treatments to add hydrophobicity to soil surfaces, can potentially be used to decrease the negative effects of adsorption on enzymatic activity, at least to some extent. Finally, the high amount of urease activity in hydrophobic soils containing more than 50% hydrophobic-coated sands suggests that biocementation could be successfully used to stabilize soils exposed to wildfires and/or that have experienced long periods of severe drought, since such soils are associated with higher degrees of hydrophobicity.

5. CONCISE SUMMARY

In the bacterial enzyme-induced calcite precipitation (BEICP) technique for biocementation, the spatial distribution of adsorbed and catalytically active urease dictates the location where calcium carbonate precipitation and resulting cementation will occur. This study investigated the relationships between the amount of urease and total bacterial proteins adsorbed, the retained enzymatic activity of adsorbed urease, and the overall loss of activity upon adsorption, and how these relationships are influenced by changes in soil surface chemistry. In soils with hydrophobic contents higher than 20% (*w*/*w*) ratio, urease was preferentially adsorbed compared to the total amount of proteins present in the crude bacterial protein extract. Conversely, adsorption of urease onto silica sand and soil mixtures, including iron-coated sand, was much lower compared to the total proteins. Higher levels of urease activity were retained in hydrophobic-containing samples, with urease activity decreasing with lower hydrophobic content. These observations suggest that the surface manipulation of soils, such as treatments to add hydrophobicity to soil surfaces, can potentially be used to increase the activity of adsorbed urease to improve biocementation outcomes.

REFERENCES

1. Meyer, F.D.; Bang, S.; Min, S.; Stetler, L.D. Microbiologically-Induced Soil Stabilization: Application of Sporosarcina pasteurii for Fugitive Dust Control. In Proceedings of the Geo-Frontiers 2011: Advances in Geotechnical Engineering, Dallas, TX, USA, 13–16 March 2011; pp. 4002–4011. [**Google Scholar**] [**CrossRef**][**Green Version**]

2. Dhami, N.K.; Ereddy, M.S.; Mukherjee, A. Biomineralization of calcium carbonates and their engineered applications: A review. *Front. Microbiol.* **2013**, *4*, 314. [**Google Scholar**] [**CrossRef**] [**PubMed**][**Green Version**]

3. Whiffin, V.S.; van Paassen, L.; Harkes, M.P. Microbial Carbonate Precipitation as a Soil Improvement Technique. *Geomicrobiol. J.* **2007**, *24*, 417–423. [**Google Scholar**] [**CrossRef**]

4. Van Paassen, L.A.; Ghose, R.; Van Der Linden, T.J.M.; Van Der Star, W.R.L.; Van Loosdrecht, M.C.M. Quantifying Biomediated Ground Improvement by Ureolysis: Large-Scale Biogrout Experiment. *J. Geotech. Geoenviron. Eng.* **2010**, *136*, 1721–1728. [**Google Scholar**] [**CrossRef**]

5. DeJong, J.T.; Mortensen, B.M.; Martinez, B.C.; Nelson, D.C. Bio-mediated soil improvement. *Ecol. Eng.* **2010**, *36*, 197–210. [Google Scholar] [CrossRef]

6. Cetin, B.; Aydilek, A.H.; Li, L. Trace Metal Leaching from Embankment Soils Amended with High-Carbon Fly Ash. *J. Geotech. Geoenviron. Eng.* **2014**, *140*, 1–13. [Google Scholar] [CrossRef]

7. Komonweeraket, K.; Cetin, B.; Aydilek, A.H.; Benson, C.H.; Edil, T.B. Geochemical Analysis of Leached Elements from Fly Ash Stabilized Soils. *J. Geotech. Geoenviron. Eng.* **2015**, *141*, 04015012. [Google Scholar] [CrossRef]

8. Komonweeraket, K.; Cetin, B.; Aydilek, A.H.; Benson, C.H.; Edil, T.B. Effects of pH on the leaching mechanisms of elements from fly ash mixed soils. *Fuel* **2015**, *140*, 788–802. [Google Scholar] [CrossRef]

9. Qin, Y.; Cabral, J.M. Review Properties and Applications of Urease. *Biocatal. Biotransform.* **2002**, *20*, 1–14. [Google Scholar] [CrossRef]

10. Vilar, R.P.; Ikuma, K. Adsorption of urease as part of a complex protein mixture onto soil and its implications for enzymatic activity. *Biochem. Eng. J.* **2021**, *171*, 108026. [Google Scholar] [CrossRef]

11. Carmona, J.P.; Oliveira, P.V.; Lemos, L. Biostabilization of a Sandy Soil Using Enzymatic Calcium Carbonate Precipitation. *Procedia Eng.* **2016**, *143*, 1301–1308. [Google Scholar] [CrossRef][Green Version]

12. Kavazanjian, E.; Hamdan, N. Enzyme induced carbonate precipitation (eicp) columns for ground improvement. In Proceedings of the IFCEE 2015, Antonio, TX, USA, 17–21 March 2015; pp. 2252–2261. [Google Scholar] [CrossRef]

13. Neupane, D.; Yasuhara, H.; Kinoshita, N.; Unno, T. Applicability of Enzymatic Calcium Carbonate Precipitation as a Soil-Strengthening Technique. *J. Geotech. Geoenviron. Eng.* **2013**, *139*, 2201–2211. [Google Scholar] [CrossRef]

14. Yasuhara, H.; Neupane, D.; Hayashi, K.; Okamura, M. Experiments and predictions of physical properties of sand cemented by enzymatically-induced carbonate precipitation. *Soils Found.* **2012**, *52*, 539–549. [Google Scholar] [CrossRef][Green Version]

15. Larson, A.D.; Kalion, R.E. Purification and properties of bacterial urease. *J. Bacteriol.* **1954**, *68*, 67–73. [Google Scholar] [CrossRef] [PubMed][Green Version]

16. Hoang, T.; Alleman, J.; Cetin, B.; Ikuma, K.; Choi, S.-G. Sand and silty-sand soil stabilization using bacterial enzyme–induced calcite precipitation (BEICP). *Can. Geotech. J.* **2019**, *56*, 808–822. [Google Scholar] [CrossRef]

17. Hoang, T.; Alleman, J.; Cetin, B.; Choi, S.-G. Engineering Properties of Biocementation Coarse- and Fine-Grained Sand Catalyzed By Bacterial Cells

and Bacterial Enzyme. *J. Mater. Civ. Eng.* **2020**, *32*, 04020030. [Google Scholar] [CrossRef]

18. Fusi, P.; Ristori, G.; Calamai, L.; Stotzky, G. Adsorption and binding of protein on "clean" (homoionic) and "dirty" (coated with Fe oxyhydroxides) montmorillonite, illite and kaolinite. *Soil Biol. Biochem.* **1989**, *21*, 911–920. [Google Scholar] [CrossRef]
19. Gianfreda, L.; Rao, M.A.; Violante, A. Invertase β-fructosidase)- Effects of montmorillonite, AL-hydroxide and AL(OH)x-montmorillonite on activity and kinetics properties. *Soil Biol. Biochem.* **1991**, *23*, 581–587. [Google Scholar] [CrossRef]
20. Absolom, D.R.; Zingg, W.; Neumann, A.W. Protein adsorption to polymer particles: Role of surface properties. *J. Biomed. Mater. Res.* **1987**, *21*, 161–171. [Google Scholar] [CrossRef]
21. Azioune, A.; Chehimi, M.M.; Miksa, B.; Basinska, T.; Slomkowski, S. Hydrophobic Protein-Polypyrrole Interactions: The Role of van der Waals and Lewis Acid-Base Forces As Determined by Contact Angle Measurements. *Langmuir* **2002**, *18*, 1150–1156. [Google Scholar] [CrossRef]
22. Huang, Q.; Jiang, M.; Li, X. Adsorption and Properties of Urease Immobilized on Several Iron and Aluminum Oxides (Hydroxides) and Kaolinite. In *Effect of Mineral-Organic-Microorganism Interactions on Soil and Freshwater Environments*; Springer: Berlin/Heidelberg, Germany, 1999; pp. 167–168. [Google Scholar]
23. Kim, J.; Somorjai, G.A. Molecular Packing of Lysozyme, Fibrinogen, and Bovine Serum Albumin on Hydrophilic and Hydrophobic Surfaces Studied by Infrared-Visible Sum Frequency Generation and Fluorescence Microscopy. *J. Am. Chem. Soc.* **2003**, *125*, 3150–3158. [Google Scholar] [CrossRef]
24. Tilton, R.; Robertson, C.R.; Gast, A.P. Manipulation of hydrophobic interactions in protein adsorption. *Langmuir* **1991**, *7*, 2710–2718. [Google Scholar] [CrossRef]
25. Tangpasuthadol, V.; Pongchaisirikul, N.; Hoven, V.P. Surface modification of chitosan films.: Effects of hydrophobicity on protein adsorption. *Carbohydr. Res.* **2003**, *338*, 937–942. [Google Scholar] [CrossRef]
26. Wang, W.; Chen, L.; Zhang, Y.; Liu, G. Adsorption of bovine serum albumin and urease by biochar. In *IOP Conference Series: Earth and Environmental Science*; IOP Publishing: Bristol, UK, 2017; Volume 61, pp. 8–13. [Google Scholar] [CrossRef][Green Version]
27. Lu, J.; Su, T.; Thirtle, P.; Thomas, R.; Rennie, A.; Cubitt, R. The Denaturation of Lysozyme Layers Adsorbed at the Hydrophobic Solid/Liquid Surface Studied by Neutron Reflection. *J. Colloid Interface Sci.* **1998**, *206*, 212–223. [Google Scholar] [CrossRef] [PubMed]

28. Onweremadu, E.U. Hydrophobicity of soils formed over different lithologies. *Malasyan J. Soil Sci.* **2008**, *12*, 19–30. [Google Scholar]

29. Hallett, P.; Baumgartl, T.; Young, I. Subcritical Water Repellency of Aggregates from a Range of Soil Management Practices. *Soil Sci. Soc. Am. J.* **2001**, *65*, 184–190. [Google Scholar] [CrossRef]

30. Dekker, L.W.; Ritsema, C.J. How water moves in a water repellent sandy soil: 1. Potential and actual water repellency. *Water Resour. Res.* **1994**, *30*, 2507–2517. [Google Scholar] [CrossRef]

31. Dekker, L.W.; Doerr, S.H.; Oostindie, K.; Ziogas, A.K.; Ritsema, C.J. Water Repellency and Critical Soil Water Content in a Dune Sand. *Soil Sci. Soc. Am. J.* **2001**, *65*, 1667–1674. [Google Scholar] [CrossRef]

32. Jaramillo, D.F.; Dekker, L.W.; Ritsema, C.J.; Hendrickx, J.M.H. Soil water repellency in arid and humid climates. *Soil Water Repel. Occur. Conseq. Amelior.* **2003**, *232*, 93–98. [Google Scholar] [CrossRef]

33. Roberts, F.; Carbon, B. Water repellence in sandy soils of South-Western Australia. II. Some chemical characteristics of the hydrophobic skins. *Soil Res.* **1972**, *10*, 35–42. [Google Scholar] [CrossRef]

34. Jaramillo, D.F.; Herrón, F.E. Evaluacion de la repelencia al agua de algunos andisols de antioquia bajo cobertura de Pinus patula. *Acta Agron.* **1991**, *4*, 79–85. [Google Scholar]

35. Debano, L.F.; Krammes, J.S. Water repellent soils and their relation to wildfire temperatures. *Int. Assoc. Sci. Hydrol. Bull.* **1966**, *11*, 14–19. [Google Scholar] [CrossRef]

36. Doerr, S.; Shakesby, R.; Walsh, R. Soil water repellency: Its causes, characteristics and hydro-geomorphological significance. *Earth-Sci. Rev.* **2000**, *51*, 33–65. [Google Scholar] [CrossRef]

37. McHale, G.; Newton, M.I.; Shirtcliffe, N.J. Water-repellent soil and its relationship to granularity, surface roughness and hydrophobicity: A materials science view. *Eur. J. Soil Sci.* **2005**, *56*, 445–452. [Google Scholar] [CrossRef] [Green Version]

38. Osborn, J.F.; Pelishek, R.E.; Krammes, J.S.; Letey, J. Soil Wettability as a Factor in Erodibility. *Soil Sci. Soc. Am. J.* **1964**, *28*, 294–295. [Google Scholar] [CrossRef]

39. Jungerius, P.; van der Meulen, F. Erosion processes in a dune landscape along the Dutch coast. *Catena* **1988**, *15*, 217–228. [Google Scholar] [CrossRef]

40. Lowe, M.-A.; McGrath, G.; Leopold, M. The Impact of Soil Water Repellency and Slope upon Runoff and Erosion. *Soil Tillage Res.* **2021**, *205*, 104756. [Google Scholar] [CrossRef]

41. Almajed, A.; Lemboye, K.; Arab, M.G.; Alnuaim, A. Mitigating wind erosion of sand using biopolymer-assisted EICP technique. *Soils Found.* **2020**, *60*, 356–371. [Google Scholar] [CrossRef]

42. Chae, S.H.; Chung, H.; Nam, K. Evaluation of microbially Induced calcite precipitation (MICP) methods on different soil types for wind erosion control. *Environ. Eng. Res.* **2020**, *26*, 123–128. [Google Scholar] [CrossRef]

43. Jiang, X.; Rutherford, C.; Cetin, B.; Ikuma, K. Reduction of Water Erosion Using Bacterial Enzyme Induced Calcite Precipitation (BEICP) for Sandy Soil. In Proceedings of the Geo-Congress 2020 Biogeotechnics, Reston, VA, USA; 2020; pp. 104–110. [Google Scholar] [CrossRef]

44. Jiang, N.-J.; Soga, K. The applicability of microbially induced calcite precipitation (MICP) for internal erosion control in gravel–sand mixtures. *Géotechnique* **2017**, *67*, 42–53. [Google Scholar] [CrossRef][Green Version]

45. Gianfreda, L.; Rao, M.A.; Violante, A. Adsorption, activity and kinetic properties of urease on montmorillonite, aluminium hydroxide and AL(OH) x-montmorillonite complexes. *Soil Biol. Biochem.* **1992**, *24*, 51–58. [Google Scholar] [CrossRef]

46. Artioli, Y. Adsorption. In *Encyclopedia of Ecology*; Elsevier: Amsterdam, The Netherlands, 2008; pp. 60–65. [Google Scholar]

47. Hoeve, C.A.J.; DiMarzio, E.A.; Peyser, P. Adsorption of Polymer Molecules at Low Surface Coverage. *J. Chem. Phys.* **1965**, *42*, 2558–2563. [Google Scholar] [CrossRef]

48. Larsericsdotter, H.; Oscarsson, S.; Buijs, J. Thermodynamic Analysis of Proteins Adsorbed on Silica Particles: Electrostatic Effects. *J. Colloid Interface Sci.* **2001**, *237*, 98–103. [Google Scholar] [CrossRef] [PubMed]

49. Norde, W.; Favier, J.P. Structure of adsorbed and desorbed proteins. *Colloids Surfaces* **1992**, *64*, 87–93. [Google Scholar] [CrossRef]

50. Norde, W.; Giacomelli, C. BSA structural changes during homomolecular exchange between the adsorbed and the dissolved states. *J. Biotechnol.* **2000**, *79*, 259–268. [Google Scholar] [CrossRef]

51. Norde, W.; Giacomelli, C.E. Conformational changes in proteins at interfaces: From solution to the interface, and back. *Macromol. Symp.* **1999**, *145*, 125–136. [Google Scholar] [CrossRef]

52. Gianfreda, L.; Rao, M.A.; Violante, A. Formation and Activity of Urease-Tannate Complexes Affected by Aluminum, Iron, and Manganese. *Soil Sci. Soc. Am. J.* **1995**, *59*, 805–810. [Google Scholar] [CrossRef]

53. Gianfreda, L.; De Cristofaro, A.; Rao, M.A.; Violante, A. Kinetic Behavior of Synthetic Organo-and Organo-Mineral-Urease Complexes. *Soil Sci. Soc. Am. J.* **1995**, *59*, 811–815. **[Google Scholar]** **[CrossRef]**

54. He, S.; Feng, Y.; Ren, H.; Zhang, Y.; Gu, N.; Lin, X. The impact of iron oxide magnetic nanoparticles on the soil bacterial community. *J. Soils Sediments* **2011**, *11*, 1408–1417. **[Google Scholar]** **[CrossRef]**

55. Rabe, M.; Verdes, D.; Seeger, S. Understanding protein adsorption phenomena at solid surfaces. *Adv. Colloid Interface Sci.* **2011**, *162*, 87–106. **[Google Scholar]** **[CrossRef][Green Version]**

56. Bordbar, A.-K.; Sohrabi, N.; Hojjati, E. The estimation of the hydrophobic and electrostatic contributions to the free energy change upon cationic surfactants binding to Jack bean urease. *Colloids Surf. B Biointerfaces* **2004**, *39*, 171–175. **[Google Scholar]** **[CrossRef]**

57. Tipping, E.; Jones, M.N.; Skinner, H.A. Enthalpy of interaction between some globular proteins and sodium n-dodecyl sulphate in aqueous solution. *J. Chem. Soc. Faraday Trans. 1 Phys. Chem. Condens. Phases* **1974**, *70*, 1306–1315. **[Google Scholar]** **[CrossRef]**

58. Bordbar, A.-K. Thermodynamic Analysis for Cationic Surfactants Binding to Bovine Serum Albumin. *J. Phys. Theor. Chem.* **2006**, *2*, 197–204. **[Google Scholar]** **[CrossRef]**

59. Takishima, K.; Suga, T.; Mamiya, G. The structure of jack bean urease. The complete amino acid sequence, limited proteolysis and reactive cysteine residues. *JBIC J. Biol. Inorg. Chem.* **1988**, *175*, 151–157. **[Google Scholar]** **[CrossRef]**

60. Kuscu, I.S.K.; Cetin, M.; Yigit, N.; Savaci, G.; Sevik, H. Relationship between Enzyme Activity (Urease-Catalase) and Nutrient Element in Soil Use. *Pol. J. Environ. Stud.* **2018**, *27*, 2107–2112. **[Google Scholar]** **[CrossRef]**

61. Christians, S.; Kaltwasser, H. Nickel-content of urease from Bacillus pasteurii. *Arch. Microbiol.* **1986**, *145*, 51–55. **[Google Scholar]** **[CrossRef]**

62. Kuo, J.F.; Angeles, L. Further Investigation of the Surface Charge Properties of Oxide Surfaces in Oil-Bearing Sands and Sandstones. *J. Colloid.* **1987**, *115*, 9–16. **[Google Scholar]** **[CrossRef]**

63. Xia, J.; Wishart, D.S. Metabolomic data processing, analysis, and interpretation using MetaboAnalyst. *Curr. Protoc. Bioinform.* **2011**, *34*. **[Google Scholar]** **[CrossRef]**

64. Garwood, G.; Mortland, M.; Pinnavaia, T. Immobilization of glucose oxidase on montmorillonite clay: Hydrophobic and ionic modes of binding. *J. Mol. Catal.* **1983**, *22*, 153–163. **[Google Scholar]** **[CrossRef]**

65. Quiquampoix, H.; Abadie, J.; Baron, M.H.; Leprince, F.; Matumoto-Pintro, P.T.; Ratcliffe, R.G.; Staunton, S. Mechanisms and Consequences of Protein Adsorption on Soil Mineral Surfaces. In *Proteins at Interfaces II*; ACS Symposium Series; American Chemical Society, ACS: Washington, DC, USA, 1995; pp. 321–333. Available online: **https://pubs.acs.org/doi/abs/10.1021/bk-1995-0602.ch023** (accessed on 9 January 2022).

66. Lahari, C.; Jasti, L.S.; Fadnavis, N.W.; Sontakke, K.; Ingavle, G.; Deokar, S.; Ponrathnam, S. Adsorption Induced Enzyme Denaturation: The Role of Polymer Hydrophobicity in Adsorption and Denaturation of α-Chymotrypsin on Allyl Glycidyl Ether (AGE)-Ethylene Glycol Dimethacrylate (EGDM) Copolymers. *Langmuir* **2010**, *26*, 1096–1106. **[Google Scholar]** **[CrossRef]**

67. Hirai, M.; Kawai-Hirai, R.; Hirai, T.; Ueki, T. Structural change of jack bean urease induced by addition surfactants studied with synchrotron-radiation small-angle X-ray scattering. *JBIC J. Biol. Inorg. Chem.* **1993**, *215*, 55–61. **[Google Scholar]** **[CrossRef]**

68. Contaxis, C.; Reithel, F. Studies on Protein Multimers: II. A Study of the Mechanism OF Urease Dissociation in 1, 2-Propanediol: Comparative Studies with Ethylene Glycol and Glycerol. *J. Biol. Chem.* **1971**, *246*, 677–685. **[Google Scholar]** **[CrossRef]**

69. dos Santos, E.A.; Farina, M.; Soares, G.A.; Anselme, K. Surface energy of hydroxyapatite and β-tricalcium phosphate ceramics driving serum protein adsorption and osteoblast adhesion. *J. Mater. Sci. Mater. Med.* **2008**, *19*, 2307–2316. **[Google Scholar]** **[CrossRef]**

70. Dufrêne, Y.F.; Marchal, T.G.; Rouxhet, P.G. Influence of Substratum Surface Properties on the Organization of Adsorbed Collagen Films: In Situ Characterization by Atomic Force Microscopy. *Langmuir* **1999**, *15*, 2871–2878. **[Google Scholar]** **[CrossRef]**

71. Benjamin, M.M.; Sletten, R.S.; Bailey, R.P.; Bennett, T. Sorption and filtration of metals using iron-oxide-coated sand. *Water Res.* **1996**, *30*, 2609–2620. **[Google Scholar]** **[CrossRef]**

72. Olorunfemi, I. Soil Hydrophobicity: An Overview. *J. Sci. Res. Rep.* **2014**, *3*, 1003–1037. **[Google Scholar]** **[CrossRef]**

73. Rahmatullah, M.; Boyde, T. Improvements in the determination of urea using diacetyl monoxime; methods with and without deproteinisation. *Clin. Chim. Acta* **1980**, *107*, 3–9. **[Google Scholar]** **[CrossRef]**

74. Kandhavelu, J.; Demonte, N.L.; Namperumalsamy, V.P.; Prajna, L.; Thangavel, C.; Jayapal, J.M.; Kuppamuthu, D. Data set of Aspergillus flavus induced alterations in tear proteome: Understanding the pathogen-induced host response to fungal infection. *Data Brief* **2016**, *9*, 888–894. [**Google Scholar**] [**CrossRef**]

75. UniProt Consortium. UniProt: A worldwide hub of protein knowledge. *Nucleic Acids Res.* **2019**, *47*, D506–D515. [**Google Scholar**] [**CrossRef**][**Green Version**]

CHAPTER-12

METAGENOMIC APPROACHES AS A TOOL TO UNRAVEL PROMISING BIOCATALYSTS FROM NATURAL RESOURCES: SOIL AND WATER

1. INTRODUCTION: AN OVERVIEW

In the pursuit of sustainable solutions for industrial processes and environmental challenges, the exploration of natural resources has become a focal point. Metagenomics, a powerful approach in microbial ecology, offers a unique lens into the vast and unexplored genetic reservoirs present in environments such as soil and water. This study embarks on a journey to unravel promising biocatalysts through metagenomic approaches, specifically focusing on the rich microbial diversity within soil and water ecosystems.

The microbial communities dwelling in these environments harbor an immense genetic diversity, often encoding enzymes with unique catalytic properties. Biocatalysts, including enzymes, derived from such sources have the potential to revolutionize various biotechnological applications, ranging from industrial processes to environmental remediation. Metagenomic techniques enable the direct examination of the collective genetic material within a given environment, bypassing the need for traditional isolation and cultivation of individual microorganisms. This not only provides access to previously unculturable microbes but also unveils a treasure trove of enzymatic diversity.

In this exploration of metagenomic approaches, our aim is to harness the genetic richness inherent in soil and water ecosystems for the identification and characterization of novel biocatalysts. The implications of this research extend beyond the laboratory, offering sustainable alternatives for biotechnological applications and contributing to the growing field of environmental biotechnology. As we delve into the vast genetic landscapes of these natural resources, we anticipate uncovering biocatalysts that hold the key to addressing current challenges and advancing the frontiers of biotechnology.

The natural resources available on Earth have played an important role in the history of human civilization since they have provided the necessary materials and energy for the preservation and proliferation of life. Besides the basic sustenance, proper use of these resources can also contribute to the improvement of our comfort, protection and well-being [1]. Over the centuries, we have witnessed the continuous exploitation of some finite natural resources, which resulted in the degradation of important ecosystems, subsequently creating severe environmental, economic and technological issues [2,3]. This overconsumption of materials and energy will eventually lead to inevitable resource depletion. Consequently, there is a need to moderate the demand for finite natural resources and simultaneously search for efficient and sustainable ways to extract and convert energy/materials from renewable resources [4]. In this context, biocatalysts can play a crucial role. Due to their high specificity and selectivity, the biocatalysts can generally ensure the effective conversion of substrates, minimizing the formation of undesirable side-products and reducing the energetic costs associated with the process. Hence, the use of suitable and robust biocatalysts can greatly contribute to the implementation of greener and sustainable bioprocesses that efficiently compete with the classical chemical routes. Currently, the use of biocatalysts for the valorization of alternative non-finite resources, under the concept of bio economy and the EU Green Deal, has gained increased attention.

Unexplored or slightly explored environments, such as soil and water, are interesting sources of novel and promising biocatalysts. Despite the clear differences at the physicochemical level, soil and water are both regarded as natural bio-reservoirs with great microbial diversity. For this reason, these environments have been the focus of several microbial studies in the last few decades. Microorganisms are considered important suppliers of various bioproducts with applications in several industrial areas, such as enzymes. In fact, in the last decade, we have seen a significant increase in the demand for enzymes [5], which is easily explained by their great biotechnological potential. However, the presence of a significant number of unculturable microorganisms both in the soil and water can limit or make unfeasible some microbial studies to find novel biocatalysts. In this context, metagenomics can play a crucial role.

Metagenomics has emerged as a culture-independent technique that allows exploring the genetic material of whole microbial communities present in a given environment [6]. This technique has been successfully used to identify novel enzymes with promising catalytic activities and some of them have been patented and already translated to the market [7]. Two different metagenomic approaches have been described, namely, sequence-based or function-based metagenomics. In both cases, an initial step of DNA extraction from an environmental sample is needed [8]. The sequence-based studies allow the identification of candidate genes, while the function-based screenings include the detection and isolation of clones from metagenomic libraries with a positive response to the desired phenotype [9]. The construction of a metagenomic library requires the selection of the most suitable expression vector, in which the environmental DNA fragment will be inserted. In addition to other aspects, such as the quality and size of the environmental DNA, this selection depends on the purpose of the functional screening. Plasmids can be used when DNA fragments are small (≤15 kb insert size) and contain only individual genes. On the other hand, some expression vectors, such as fosmids and cosmids (<40 kb insert size), or bacterial artificial chromosomes (BACs) and yeast artificial chromosomes (YACs) (>40 kb insert size), allow the recovery of large biosynthetic gene clusters that encode the production of one or more specialized metabolites. Besides the expression vector, expression systems should be selected in such a way that gene expression and target gene detection are maximized. *Escherichia coli* has been widely used due to the extensive genetic knowledge on this microorganism that makes it suitable for effective and profitable cloning and protein expression [10,11]. However, the demand for robust biocatalysts requires functional screenings at temperatures other than the growth temperatures of mesophilic hosts. Therefore, other expression hosts, such as Thermus thermophilus, have already been shown to be good candidates as expression hosts in the functional metagenome analysis [12].

This review discusses the most significant works reported in the last decade about metagenomic studies performed in soil and water environments for the discovery of novel and interesting enzymes. The bibliographic survey was carried out in the international database Scopus between 2010 and 2021 using several search criteria as illustrated in **Figure 1**. Briefly, in the first criteria the keywords used were: "metagenomic" and "enzyme" and "soil" or "water". After that, for each natural resource, the search was refined according to the type and origin of the samples. Finally, the different groups of enzymes were considered in the search.

Figure 1. Search criteria defined for the bibliographic survey carried out in the international database Scopus. The first research criteria consisted of a generic search for metagenomic studies involving enzymes; the second criteria restricted the search to different types of samples and their origins; and the last criteria focused on the different groups of enzymes. In each case, the established keywords were searched in the title, abstract or keywords of the article.

Based on the results obtained in the bibliographic survey, the studies were divided into three categories according to the type of sample (e.g., raw resources, human manipulated resources and unspecified resources) as illustrated in **Figure 2**. Within the category of raw resources, differences were established according to the temperature of each specific environment. In the case of human manipulated resources, three conditions were considered for soil, namely, samples from polluted environments, composting processes and agricultural lands and grassland. For the water samples resulting from human manipulation two conditions were defined, namely, contaminated groundwater/freshwater and coastal water. In the category of unspecific resources, both for soil and water, all studies that did not present enough information to classify them according to the categories mentioned above were included. Overall, for each study, the environmental conditions (temperature and pH), the type of metagenomic approach and its main characteristics, as well as the catalytic activity, was revised. Furthermore, the number of citations found in the international database Scopus was also presented as an indicator of the impact of the study. All this information was summarized in the tables presented in each specific subsection.

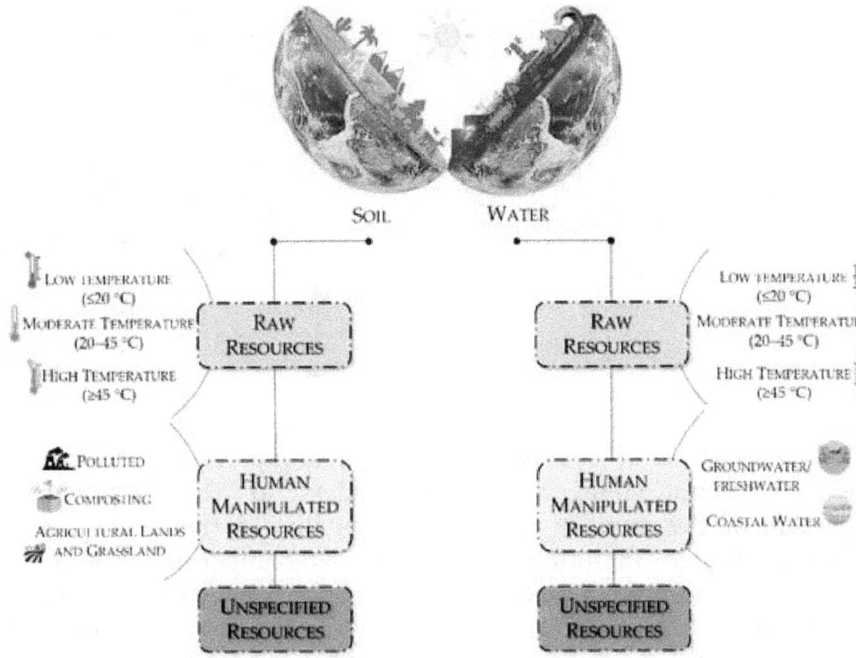

Figure 2. Classification of the different soil and water sources used in metagenomic studies conducted to identify promising biocatalysts.

2. SOIL

Soil is more than just land; it represents one of the most important natural resources available on Earth. It is considered the basis of agriculture, an important water filter, a natural reserve of carbon and water and also the regular habitat for several living organisms. Soils originate from rocks through complex chemical, mechanical and biological processes which result in the formation of small particles and grains [13]. In terms of composition, soils are generally constituted by solid, liquid and gas fractions in a 2:1:1 ratio (volume basis). The solid fraction is mainly composed of inorganic materials but also contains a small portion of organic matter (around 1–5%) [14]. The liquid phase is generally an aqueous solution of electrolytes (e.g., micronutrients and macronutrients). On the other hand, the main gases present in soil are water vapor, CO_2, O_2 and N_2 [13]. Depending on its origin and composition, different types of soil can be defined. Microbial diversity and functionality are strongly dependent on the specific characteristics of each type of soil. The classification established for the different soil samples considered in this review is presented in **Figure 3**a. A total of 173 metagenomic studies were performed, considering soil samples aiming to explore promising enzymes.

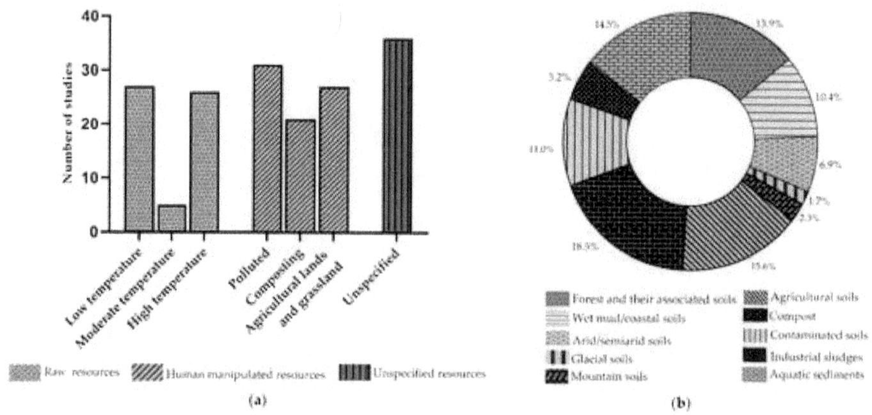

Figure 3. Classification of soil sampling sources explored to find promising biocatalysts through metagenomic approaches in the last decade. (a) Representation of the number of studies according to the defined categories (raw resources, human manipulated resources and unspecified resources). (b) Assignment of the soil type to each collected sample from different environments (forest and their associated soils, wet mud/coastal soils, arid/semiarid soils, glacial soils, mountain soils, agricultural soils, compost, contaminated soils, industrial sludge and aquatic sediments).

As shown in Figure 3b, in the last decade, the origin of the soil samples was mostly natural or controlled composting (18.5%), followed by agriculture (15.6%) and aquatic environments (14.5%). Forests and environments associated with them, e.g., groves (13.9%), were also moderately studied. The lowest studied environments included contaminated areas (11.0%), wetlands/coastal areas (10.4%), arid/semiarid areas (6.9%), industrial sludges (5.2%), mountainous areas (2.3%) and, finally, glacial areas (1.7%), probably due to the difficulty in accessing them.

2.1. Raw Resources

Depending on the geographical location, the soil may present itself in various forms at the Earth's surface as the result of environmental conditions, organic matter content and type of vegetation [15]. An abiotic factor with high importance in any ecosystem is temperature since it can directly or indirectly affect microbial activity and, consequently, the composition of microbial communities. Each microbial species has a different temperature tolerance range and is capable of producing distinct types of biocatalysts at different production rates. Still, certain microorganisms can thrive and function well metabolically in adverse conditions, notably at extreme temperatures [16,17].

Therefore, the sampling locations in which the soil was directly collected and evaluated "as found in the nature" (raw resources) were classified according to the recorded temperature: low temperature (location temperature below 20°C); moderate temperature (location temperature between 20 and 45°C); and high temperature (location temperature above 45°C) [17].

As shown in **Figure 3**a, a greater number of studies were accomplished in low-temperature environments (27 studies), followed by high-temperature environments (26 studies) and, finally, moderate temperature environments (5 studies). The unique characteristics of extreme environments, namely, extreme temperatures, make them more interesting places to find promising and robust enzymes. This fact is probably the main reason for the reduced number of studies accomplished in moderate temperature environments. Furthermore, there is an increasing need to find highly stable biocatalysts with efficient activities capable of acting in various industrial processes that make use of severe conditions, similar to what occurs in hot and cold environments [**18**]. **Table 1** compiles all the data considered in the category of raw resources.

Table 1. Metagenomic studies reported in the last decade for soil samples classified in the category of raw resources.

Sample Source	Source Properties pH	Source Properties T (°C)	Metagenomic Approach	Vector/ Host	No. Positive Clones/ Total Clones	Enzyme	No. Citations *	Reference
Astaka region, Kargil, Northwestern Himalayas (India)	-	-	Functional	Cosmid/*E. coli*	1/35,000	Amylase	42	**[19]**
Deep-sea, South China Sea (Pacific Ocean)	-	-	Functional	Fosmid/*E. coli*	1/20,000	α-amylase	32	**[20]**
Ikaite tufa columns, Ikka Fjord (Greenland)	10.4	4.0	Sequence-based and functional	BAC/*E. coli*	3/2843 2/2843	α-amylase β-galactosidases	70	**[21]**
Hot sulfur springs, Nubra Valley, Leh, Northern Himalayas (India)	7.5–8.5	60.0–80.0	Functional	Plasmid/*E. coli*	1/10,000	β-D-galactosidase	24	**[22]**
Mountain of Flames, Turpan Basin, Xinjiang Uygur Autonomous Region (China)	-	76.0	Functional	Plasmid/*E. coli*	1/8000	β-galactosidase	21	**[23]**
Composting, EXPO Park, Osaka (Japan)	7.5	67.0	Functional	Fosmid/*E. coli*	10/6000	Cellulase	15	**[24]**
Farm-made composting, Khon Kaen (Thailand)	4.2–6.8	30.0–50.0	Functional	Fosmid/*E. coli*	10/251	Endocellulase	13	**[25]**
Composting, Tainan (Taiwan)	-	-	Functional	Lambda phage vector/*E. coli*	1/2739	Endoglucanase	30	**[26]**
Hydrothermal vent, Vulcano island (Italy)	5.9	100.0	Sequence-based	-	-	Endo-β-glucanase	14	**[27]**
Caldeirão hot spring, Furnas Valley, São Miguel Island, Azores (Portugal)	6.0–7.0	60.0–70.0	Functional	Plasmid/*E. coli*	-	β-glucosidase	70	**[28]**
Turpan Basin, Xinjiang Uygur Autonomous Region (China)	-	82.0	Functional	Plasmid/*E. coli*	5/50,000	β-glucosidase	68	**[29]**

Sample Source	Source Properties		Metagenomic Approach	Vector/ Host	No. Positive Clones/ Total Clones	Enzyme	No. Citations *	Reference
	pH	T (°C)						
Forest, National Institute of Advanced Industrial Science and Technology, Tsukuba (Japan)	-	15.0	Functional	Fosmid/*E. coli*	1/50,000	β-glycosidase	22	**[30]**
Suruga Bay, Shizuoka, Japan (Pacific Ocean)	-	-	Sequence-based	-	-	β-glucosidases Endomannanase Endoxylanases β-xylosidases	17	**[31]**
Tattapani thermal spring, Chhattisgarh (India)	7.5–8.3	55.0–98.0	Sequence-based	-	-	β-glucosidase Xylanase	24	**[32]**
Thermophilic composting, Viçosa and Urucânia, Minas Gerais (Brazil)	-	55.0–65.0	Functional	Fosmid/*E. coli*	159/6720 9/6720 14/6720	Cellulase Xylanase β-glucosidase	11	**[33]**
Solfatara-Pisciarelli hydrothermal pool, Agnano, Naples (Italy)	1.5	92.0	Sequence-based	-	-	β-mannanase/ β-1,3-glucosidase β-*N*-acetyl-glucosaminidase/β-glucosidase	11	**[34]**
Hot spring compost-soil, Fukuoka (Japan)	3.0–4.5	-	Functional	Plasmid/*E. coli*	-	Xylanase	73	**[35]**
Lobios hot spring, Ourense, Galicia (Spain)	5.9	76.0	Sequence-based and functional	Fosmid/*E. coli*	1/150,000	Xylanase	15	**[36]**
Forest, National Institute of Advanced Industrial Science and Technology, Tsukuba (Japan)	-	15.0	Functional	Fosmid/*E. coli*	1/50,000	α-xylosidase	21	**[37]**
Haiyan wetland, Qinghai-Tibetan Plateau (China)	-	-	Sequence-based	-	-	Chitinase	1	**[38]**
Cold desert, McMurdo Dry Valleys, South Victoria Land (Antarctica)	-	3.0	Functional	Fosmid/*E. coli*	-/10,000	Esterase	38	**[39]**

Sample Source	Source Properties pH	Source Properties T (°C)	Metagenomic Approach	Vector/ Host	No. Positive Clones/ Total Clones	Enzyme	No. Citations *	Reference
Lake Arreo, Álava (Spain)	8.0	6.9	Functional	Fosmid/*E. coli*	10/11,520	Esterases/lipases	39	[40]
Apharwat mountain, Jammu and Kashmir, North-western Himalayas (India)	-	-	Functional	Plasmid/*E. coli*	1/10,000	Thioesterase	1	[41]
Composting, EXPO Park, Osaka (Japan)	7.5	67.0	Functional	Fosmid/*E. coli*	19/6000	Esterase	9	[42]
Hot spring, Furnas Valley, São Miguel Island, Azores (Portugal)	2.0–7.0	60.0–62.0	Sequence-based and functional	Fosmid/*E. coli* Fosmid/ *T. thermophilus*	1/6048 5/6048	Esterase	50	[43]
Apharwat mountain, Jammu and Kashmir, North-western Himalayas (India)	-	-	Functional	Plasmid/*E. coli*	3/10,000	Esterase	3	[44]
Karuola glacier, Tibetan Plateau (China)	-	-	Functional	Fosmid/*E. coli*	5/10,000	Esterase	10	[45]
Solnechny hot spring, Uzon Caldera, Kronotsky reserve, Kamchatka (Russia)	5.8–6.0	61.0–64.0	Sequence-based	-	-	Esterase	39	[46]
Permafrost, Kolyma Lowland, North-eastern Siberia (Russia)	-	-	Functional	Fosmid/*E. coli*	7/5000	Esterase	30	[47]
Deep-sea, Barents Sea (Arctic Ocean)	-	-	Functional	Fosmid/*E. coli*	19/3884	Esterase	24	[48]
Deep-sea, Svalbard e Jan Mayen (Arctic Ocean)	-	-	Functional	Fosmid/*E. coli*	19/3884	Esterase		
Yellow Sea, China (Pacific Ocean)	-	-	Functional	Fosmid/*E. coli*	34/40,000	Esterase	42	[49]

Sample Source	Source Properties pH	T (°C)	Metagenomic Approach	Vector/ Host	No. Positive Clones/ Total Clones	Enzyme	No. Citations *	Reference
Acidic pool, Vulcano island (Italy)	-	≈25.0	Functional	Fosmid/*E. coli*	44/1920	Carboxylesterases	44	[50]
Deep-sea (Atlantic Ocean)	-	-	Functional	Fosmid/*E. coli*	6/17,000	Esterase	7	[51]
Deep-sea (Atlantic Ocean)	-	-	Functional	Fosmid/*E. coli*	6/17,000	Esterase		
Solar saltern, Ribandar, Goa (India)	7.5	35.0	Functional	Fosmid/*E. coli*	1/5100	Esterase	14	[52]
Qiongdongnan Basin, South China Sea (Pacific Ocean)	-	-	Functional	Fosmid/*E. coli*	19/40,000	Esterase	68	[53]
Qiongdongnan Basin, South China Sea (Pacific Ocean)	-	-	Functional	Fosmid/*E. coli*	1/200,000	Esterase	52	[54]
Deep-sea, South China Sea (Pacific Ocean)	-	-	Functional	Plasmid/*E. coli*	15/60,000	Esterase	43	[55]
Kongsfjorden seashore, Ny-Alesund, Svalbard Archipelago (Norway)	-	-	Functional	Fosmid/*E. coli*	2/60,000	Esterase	61	[56]
Turpan Basin, Xinjiang Uygur Autonomous Region (China)	-	82.0	Functional	Plasmid/*E. coli*	3/21,000	Pyrethroid-hydrolysing esterase	49	[57]
Deep-Sea (Pacific Ocean)	-	1.5	Functional	Fosmid/*E. coli*	12/20,000	Esterase	43	[58]
Composting, EXPO Park, Osaka (Japan)	7.5	67.0	Functional	Fosmid/*E. coli*	19/6000	Cutinase	180	[59]
Taptapani hot spring, Odisha (India)	8.5	50.0	Functional	Plasmid/*E. coli*	7/13,298	Lipase	21	[60]
Taptapani hot spring, Odisha (India)	8.5	50.0	Functional	Plasmid/*E. coli*	7/13,000	Lipase	4	[61]

Sample Source	Source Properties pH	Source Properties T (°C)	Metagenomic Approach	Vector/ Host	No. Positive Clones/ Total Clones	Enzyme	No. Citations *	Reference
Qiongdongnan Basin, South China Sea (Pacific Ocean)	-	-	Functional	Plasmid/*E. coli*	1/60,000	Laccase	28	[62]
Hot spring, Jiuqu, Guangxi (China)	-	65.0–68.0	Sequence-based	-	-	Aldehyde dehydrogenase	6	[63]
Ikahama hot spring, Tonami, Toyama (Japan)	-	-	Sequence-based	Plasmid/*E. coli*	240/2800	Alcohol dehydrogenases	18	[64]
Farm composting, Toyama (Japan)	-	35.0–45.0	Sequence-based and functional	Plasmid/*E. coli*	1200/2000	Alcohol dehydrogenases	15	[65]
Death Valley desert dune, California (USA)	-	-	Functional	Plasmid/*E. coli*	1/30,000	Serine proteases	48	[66]
Gobi Desert dune, Ulaanbadrakh (Mongolia)				Fosmid/*E. coli*	16/17,000			
Chumathanghotspring,Ladakhregion, North-western Himalayas (India)	-	-	Functional	Plasmid/*E. coli*	1/9000	Protease	10	[67]
Creek soil, Ardley Island, FILDES Peninsula (Antarctica)	-	-	Functional	Fosmid/*E. coli*	7/-	Protease	0	[68]
Composting, EXPO Park, Osaka (Japan)	7.0	50.0	Functional	Plasmid/*E. coli*	17/300,000	RNase H	10	[69]
Ladakh region, North-western Himalayas (India)	-	-	Functional	Plasmid/*E. coli*	1/8500	Rhodanase	4	[70]
Caatinga biome, João Câmara, Rio Grande do Norte (Brazil)	-	21.0–32.0	Functional	Plasmid/*E. coli*	1/2688	Exonuclease	12	[71]
Deep-sea (Arctic Ocean)	-	-	Sequence-based	Plasmid/*E. coli*	1/2750	Chitin deacetylase	11	[72]
Tattapani thermal spring, Chhattisgarh (India)	7.7	98.0	Sequence-based	-	-	Trehalose synthase	0	[73]

* Obtained from the international database Scopus (21 March 2022).

2.1.1. Low-Temperature Environments

Cold soils, in addition to being exposed to very low temperatures, are also subject to other harsh conditions, such as freeze-thaw cycles, UV radiation and restricted availability of water and nutrients. Indeed, arid/semiarid regions, aquatic environments, polar regions, mountainous and forest areas are examples of cold habitats that have been the focus of metagenomic studies. Although they usually have low biodiversity, the microorganisms that inhabit them have acquired survival machinery and, for this reason, these types of habitats constitute stimulating reservoirs of biotechnological molecules [74].

Deep-sea sediments, mainly sands and clays from several aquatic environments, are the type of samples that most contribute to the exploration of enzymes in these cold habitats. The South China Sea is a marginal sea and one of the most studied environments since it constitutes an important reservoir of sediments rich in organic materials [75]. Other regions of the Pacific Ocean, such as Suruga Bay in Japan [31], as well as the depths of the Atlantic Ocean [51] and the Arctic Ocean, notably in the Barents Sea area [48], have also been studied to search not only for lipolytic enzymes but also enzymes belonging to the glycosyl hydrolases class. Additionally, the submarine tufa columns of the Ikka Fjord in Greenland [21] and the karstic lake in Spain (Lake Arreo) [40], which, in addition to being permanently cold, are alkaline, and valuable sources of microorganisms adapted to these environments. In addition to aquatic sediments, other types of samples with metagenomic exploitation potential come from arid/semiarid and mountain soils, for example in the Ladakh region [70] and the Apharwat mountain [41,44], respectively, both in the north-western Himalayas that have distinct geo-climatic characteristics like extremely cold and dry weather, high altitude and glacial and permafrost soils. Relevant examples of this are the Karuola glacier in Tibetan Plateau [45] and the Kolyma Lowland permafrost in north-eastern Siberia [47], which are extremely hostile environments and inhabited by unique microbial communities.

2.1.2. Moderate Temperature Environments

There are few studies in which sampling sources are identified as moderate temperature environments. Still, certain environments of this nature have other characteristics that also make them interesting for metagenomic purposes. An example is the Caatinga biome of João Câmara (Brazil) which presents sandy loam soil and constitutes an ecosystem of high biological relevance due to the features of the area, such as the semiarid climate, the high exposure to UV radiation and the long periods of drought [71,76]. Another example is the solar saltern of Goa, which differs by its high salinity and represents an important source for metagenomic studies given the difficulty in cultivating halophilic microorganisms through conventional techniques [52].

2.1.3. High-Temperature Environments

High-temperature environments have also proven to be important sources of very useful thermostable enzymes with applications in various industrial fields, such as food and chemical synthesis industries. In addition to geothermally heated environments, such as hot springs and hydrothermal vents, arid/semiarid regions and environments subject to natural composting processes are often good targets for the application of metagenomic tools [77]. Compost samples from Expo Park in Japan, produced from leaves and branches, are a good example of natural composting since they have been studied over a few years [24,42,59,69]. Once thermophilic composting reaches high temperatures, there is a greater predominance of microorganisms capable of degrading complex molecules, with this type of environment being a potential source of lignocellulose-degrading enzymes and, for this reason, an interesting subject of study [78]. Several arid/semiarid regions have been explored given the typical characteristics of these environments, including deserts [66] and also other sites—more specifically, the Turpan Basin, which represents China's hottest place and has proven to be a valuable source of different types of highly thermostable enzymes [57]. Hot springs and hydrothermal vents from different portions of the planet, e.g., Caldeirão hot spring in Portugal [28], Solnechny hot spring in Russia [46] and Solfatara-Pisciarelli hydrothermal pool in Italy [34] have also contributed to finding robust enzymes through the construction of metagenomic libraries using DNA extracted from wet mud and/or sediments collected from these places. For all other metagenomic studies, the expression host used was *E. coli*, except for the metagenomic library constructed from sediments of a hot spring in the Azores, Portugal, which used the *T. thermophilus* as the host. Using this thermophilic host, Leis and co-workers intended to increase the probability of detecting genes derived from extreme environments that would encode for new thermostable biocatalysts and allow the screening of phenotypes that are not observable in *E. coli* [43].

In some studies performed from these raw resources, an additional enrichment step was performed to provide favourable growth conditions for certain microorganisms of interest, often present in small abundance, to the detriment of others [79]. These enrichments were implemented by introducing specific substrates, such as cellulose, xylan, chitin, starch and glucose [31,33,38], and even olive oil [47], that stimulate specific microbial activities. On the other hand, culture enrichment also occurred by controlling environmental conditions, in particular the temperature, which is generally in agreement with the temperature of the sampling locations [26].

Over the past decade, in addition to function-based metagenomic screenings, sequence-based metagenomic screenings have also been performed. Sequential metagenomics showed that the phyla that predominate high-temperature environments are *Crenarchaeota, Thaumarchaeota, Acidobacteria* and *Proteobacteria* capable of mineral-based metabolism and generally associated with soil, found more specifically in sediments from hot springs and hydrothermal vents [21,31,34,36,43].

2.2. Human Manipulated Resources

Soil, in addition to being "discovered" as it is exposed in nature, without any kind of alteration, can be studied in a variety of scenarios, including contaminated/polluted, agricultural and controlled composting as a consequence of human manipulations. According to **Figure 3**a, 31 metagenomic studies were performed in polluted soil samples, 27 studies were conducted in agricultural fields and grasslands and 21 studies were performed on samples provided by composting facilities. **Table 2** comprises all the data considered in the category of human manipulated resources.

2.2.1. Polluted Environments

The intensification of industrialization, urbanization and mining have negatively affected the soil as a natural source. It has been observed that soil has been contaminated by different factors, namely, industrial sewages, solid wastes and urban activities. Some organic and inorganic pollutants have been responsible for soil contamination, such as heavy metals, alkaline or acidic constituents, toxins, oil contaminants and others [146].

It was found that the following categories of polluted samples have been used as the object of metagenomic studies: soils contaminated by oil and its constituents (such as polycyclic aromatic hydrocarbons (PAHs)), fertilizers and other alkaline pollutants, industrial sludge and sediments. Oil production sites [80], soils where oil spills or runoffs have occurred [50,101,111,139], soils near industrial areas [50,87,117,140] and soils treated with fertilizers [136] were particularly analyzed. Since pollutants are rich in toxic compounds, they affect the activity and diversity of microbial communities present in these adulterated soils. Therefore, metagenomic studies have been developed in this type of compromised environment to unravel the gene clusters that encode enzymes involved in the biodegradation of the various pollutants already mentioned. Only the microorganisms that transport machinery capable of resisting and degrading these types of recalcitrant compounds can survive in such environments. The toxic compounds can even act as substrates and enrich some specific microorganisms [147].

Table 2. Metagenomic studies reported in the last decade for soil samples classified in the category of human manipulated resources.

Sample Source	Source Properties		Metagenomic Approach	Vector/ Host	No. Positive Clones/Total Clones	Enzyme	No. Citations *	Reference
	pH	T (°C)						
Daqing oil field, Heilongjiang (China)	-	-	Functional	Plasmid/*E. coli*	3/12,000	β-galactosidase	47	**[80]**
Agricultural corn field, Elora, Ontario (Canada)	7.8	-	Functional	Cosmid/*E. coli*	161/79,060	β-galactosidase	28	**[81]**
Grover Soil Solutions compost facility, California (USA)	-	-	Sequence-based	-	-	Cellulase	139	**[82]**
Sugarcane land field, São Carlos (Brazil)	-	-	Functional	Plasmid/*E. coli*	1/26,900	Cellulase	43	**[83]**
Pollutants- contaminated stream ground surface, Guangxi (China)	9.5	-	Functional	Plasmid/*E. coli*	2/30,000	β-glucosidases	40	**[84]**
Yingtan Red Soil Ecological Station, Jiangxi (South China)	-	-	Functional	Plasmid/*E. coli*	1/3024	Endo-β-1,4-glucanase	54	**[85]**
Straw stook, Jiangxia, Wuhan (China)	-	-	Functional	Plasmid/*E. coli*	1/24,000	Endoglucanase	12	**[86]**
Agricultural fields irrigated with effluents of paper and pulp mill, Uttarakhand (India)	-	-	Functional	Fosmid/*E. coli*	1/7500	β-1,4-endoglucanase	6	**[87]**
Shek Wu Hui Sewage Treatment Works, Hong Kong (China)	-	55.0	Sequence-based	-	-	Endo-β-1,4-glucanases	50	**[88]**
Composting, State University of Paraíba composting cells, Campina Grande (Brazil)	-	55.0–70.0	Functional	Cosmid/*E. coli*	1/10,000	Endoglucanase	16	**[89]**

Sample Source	Source Properties		Metagenomic Approach	Vector/ Host	No. Positive Clones/Total Clones	Enzyme	No. Citations *	Reference
	pH	T (°C)						
Sugarcane land field, São Carlos (Brazil)	-	-	Functional	Plasmid/*E. coli*	1/26,900	Endoglucanase	15	**[90]**
Yonghyeon Nonghyup compost factory, Sacheon (South Korea)	8.9–9.2	40.0–73.0	Functional	Fosmid/*E. coli*	2/12,380	Endo-β-1,4-glucanases	10	**[91]**
Grassland rhizosphere, Teagasc Oak Park research facility, Carlow (Ireland)	-	-	Functional	Fosmid/*E. coli*	1/45,000	Endo-β-1,4-glucanase	14	**[92]**
Municipal compost platform, Bailly (France)	-	50.0	Functional	Fosmid/*E. coli*	74/48,000	Endoglucanases	2	**[93]**
Paddy soil, Liaoning (China)	-	-	Functional	Fosmid/*E. coli*	-/25,000	β-glucanase	26	**[94]**
University of Agricultural Sciences field, Uppsala (Sweden)	6.9	-	Sequence-based	Fosmid/*E. coli*	-/7800	Chitinase	64	**[95]**
University of Agricultural Sciences field, Uppsala (Sweden)	6.9	-	Functional	Fosmid/*E. coli*	1/7800	Chitinase	57	**[96]**
Chitin-treated agricultural field, Experimental farm "Vredepeel" (The Netherlands)	5.7	-	Sequence-based and functional	Fosmid/*E. coli*	5/145,000	Chitinase	17	**[97]**
Chitin- contaminated soil, Mahtani Chitosan, Gujarat (India)	-	-	Sequence-based	-	-	Chitinase	13	**[98]**
Fat-contaminated soil, industrial wastewater treatment plant lagoon, Paraná (Brazil)	-	30.0	Functional	Fosmid/*E. coli*	15/500,000	Chitinase	15	**[99]**
Carnoulès acid- mine drainage, Gard (France)	3.8	15.1	Functional	Plasmid/*E. coli*	28/80,000	Amylases	16	**[100]**

Sample Source	Source Properties		Metagenomic Approach	Vector/ Host	No. Positive Clones/Total Clones	Enzyme	No. Citations *	Reference
	pH	T (°C)						
Oil-contaminated mangrove site, Bertioga, São Paulo (Brazil)	-	-	Functional	Fosmid/*E. coli*	1/12,960	β-*N*-acetylhexosaminidase	8	**[101]**
Yonghyeon Nonghyup compost factory, Sacheon (South Korea)	8.9 – 9.2	40.0 – 73.0	Functional	Fosmid/*E. coli*	5/12,380	Xylanase	30	**[102]**
Sugarcane land field, São Carlos (Brazil)	-	-	Functional	Plasmid/*E. coli*	1/26,900	Endoxylanase	20	**[103]**
Ämmässuo composting plant (Finland)	-	-	Functional	Fosmid/*E. coli* Plasmid/*E. coli*	21/43,000 18/40,000 1/760,000	Xylanases	8	**[104]**
Jiaozuo Ruifeng Paper Co., Ltd., Jiaozuo, Henan province (China)	-	-	Sequence-based	-	-	Xylanase	2	**[105]**
Commercial compost production facility, Western Cape (South Africa)	-	70.0	Functional	Fosmid/*E. coli*	26/20,000	β-xylosidase	0	**[106]**
Yonghyeon Nonghyup compost factory, Sacheon (South Korea)	8.9 – 9.2	40.0 – 73.0	Functional	Fosmid/*E. coli*	2/12,380 5/12,380	Cellulase Xylanases	16	**[107]**
Grover Soil Solutions and the Jepson Prairie Organics compost factories, California (USA)	-	-	Sequence-based	-	-	Endoglucanase Xylanase	76	**[108]**
Grover Soil Solutions compost facility, California (USA)	-	-	Sequence- based	-	-	β-xylosidase/ α-arabinofuranosidase Endoxylanases α-fucosidase	41	**[109]**

Sample Source	Source Properties		Metagenomic Approach	Vector/ Host	No. Positive Clones/Total Clones	Enzyme	No. Citations *	Reference
	pH	T (°C)						
Industrial bagasse collection site, Phu Khieo Bio-Energy, Chaiyapoom (Thailand)	-	49.0–52.0	Functional	Fosmid/*E. coli*	7/100,000	Endoglucanase Endoxylanase	34	**[110]**
Oil- contaminated soils, Ile des Petrels, Terre Adélie (Antarctica)	-	12.0	Functional	BAC/*E. coli*	14/113,742 14/113,742 3/113,742 11/113,742	Lipases/esterases Amylases Proteases Cellulases	32	**[111]**
Experimental field luvisoil, Gembloux (Belgium)	6.5–7.0	-	Functional	Yeast episomal shuttle vector/ *E. coli*	7/500,000 2/500,000	β-glycosidases Glycosyltransferas-es	7	**[112]**
Swine wastewater treatment facility, Tainan (Taiwan)	7.6	30.0	Sequence-based and functional	Plasmid/*E. coli*	13/3818	Esterases	37	**[113]**
Yonghyeon Nonghyup compost factory, Sacheon (South Korea)	8.9–9.2	40.0–73.0	Functional	Fosmid/*E. coli*	19/23,400	Esterase	52	**[114]**
Plants rhizosphere, Gyeongsang (South Korea)	-	-	Functional	Fosmid/*E. coli*	14/142,900	Esterase	42	**[115]**
Wheat field, Shouguang, Shandong (China)	-	-	Functional	Plasmid/*E. coli*	6/50,000	Feruloyl esterase	18	**[116]**
Petroleum hydrocarbons-contaminated region, Ribeirão Preto, São Paulo (Brazil)	-	-	Functional	Fosmid/*E. coli*	30/4224	Esterase	21	**[117]**
Commercial compost production facility, Western Cape (South Africa)	6.1	70.0	Sequence-based and functional	Fosmid/*E. coli*	25/110,592	Feruloyl esterase	12	**[118]**

Sample Source	Source Properties pH	T (°C)	Metagenomic Approach	Vector/ Host	No. Positive Clones/Total Clones	Enzyme	No. Citations *	Reference
Bioenergiezentrum GmbH compost facility, Göttingen (Germany)	-	63.3	Sequence-based and functional	Fosmid/*E. coli* Fosmid/ *T. thermophilus*	1/1920 1/1920	Esterase	50	**[43]**
Yonghyeon Nonghyup compost factory, Sacheon (South Korea)	8.9 – 9.2	40.0 – 73.0	Functional	Fosmid/*E. coli*	18/23,400	Esterase	28	**[119]**
Tembec Paper Mill, Temiscaming, Ontario (Canada)	-	25.0 – 35.0	Functional	Plasmid/*E. coli*	1/53,500	Carboxylesterases	44	**[50]**
Waste water treatment plant anaerobic digester, Evry (France)	-	33.0		Fosmid/*E. coli*	254/47,616			
Oil- contaminated harbour, Priolo Gargallo (Italy)	-	15.0		Plasmid/*E. coli*	4/118,500			
PAH contaminated soil, Michle (Czech Republic)	-	20.0 – 25.0			20/99,900			
PAH contaminated bioremediation site, Sobeslav (Czech Republic)	-	-			5/114,000			
Composting plant, Liemehna (Germany)	-	30.0 – 50.0			6/60,000			
Biodiversity Exploratories Schorfheide-Chorin forest and grassland site (Germany)	-	8.0–8.5	Functional	Plasmid/*E. coli* Fosmid/*E. coli*	28/40,000–341,000 9/4600–300,000	Esterase	15	**[120]**
Bioenergiezentrum GmbH compost facility, Göttingen (Germany)	-	55.0	Functional	Plasmid/*E. coli*	279/675,200	Esterases	6	**[121]**

Sample Source	Source Properties		Metagenomic Approach	Vector/ Host	No. Positive Clones/Total Clones	Enzyme	No. Citations *	Reference
	pH	T (°C)						
Cotton field (China)	-	-	Functional	Plasmid/*E. coli*	1/92,000	Esterase	0	**[122]**
Cornfield, Shangqiu, Henan province (China)	-	-	Functional	Fosmid/*E. coli*	1/30,000	Carboxylesterase	1	**[123]**
Waste contaminated farmland, Nanjing, Jiangsu province (China)	-	-	Functional	Fosmid/*E. coli*	9/60,000	Carboxylesterase	2	**[124]**
Yonghyeon Nonghyup compost factory, Sacheon (South Korea)	8.9 – 9.2	40.0 – 73.0	Functional	Fosmid/*E. coli*	19/23,400	Esterase	0	**[125]**
Yonghyeon Nonghyup compost factory, Sacheon (South Korea)	8.9 – 9.2	40.0 – 73.0	Functional	Fosmid/*E. coli*	19/23,400	Esterases	1	**[126]**
Oil-contaminated Mud Flat, Taean (South Korea)	-	-	Functional	Plasmid/*E. coli*	8/3000	Esterases	2	**[127]**
Teagasc Oak Park research facility organic field trial site, Carlow (Ireland)	-	-	Functional	Fosmid/*E. coli*	1/14,000	Esterase/lipase	9	**[128]**
Fat-contaminated soil, Paraná (Brazil)	-	30.0	Functional	Fosmid/*E. coli*	32/500,000	Lipase	129	**[129]**
Oil-contaminated soil, Qingshan Branch oil field, Anhui (China)	-	-	Functional	Plasmid/*E. coli*	6/20,000	Lipase	17	**[130]**
Oil-contaminated mud flat, Taean (South Korea)	-	-	Functional	Plasmid/*E. coli*	9/3000	Lipase	25	**[131]**
Cotton field (China)	-	-	Functional	Plasmid/*E. coli*	1/92,000	Tannase	28	**[132]**

Sample Source	Source Properties pH	T (°C)	Metagenomic Approach	Vector/ Host	No. Positive Clones/Total Clones	Enzyme	No. Citations *	Reference
Biodiversity Exploratories Hainich-Dün forest and grassland site (Germany)	6 . 5 – 8.0	-	Functional	Plasmid/*E. coli* Fosmid/*E. coli*	28/40,000–341,000 9/4600–300,000	Lipolytic enzymes		**[133]**
Biodiversity Exploratories Schwäbische Alb forest and grassland site (Germany)	6 . 0 – 7.0						54	
Biodiversity Exploratories Schorfheide-Chorin forest and grassland site (Germany)	8 . 0 – 8.5							
Experimental field luvisoil, Gembloux (Belgium)	6 . 5 – 7.0	-	Functional	Yeast episomal shuttle vector/*E. coli*	19/420,000	Lipolytic enzymes	1	**[134]**
Wastewater treatment facility (Japan)	-	-	Functional	Fosmid/*E. coli*	1/100,000	Bilirubin oxidase	8	**[135]**
Contaminated rice field, Nanning, Guangxi (China)	-	-	Sequence-based	Plasmid/*E. coli*	1/32,000	D-Amino acid oxidase	7	**[136]**
Coke plant wastewater treatment facility (Japan)	-	-	Functional	Fosmid/*E. coli*	6/40,000	Oxygenases	19	**[137]**
Pesticide industry effluent treatment plant (India)	-	-	Functional	Plasmid/*E. coli*	2/40,000	Flavin monooxygenases	22	**[138]**
PAH-contaminated wetland, Chambéry (France)	-	-	Sequence-based	-	-	Dioxygenases	34	**[139]**
Polychlorinated biphenyl-contaminated, Fushun, Liaoning (China)	8.3	-	Functional	Plasmid/*E. coli*	1/-	2,4-dichlorophenol hydroxylase	23	**[140]**

Sample Source	Source Properties pH	T (°C)	Metagenomic Approach	Vector/Host	No. Positive Clones/Total Clones	Enzyme	No. Citations *	Reference
Compost-producing company, Toyama (Japan)	-	50.0–80.0	Sequence-based and functional	Plasmid/*E. coli*	1200/2000	Alcohol dehydrogenases	15	**[65]**
Farm, Knockbeg (Ireland)	-	-	Functional	Fosmid/*E. coli*	28/14,400	Phytases	14	**[141]**
Paddy field, Yuseong, Daejeon (South Korea) Plants rhizosphere, Gyeongsang (South Korea)	-	-	Functional	Fosmid/*E. coli*	157/326,200	Wax ester synthase	6	**[142]**
Lop Nur, Xinjiang Uigur Autonomous Region (China)	-	-	Functional	Plasmid/*E. coli*	1/85,000	Trehalose synthase	34	**[143]**
Cornfield, Turpan Basin, Xinjiang Uygur Autonomous Region (China)	-	-	Functional	Plasmid/*E. coli*	1/700,000	Glycosyltransferase	14	**[144]**
Municipal sewage plant, Garmerwolde (The Netherlands)	6.4	-	Functional	Plasmid/*E. coli*	6/1,250,000 8/1,250,000	Sulfatases Phosphotriesterases	146	**[145]**
Urban composting facility, Groningen (The Netherlands)	7.9							
Local agricultural field (The Netherlands)	8.2							

* Obtained from the international database Scopus (21 March 2022).

The acid mine drainage in Carnoulès (France) is considered an interesting reservoir of enzymes capable of degrading polymers and pollutants simultaneously producing antimicrobial agents, since this polyextreme environment, in addition to being highly acidic, presents high concentrations of heavy metals, such as iron and arsenic, as a consequence of mining [100]. Another example is the saline-alkali soil of Lop Nur (China) which is characterized by extreme aridity and is a location that suffers from severe human manipulation. Since it serves as a basis for monitoring and verification of nuclear tests [148], microorganisms present in this soil are certainly subject to a high degree of stress and, for this reason, it may present an interesting microbial diversity and functionality [143].

Other areas exposed to other components such as fats [99,129] or chitin [97,98] have also been the focus of metagenomic studies as they are potential sources of new genes encoding groups of specific enzymes (lipolytic and chitinolytic enzymes, respectively). Activated sludge from different municipal [50,88,135,145] or industrial effluents, such as pesticide [138], swine [113] and paper and pulp [50] industries, are also a rich source of microorganisms producing enzymes capable of degrading protein, lipids and other pollutants. Of the thirty-one studies, two of them performed the analysis of the 16S rRNA gene libraries constructed, one of activated sludge from a swine wastewater treatment facility and the other one from soil contaminated and enriched with chitin. In both studies, it was found that the most predominant phylum was *Proteobacteria* [98,113]. Nevertheless, different samples can unravel other dominant ones, since the composition of the sources (activated sludge from industrial or municipal wastewater treatment plants or treated soils) and the type of treatment accomplished may influence the bacterial diversity.

2.2.2. Agricultural Lands and Grassland

Land use and management have a great influence on the functioning of the soil ecosystem. Microbial diversity and functionality are sensitive to land use considering the important role of soil microorganisms in soil formation processes and nutrient cycling [149].

Hence, several metagenomic studies were implemented in fields designed for agriculture and/or grasslands, many of them subject to the ploughing and cultivation of different crops. Rhizosphere soils of, for example, red pepper plants and strawberry plants represent a complex but interesting ecosystem due to the symbiosis and parasitism interactions that happen between plants and microorganisms in these soil regions [92,115,142]. Cotton [132], wheat [116,141], sugarcane [83,90,103], corn [81,144], straw [86] and paddy [94,142] fields are examples of agricultural environments from which samples were collected, in particular from topsoil, to be analysed through metagenomic approaches. The selection of topsoil samples and not samples at higher depths is due to the presence of a higher soil microbial biomass on the surface since there is larger evidence of litter composition and root turnover rates in this type of land [149]. Another important fact is that decomposition of a variety of lignocellulosic residues occurs in these environments making them attractive to isolate lignocellulose-degrading enzymes relevant to several industrial applications.

Three large-scale research landscapes [133] in Germany (Hainich-Dün, Schorfheide-Chorin and Schwäbische Alb) are defined as exploratory environments. They present different geological and climatic conditions and are characterized by the different intensities of use and management of agricultural fields and grasslands. Therefore, they certainly have great microbial diversity. In these environments, 37 novel lipolytic enzymes, the vast majority belonging to the hormone-sensitive lipase family, were reported (**Table 2**). These exploratory environments, together with the Oak Park research facility in Ireland [128], are valuable sources of promising biocatalysts, notably lipases/esterases, as their land is essentially fertilized with compost and/or manure and is subject to crop rotation. The crop rotation system benefits certain chemical and physical properties of the soil, which is very important and favourable for soil microorganisms [150].

2.2.3. Industrial Composting

Among the different manipulated sources, composting is considered one of the most important bio-reactions for renewable bioenergy on the planet due to the huge variety of microorganisms capable of degrading lignocellulosic biomass. Composting is a sustainable and efficient microbiological process in which the stabilization of the organic matter occurs due to the passage through a thermophilic phase promoted by the proliferation of thermophilic microorganisms [78,151].

The great contribution of composting to the circular economy has led to an increase in the number of composting facilities that are responsible for the production of compost, rich in humic substances, with high agronomic value in organic fertilization of agricultural soils. Different parameters are controlled and adjusted throughout the industrial composting process, such as temperature, pH, humidity, nature of organic materials, particle size and C/N ratio [151].

Some raw materials used in industrial composting are agricultural and agro-industrial residues, including animal faeces [91,102,107,114,118,119], household wastes [108,121], residues from crop harvesting [82,89,91,102,107,109,110,114,119], green wastes [82,104,108,109,118,121], wood chips and sawdust [43,118] and the organic putrescible fraction of municipal solid waste [93,145].

For the metagenomic studies in which composting samples were used, the sample collection essentially occurred during the thermophilic phase of composting that reaches high temperatures (above 45°C) due to microbial metabolic activity. Certain studies refer to microbial diversity and confirm the prevalence of thermophilic microorganisms, namely, *Actinobacteria, Bacteroidetes, Firmicutes* and *Proteobacteria*, producing enzymes able to degrade the complex molecules that compose the lignocellulosic biomass [**43,82,108,118**]. Additionally, some studies also report enrichment with lignocellulosic substrates, namely, switchgrass, steam-exploded spruce, cutter chips, Whatman filter paper, pre-treated *Miscanthus giganteus* and wheat straw. These substrates are firstly incubated with the composting samples at a constant temperature and pH-defined according to the phase at which they were collected to increase the abundance of thermophilic microorganisms and develop a suitable consortium able to degrade lignocellulosic biomass. Over the years, several lignocellulose degrading enzymes have been reported through the analysis of industrial composting metagenomes [**82,93,104,108,109**].

2.3. Unspecified Resources

The designation "unspecified" includes studies from soil resources that do not fall into either the raw resources section or the section of resources subject to human manipulation, since neither the temperature of the sample collected, the sampling site nor the existence of any human activity which may alter the natural properties of the sample is mentioned.

A large part of the unspecified resources are forests that are made up of microorganisms responsible for mediating biogeochemical cycles in terrestrial ecosystems. Additionally, forest soils have a high microbial diversity due to the great accumulation of organic matter [**15**]. Several metagenome studies have been developed, notably in soil samples from the Amazon rainforest [**152**], mangrove forests [**153,154,155,156,157**], peat-swamp forests [**158,159**], beech forests [**160**] and chestnut groves [**161**], to assess microbial diversity, as well as the metabolic capacities of these communities to decompose natural biomass. A metagenome from the *Eucalyptus* sp. forest in Brazil revealed that two of the most abundant phyla are *Actinobacteria* and *Firmicutes*, commonly associated with forest soils [**15,162**].

This category also includes samples from soils with high moisture content, such as alluvial soils from Eulsukdo Island (South Korea) [**142,163**], soils collected from lakes [**64,145**] and even soils with a significant salinity content [**145,164**], mountain soils such as an acid peatland site in Germany [**159**] and arid areas, namely, soils of the Cerrado region of Brazil [**165,166**] that present a high clay content, low pH and high iron levels. All these environments have interesting characteristics that give them unique ecosystems with high enzymatic potential. **Table 3** comprises all the data considered in the category of unspecified resources.

Table 3. Metagenomic studies reported in the last decade for soil samples classified in the category of unspecified resources.

Sample Source	Source Properties pH	T (°C)	Metagenomic Approach	Vector/ Host	No. Positive Clones/ Total Clones	Enzyme	No. Citations *	Reference
Shenzhen Mangrove Reserve, Guangdong (China)	-	-	Functional	Plasmid/*E. coli*	1/30,000	β-glucosidase	64	**[155]**
Amazon forest, Moju, Pará (Brazil)	5.5	-	Functional	Fosmid/*E. coli*	5/97,500	β-glucosidase	18	**[152]**
Eucalyptus sp. forest, UNESP campus, São Paulo (Brazil)	-	-	Sequence-based	-	-	β-glucosidase	16	**[162]**
Mangrove Reserve, Sanya, Hainan (China)	-	-	Functional	Fosmid/*E. coli*	1/100,000	Endo-β-1,4-glucanase	24	**[156]**
Junggar Basin, Xinjiang (China)	-	-	Functional	Plasmid/*E. coli*	1/7200	Cellulase	4	**[167]**
Garden, Zwingenberg (Germany)	-	-	Functional	Plasmid/*E. coli*	6/1,335,000	β-galactosidase	39	**[168]**
RGS 90 A/S environmental company, Copenhagen (Denmark)	-	-	Functional	Fosmid/*E. coli*	7/100,000	α-fucosidase	43	**[169]**
Forest, Daejeon (South Korea)	-	-	Functional	Plasmid/*E. coli*	1/19,626	Lichenase	6	**[170]**
Mangrove Reserve, Sanya, Hainan (China)	-	-	Functional	Fosmid/*E. coli*	1/100,000	β-agarase	18	**[157]**
Sathyamangalam forest, Erode, Tamilnadu (India)	-	-	Functional	Plasmid/*E. coli*	9/2000	Pectin degrading glycosyl hydrolase	13	**[171]**
Coastal saline area, Tianjin (China)	-	-	Sequence-based	Plasmid/*E. coli*	12/-	Xylanase	12	**[164]**
Backyard, Kyushu University, Fukuoka (Japan)	-	-	Functional	Plasmid/*E. coli*	5/150,000 1/150,000	Cellulase Xylanase	24	**[172]**

Sample Source	Source Properties		Metagenomic Approach	Vector/ Host	No. Positive Clones/ Total Clones	Enzyme	No. Citations *	Reference
	pH	T (°C)						
Forest, Yunnan (China)	-	-	Functional	Cosmid/*E. coli*	1/150,000	Cellulase/ Hemicellulase	8	**[173]**
Mangrove forest, Pontal do Paraná (Brazil)	-	-	Functional	Fosmid/*E. coli*	1/2400	Lipase	41	**[153]**
Atlantic forest, Paraná (Brazil)	3.7–4.4	-	Functional	Fosmid/*E. coli*	1/34,560	Lipase	25	**[174]**
Cerrado *sensu stricto* area, Brazilian Institute of Geography and Statistics (Brazil)	4.7	-	Functional	Plasmid/*E. coli*	3/6720	Lipase	3	**[165]**
Mt. Jumbong reed marsh, Gangwon province (South Korea)	-	-	Functional	Plasmid/*E. coli*	46/112,500	Lipase	0	**[175]**
Sirinthon Peat-Swamp Forest, Narathiwat (Thailand)	5.0	-	Functional	Fosmid/*E. coli*	6/15,000	Esterase	34	**[158]**
Alluvium, Eulsukdo Island, Saha-Gu, Busan (South Korea)	-	-	Functional	Fosmid/*E. coli*	50/45,300	Esterases	20	**[163]**
Mount Fanjing, Tongren, Guizhou (China)	-	-	Functional	Fosmid/*E. coli*	1/50,000	Esterase	35	**[176]**
Göttingen beech forest, Georg-August University, Göttingen (Germany)	-		Functional	Plasmid/*E. coli*	3/70,000	Esterase	11	**[160]**
Forest, Groenendaal (Belgium)	-	-	Functional	Plasmid/*E. coli*	3/70,000	Esterases	40	**[177]**
Chestnut grove surface, Yunnan (China)	-	-	Functional	Cosmid/*E. coli*	2/100,000	Esterases	7	**[161]**
Tree rhizosphere, Korea Expressway Corporation Arboretum, Jeonju (South Korea)	-	-	Functional	Cosmid/*E. coli*	1/7968	Carboxylesterase	10	**[178]**

Sample Source	Source Properties		Metagenomic Approach	Vector/ Host	No. Positive Clones/ Total Clones	Enzyme	No. Citations *	Reference
	pH	T (°C)						
Composting, Cheongyang (South Korea)	-	-	Functional	Fosmid/*E. coli*	14/13,000	Carboxylesterase	4	**[179]**
Shenzhen Mangrove Reserve, Guangdong (China)	-	-	Functional	Plasmid/*E. coli*	1/8000	Laccase	69	**[154]**
Lake Biwa, Otsu, Shiga (Japan)	-	-	Sequence-based	Plasmid/*E. coli*	240/2800	Alcohol dehydrogenases	15	**[65]**
Masukata Park, Uozu, Toyama (Japan)								
Cerrado *sensu stricto* area, Brazilian Institute of Geography and Statistics (Brazil)	4.7	-	Functional	Plasmid/*E. coli* Fosmid/*E. coli*	3/150,000 3/65,000	Dioxygenase	12	**[166]**
Forest, Ehime Research Institute of Agriculture, Forestry, and Fisheries, Matsuyama (Japan)	-	-	Functional	Cosmid/*E. coli*	29/208,000	Oxygenases	35	**[180]**
Schlöppnerbrunnen peatland site, Fichtelgebirge Mountains, Northeastern Bavaria (Germany)	4.0	-	Sequence-based	-	-	Phytase	10	**[159]**
Forest, Groenendaal (Belgium)	-	-	Functional	Plasmid/*E. coli*	1/70,000	Serine protease	41	**[181]**
Saline flats surface, Paesens-Moddergat (The Netherlands)	8.0	-	Functional	Plasmid/*E. coli*	6/1,250,000 8/1,250,000	Sulfatases Phosphotriesterases	146	**[145]**
Lauwersmeer lakeshore (The Netherlands)	8.5							
Alluvium, Eulsukdo Island, Saha-Gu, Busan (South Korea)	-	-	Functional	Fosmid/*E. coli*	158/326,200	Wax ester synthase	6	**[142]**
Gwangneung forest, Korea National Arboretum, Gyeonggi (South Korea)								

* Obtained from the international database Scopus (21 March 2022).

3. WATER

Water is essential in our life. Although frequently perceived as just an ordinary substance, it plays a vital role on the Earth since life as we know it could not exist without water. As a natural resource, water is fully distributed over the planet. The oceans contain the greatest fraction of water (96.5%), while continental water is composed of freshwater (2.5%) and saline groundwater (1%) [182]. Water is present in its three physical forms (liquid, solid and vapour) depending on the climate conditions. Aqueous environments (saline and freshwater) are considered important sources of microorganisms capable of inhabiting and resisting in different physical forms, such as icebergs (important spots for marine life), seas and oceans, thermal springs, glaciers, lakes and ponds. The microbial diversity and functionality found in water are strongly dependent on the environmental conditions (e.g., salinity, temperature, physical form, nutrients, pH or depth). It is known that a large part of the aquatic microbial resources remains unexplored mainly due to access limitations. Nevertheless, the use of metagenomic tools has significantly aided the discovery of novel biocatalysts from aquatic metagenomes [183]. In this review, the water samples used in the metagenomic studies were divided according to their origin (Figure 4a) and main characteristics (Figure 4b). In the last decade, a total of 26 metagenomic studies with water samples to find promising enzymes were reported (Figure 4a).

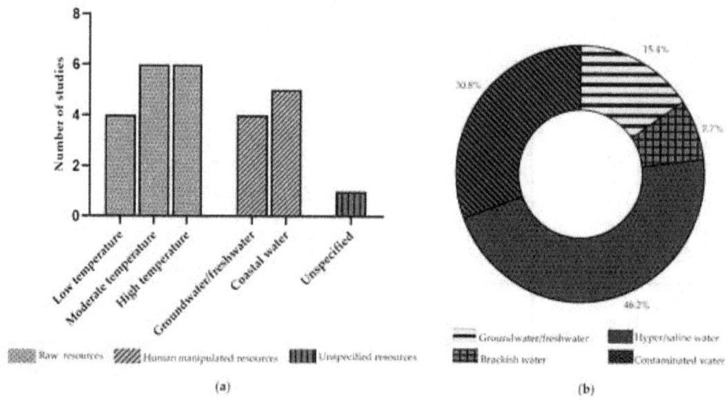

Figure 4. Classification attributed to water sampling sources explored to find promising biocatalysts through metagenomic approaches in the last decade. (a) Representation of the number of studies according to the categories defined (raw resources, human manipulated resources and unspecified resources). (b) Assignment of water type to each sample collected in different environments (groundwater/freshwater, brackish water, hyper/saline water and contaminated water).

As shown in Figure 4b, in the last decade the most studied aqueous samples were hyper/saline water (46.2%), followed by contaminated water (30.8%) and groundwater/freshwater (15.4%) and, finally, brackish water (7.7%). The first two types of water have been the most explored probably due to their extreme characteristics.

3.1. Raw Resources

As mentioned before, the microbial communities of natural water resources are strongly associated with environmental factors, one of which is temperature [184]. Indeed, in addition to the most common aquatic environments on the planet, such as seawater or lakes, extreme temperature environments also allow exploring a diversity of enzymes capable of catalysing reactions at reasonable or extreme conditions [185]. Therefore, raw water resources were classified in the same way and according to the same temperature ranges as soil resources: low temperature (location temperature below 20°C); moderate temperature (location temperature between 20 and 45°C); and high temperature (location temperature above 45°C) [17].

As shown in **Figure 4**a, the number of studies completed in the three types of environments is not very different. Nevertheless, samples were preferably collected in high-temperature (6 studies) and moderate temperature environments (6 studies) instead of low-temperature environments (4 studies). **Table 4** compiles all the data considered in the category of raw resources.

3.1.1. Low-Temperature Environments

When thinking about low-temperature aquatic environments, thoughts are immediately associated with marine environments and/or high depth environments that are intrinsically related to high pressures. The decrease in temperature and the increase in depth cause a decreased diffusion of nutrients and energy and decreased abundance of prokaryotic cells, respectively. These extreme conditions require that microorganisms found in these zones have their adapted metabolism, for example, low concentrations of nutrients, which make them fascinating targets for the bioprospection of novel microbial capabilities and, accordingly, promising enzymes [200,201].

Table 4. Metagenomic studies reported in the last decade for water samples classified in the category of raw resources.

Sample Source	Source Properties pH	Source Properties T (°C)	Metagenomic Approach	Vector/ Host	No. Positive Clones/ Total Clones	Enzyme	No. Citations *	Reference
Surface seawater, South China Sea (Pacific Ocean)	-	-	Functional	BAC/*E. coli*	6/-	β-glucosidase	95	**[186]**
Baltic Sea, Kołobrzeg, Poland (Atlantic Ocean)	-	0.8	Functional	Plasmid/*E. coli*	1/1100	β-glucosidase	47	**[187]**
Lake Poraquê, Amazon (Brazil)	7.1	29.8	Sequence-based	-	-	β-glucosidase	6	**[188]**
Atlantis II brine pool, Red Sea (Indic Ocean)	5.3	68.0	Functional	Fosmid/*E. coli*	5/10,656	Esterase	41	**[189]**
Surface seawater, South China Sea (Pacific Ocean)	-	-	Sequence-based	-	-	Esterase	25	**[190]**
Lobios hot spring, Ourense, Galicia (Spain)	>8.2	>76.0	Sequence-based and functional	Fosmid/*E. coli*	6/11,600	Esterase	36	**[191]**
Urania, deep hypersaline anoxic basin interface, Mediterranean Sea (Atlantic Ocean)	-	14.0	Functional	Plasmid/*E. coli*	41/90,800	Carboxylesterases	44	**[50]**
Surface seawater, South China Sea (Pacific Ocean)	-	-	Sequence-based	BAC/*E. coli*	-/20,000	Laccase	93	**[192]**
Atlantis II brine pool, Red Sea (Indic Ocean)	5.4	68.1	Sequence-based	-	-	Mercuric reductase	30	**[193]**
Atlantis II brine pool, Red Sea (Indic Ocean)	5.7	68.4	Sequence-based	-	-	Thioredoxin reductase	4	**[194]**
Seawater, Gulf of Mexico (Atlantic Ocean) Seawater, long-Term Ecosystem Observatory site, New Jersey, USA (Atlantic Ocean)	-	-	Sequence-based	BAC/*E. coli*	-	Ribulose 1,5-bisphosphate carboxylases/ oxygenases	14	**[195]**
Surface seawater, South China Sea (Pacific Ocean)	8.2	15.0	Sequence-based	Plasmid/*E. coli*	-/50,000	Fumarase	29	**[196]**

Sample Source	Source Properties		Metagenomic Approach	Vector/Host	No. Positive Clones/Total Clones	Enzyme	No. Citations *	Reference
	pH	T (°C)						
Atlantis II brine pool, Red Sea (Indic Ocean)	5.5	68.2	Sequence-based	-	-	Nitrilase	15	[197]
Atlantis II brine pool, Red Sea (Indic Ocean)	5.6	68.3	Sequence-based	-	-	3'-aminoglycoside phosphotransferase Beta-lactamase	11	[198]
Brackish water basin, Caspian Sea (Iran)	-	-	Sequence-based	-	-	L-asparaginases	0	[199]

* Obtained from the international database Scopus (21 March 2022).

Low-temperature water samples were collected essentially in different geographical locations in the Atlantic Ocean. Surface seawater was collected at opposite ocean regions, particularly from an ecosystem observation site in New Jersey, USA [195], and the brackish Baltic Sea in Poland [187]. The latter recorded the lowest water temperature (0.8 °C), probably due to the fact that the Baltic sea was formed as a consequence of the last glaciation that occurred 10,000–15,000 years ago and, therefore, has undergone remarkable changes in its physicochemical characteristics [202]. In these different geographical locations, different groups of enzymes were reported, namely, a ribulose 1,5-bisphosphate carboxylase/oxygenase and a cytosolic β-glucosidase with a wide range of catalytic activities. Additionally, the chemocline of the Urania basin in the Mediterranean Sea was the subject of a metagenomic study to find interesting carboxylesterases, since it is a deep-sea anoxic hypersaline basin [50]. The extreme factors that characterize the Urania basin (hypersalinity, low temperature and anoxia) together with the typical features of chemoclines make this habitat accommodate a highly diverse microbial community with pronounced microbial activities, such as CO_2 fixation and exoenzyme activities [203].

3.1.2. Moderate Temperature Environments

In moderate temperature environments, samples of different types of water are included: groundwater/freshwater, hyper/saline water and brackish water.

Hyper/saline water samples from the surface of the South China Sea were the ones that most contributed to the metagenomic studies in this category. The seasonal average water temperature that falls into the range defined for this group (20–45°C) and the unique environmental properties of the South China Sea potentially contribute to the diversity, novelty and uniqueness of genes encoding for valuable enzymes. Effectively, different groups of enzymes have already been explored in the South China Sea, including β-glucosidases, laccases and esterases [75,186,190,192].

As the lakes of the Amazon region remain unexplored, the freshwater metagenome of Lake Poraquê was functionally analysed. Being the largest hydrographic basin on Earth, the great genetic and metabolic diversity of microorganisms present in this important region may result in the discovery of new enzymes of biotechnological interest, such as enzymes involved in the degradation of plant cells walls [188].

The brackish samples of the Caspian Sea were also accessed since this environment presents a salinity and ionic concentration very similar to the human serum. In this way, there is a high probability that the secretory enzymes (more specifically, L-asparaginases) found in the Caspian Sea microbiome exhibit greater stability in the physiologic conditions of the human serum which can render them an interesting therapeutic applicability [199].

3.1.3. High-Temperature Environments

High-temperature habitats (>45 °C) are inhabited by heat-resistant microorganisms and some of these environments also combine other extreme conditions, for example, alkalinity, acidity, salinity, pressure and heavy metals [17]. Samples have been studied mainly from hyper/saline water environments and groundwater/freshwater environments.

Among the different aquatic environments, the hypersaline anoxic deep-sea basins in the Red Sea, namely, the Atlantis II deep brine pool, have received increased attention. These are characterized by a high temperature, extreme salinity, acidic pH, extremely low levels of light and oxygen and high concentrations of heavy metals. In this way, this extreme environment is expected to be an attractive location for the search for biocatalysts that can function under harsh conditions, not just those that characterize the Red Sea. An esterase capable of acting in the presence of heavy metals; a mercuric reductase extremely relevant in the detoxification system for mercuric/organomercurial species; a nitrilase useful in bioremediation processes, fine chemicals and pharmaceutics; a 3′-aminoglycoside phosphotransferase and a beta-lactamase with potential application as thermophilic selection markers; and a thioredoxin reductase important in the maintenance of the redox balance and counteracting oxidative stress inside cells are some examples of biocatalysts found in this extreme environment [189,193,194,197,198].

Another source commonly known for its high temperatures are the hot springs. A metagenomic library constructed from a groundwater sample from the Lobios hot spring in Spain was evaluated, by sequence-based and functional metagenomics approaches, given its high temperatures and alkaline pH values. Moreover, the microbial biodiversity and metagenomic potential of this source have not been sufficiently explored. This study reported a novel esterase belonging to family VIII and showed that the dominant prokaryotic phyla in this location, as in other hot springs on the planet, were *Deinococcus-Thermus*, *Proteobacteria*, *Firmicutes*, *Acidobacteria*, *Aquificae* and *Chloroflexi*. Additionally, the dominant archaeal phylum was *Thaumarchaeota* [191].

3.2. Human Manipulated Resources

Anthropogenic activities interfere negatively in many ways with the natural water cycle. Several water bodies, such as oceans, rivers and groundwater have been contaminated not only by natural events but particularly due to human interventions [204]. According to **Figure 4**a, 9 metagenomic studies were performed in contaminated water samples. This section was divided into groundwater/freshwater (4 studies) and coastal water (5 studies). **Table 5** compiles all the data considered in the category of human manipulated resources.

3.2.1. Groundwater/Freshwater

The main anthropogenic sources of water contamination are refineries, mines, factories and wastewater treatment plants, among others [204].

Over the past 10 years, different metagenomic studies have been executed in contaminated groundwater and freshwater sources for the acquisition of novel enzymes. Some examples of contaminated sources are the formation of water in a coalbed in Jharia coalfield (India) which is defined as an extreme environment [206], Eryuan Niujie hot spring in Yunnan (China), which has a high content of fats due to the wastes resulting from the livestock slaughter that occurs in the vicinity of the hot spring [207], and groundwater from an area in the Czech Republic that has been incessantly contaminated with various products from an oil industry for over 50 to 70 years [208].

The activities associated with each of the sites employ a certain continuous selective pressure on the microorganisms that live in these environments to develop enzymes capable of acting in the production of biodiesel, degradation of organophosphorus compounds and halogenated pollutants, respectively [206,207,208].

3.2.2. Coastal Water

The category of coastal waters is essentially composed of water samples from oil-contaminated harbours due to the numerous ships circulating in the waters of these areas each year and also due to the unintentional spills of hydrocarbons that may occur during the loading and unloading of petroleum-derived substances. The main pollutant responsible for the contamination of these sites is oil and the studies have mainly been focused on the Mediterranean Sea in Italy and the Barents Sea in Russia, allowing the finding of robust carboxylesterases [50].

Table 5. Metagenomic studies reported in the last decade for water samples classified in the category of human manipulated resources.

Sample Source	Source Properties pH	T (°C)	Metagenomic Approach	Vector/ Host	No. Positive Clones/ Total Clones	Enzyme	No. Citations *	Reference
Contaminated river, East China University of Science and Technology, Shanghai (China)	-	-	Functional	Fosmid/*E. coli*	6/20,400	Esterase	19	**[205]**
Coalbed water formation, Jharia coalfield, Jharkhand (India)	7.2	-	Functional	Fosmid/*E. coli*	1/208	Phosphodiesterase	5	**[206]**
Messina harbour, Mediterranean Sea, Italy (Atlantic Ocean)	-	15.0	Functional	Plasmid/*E. coli*	18/24,000	Carboxylesterases		**[50]**
Messina harbour (Int II), Mediterranean Sea, Italy (Atlantic Ocean)		15.0		Fosmid/*E. coli*	208/5760			
Milazzo, Mediterranean Sea, Italy (Atlantic Ocean)		18.0		Plasmid/*E. coli*	8/20,000		44	
Oil-contaminated coastal water, Kolguev Island, Barents Sea, Russia (Arctic Ocean)		3.0		Plasmid/*E. coli*	34/142,000			
Oil-contaminated Murmansk Port, Barents Sea, Russia (Arctic Ocean)		5.0		Plasmid/*E. coli*	43/108,000			
Eryuan Niujie hot spring, Yunnan (China)	7.0	58.0	Functional	BAC/*E. coli*	10/68,352	Lipase	11	**[207]**
Oil industry products contaminated groudwater (Czech Republic)	6.9	8.5	Sequence-based	-	-	Haloalkane dehalogenases	3	**[208]**

* Obtained from the international database Scopus (21 March 2022).

The importance of the metagenomic analysis of water samples of these types of sources is justified by the abundance of microbial species capable of degrading hydrocarbons which can be potentially applied in the bioremediation of ecosystems. To this end, additional enrichments with crude oil or specific hydrocarbons (e.g., pyrene, naphthalene and phenanthrene) are carried out to mimic the place from which they are isolated.

3.3. Unspecified Resource

The tropical underground water of the Yucatán Aquifer in Mexico was the only resource considered unspecified. Nevertheless, it is a very interesting resource of freshwater since it consists of very permeable and porous limestone that allows the infiltration of water into the deepest layers of the soil. Additionally, the Yucatán Aquifer presents cracks or interconnected spaces that allow water from distant zones and sources to move freely, carrying a collection of microorganisms from different places. Thus, although it should be a natural selection process that favours some microorganisms by eliminating others, it may also present a high microbial diversity from diverse origins [**209**].

In this way, this aquifer can represent a potential and interesting source for the acquisition of a catalogue of enzymes suitable for the degradation of natural polymers, including proteins.

4. SOIL VERSUS WATER

As previously demonstrated, soil and water have both been reported as promising sources of biocatalysts. Nevertheless, the number of metagenomic studies focused on soil have been considerably superior, as illustrated in **Figure 5**. Between 2010 and 2021, the period considered in this review, the soil was always a preferred source to search for new biocatalysts, representing more than 69% of the studies each year. Additionally, it is also important to highlight that the number of metagenomic studies for the identification of enzymes has been decreasing for both natural resources. In fact, since 2017, a significant reduction in the number of studies was observed but in 2021 this number went up again (**Figure 5**), probably as a consequence of the pandemic situation which, in general, resulted in a sharp increase in publication volume [**210**]. In the last years, we have witnessed a change in the focus of metagenomic studies from functionality to taxonomy and gene annotation. The development of more accurate software allows access to the complex data generated by the sequence-based metagenomics approach. Consequently, it is possible to explore the biosynthetic gene cluster diversity and to understand the significant role of the microorganisms in the global biogeochemical cycles.

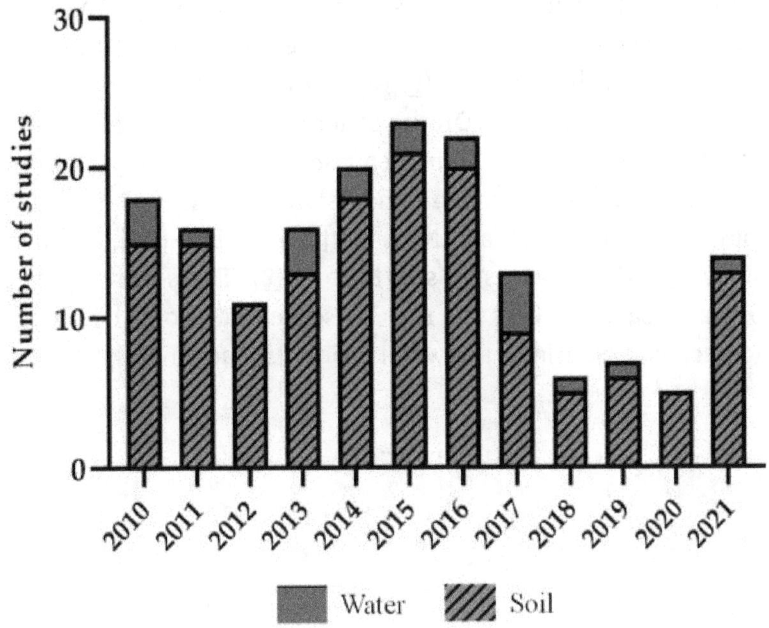

Figure 5. Number of metagenomic studies performed between 2010 and 2021 using water and soil samples to find promising enzymes.

The global distribution of the several metagenomic studies performed with soil and water samples is illustrated in **Figure 6**a. This figure shows that soil and/or water samples from five continents (America, Europe, Asia, Africa and Antarctica) and four oceans (Atlantic, Pacific, Arctic and Indic) were already explored through metagenomic studies aiming at the discovery of new enzymes. Regarding the soil samples, Asia was the continent with the higher number of studies (~52%), followed by Europe (~24%) and America (~13%) (**Figure 6**b). The samples were mostly collected from human manipulated environments, which can be justified by the higher demographic density and industrial activity attributed to these continents. The number of studies performed with soil samples from Africa or Antarctica was very low (<2%). When considering the studies performed in marine sediments, a higher percentage was found for the Pacific Ocean (~5%) (**Figure 6**b).

On the other hand, the marine water samples studied by metagenomic approaches were mostly collected in the Atlantic Ocean (~27%), Indic Ocean (~19%) and Pacific Ocean (~15%) (**Figure 6**c). In terms of continental water, samples were obtained from Asia (~15%), Europe (~8%) and America (~8%).

The main groups of enzymes identified in the metagenomic studies include lipolytic enzymes, glycosyl hydrolases, oxidoreductases, phosphatases/phytases and proteases. The global distribution of these enzymes according to the type of sample used in the metagenomic study is represented in **Figure 7**a. The soil samples from Asia, Europe and America generally resulted in the identification of lipolytic enzymes and glycosyl hydrolases (**Figure 7**b). Oxidoreductases were mostly obtained from Asian soil samples. On the other hand, some phosphatases/phytases were identified in samples from European soils. Additionally, the metagenomic studies performed with both European and Asian soil samples showed the presence of other types of enzymes such as RNase [69], rhodanase [70], trehalose synthase [143], sulfatases [145] or wax ester synthase [142]. The soil samples collected in Africa and Antarctica, as well as the marine sediments from the Pacific, Antarctic and Atlantic Oceans, essentially allowed the identification of lipolytic enzymes.

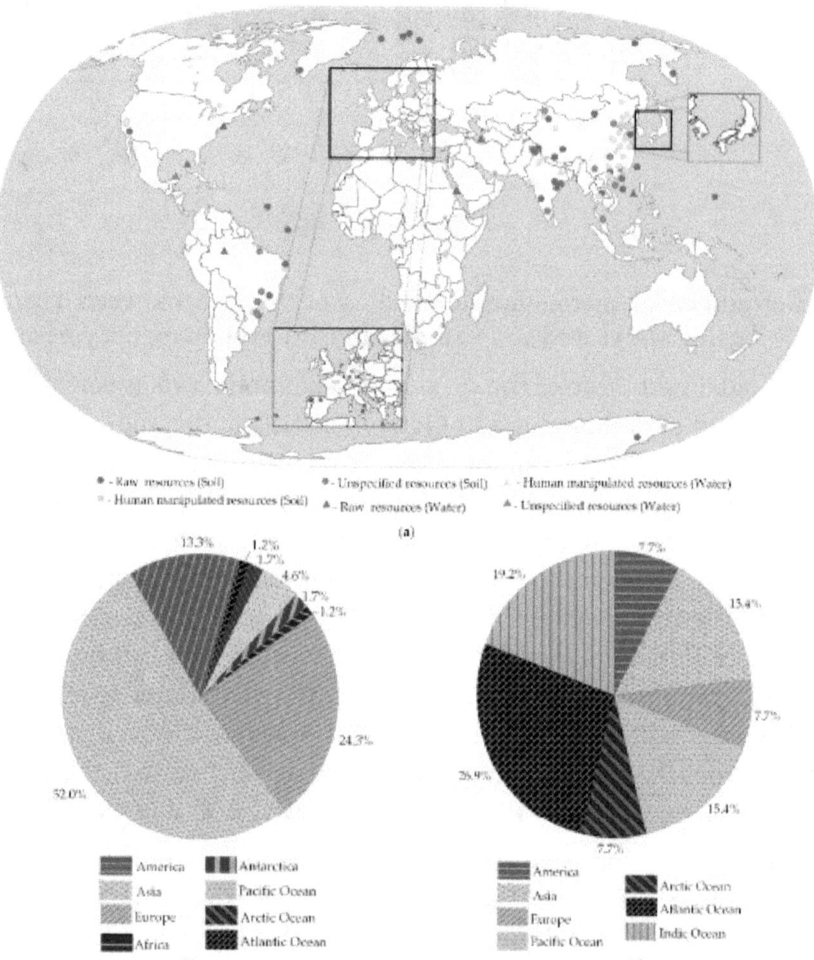

Figure 6. Distribution of the several metagenomic studies performed with soil and water samples in terms of: (a) global representation; (b) percentage of soil and sediment samples collected in each continent and ocean; (c) percentage of water samples collected in each continent and ocean.

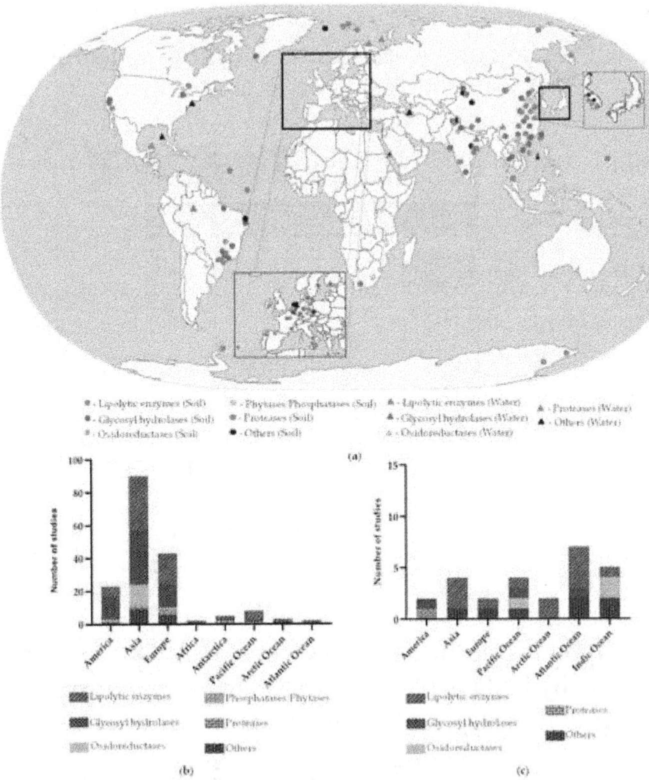

Figure 7. Distribution of the main groups of enzymes found in the metagenomic studies performed with soil and water samples in terms of: (a) global representation; (b) main groups of enzymes (%) identified in soil and sediment samples collected in each continent and ocean; (c) main group of enzymes (%) identified in water samples collected in each continent and ocean.

For the water samples, the lipolytic enzymes were also the predominant group, being identified in almost all the studied locations (except for samples from America) (Figure 7c). Nevertheless, proteases were only identified in samples collected in America. Glycosyl hydrolases were found in continental water from America and marine water from the Pacific and Atlantic Oceans. On the other hand, oxidoreductases were found in marine water samples from the Pacific and Indic Oceans. Similar to the soil, other enzymatic activities were identified in the water samples (except for samples from America and the Arctic Ocean) such as nitrilase [197], beta-lactamase [198], L-asparaginases [199], fumarase [196] or haloalkane dehalogenases [208]. However, no phosphatases/phytases were identified in the metagenomic studies of water.

Overall, it was discovered that lipolytic enzymes, glycosyl hydrolases, oxidoreductases and proteases could be found in both natural resources (Figure 8). Nevertheless, phosphatases and phytases were only identified in soil samples.

Furthermore, the enzymatic activities included in the "others" group were significantly different for soil and water samples (**Table 1**, **Table 2**, **Table 3**, **Table 4** and **Table 5**).

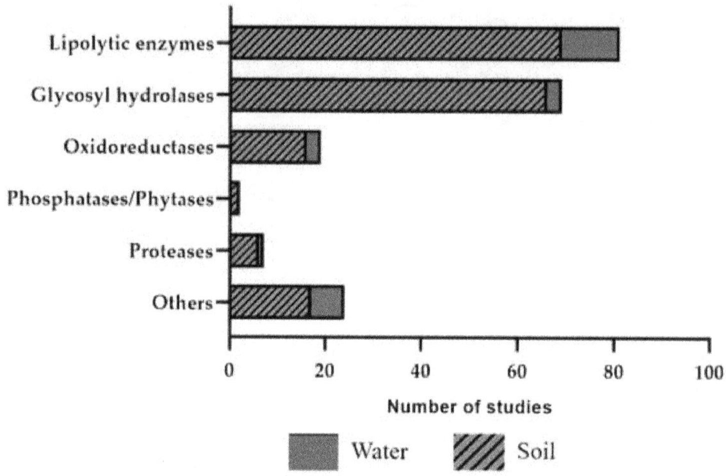

Figure 8. Number of metagenomic studies reporting the presence of lipolytic enzymes, glycosyl hydrolases, oxidoreductases, phosphatases/phytases, proteases and other types of enzymes in soil and water samples.

The five groups of enzymes mostly identified in the metagenomic studies (namely, lipolytic enzymes, glycosyl hydrolases, oxidoreductases, phosphatases/ phytases and proteases) correspond to catalytic activities with a great interest in the industry. Lipolytic enzymes, such as lipases and esterases, can catalyse either the hydrolysis or the synthesis of ester bonds. They are considered robust enzymes, which can resist the harsh conditions of some industrial processes such as high temperature and pH, and the presence of organic solvents. These enzymes have been applied in the food, cosmetic, pharmaceutical, detergent, laundry and oleo-chemical industries [211]. Lipases and esterases showing resistance to high pH values (>8.5) were mostly found in soil samples, namely, petroleum hydrocarbons-contaminated soil (esterase, optimal pH = 9.0 [117]), Brazilian cerrado soil (lipase, optimal pH = 9.0–9.5 [165]), compost units (esterase, optimal pH = 10.0 [114]), fat-contaminated soil (lipase, optimal pH = 10.0 [129]) or cold desert soil (esterase, optimal pH = 11.0 [39]). On the other hand, lipases and esterases with tolerance to high temperatures (>65 °C) were identified in hot spring soils (esterase, optimal temperature 80°C [43]; lipase, optimal temperature 65°C [61]) and compost units (esterase, optimal temperature 75°C [43,121]. Additionally, in water samples from the Red Sea (esterase, optimal temperature 65°C [189]) and the South China Sea (esterase, optimal temperature 65°C [190]), thermo and halotolerant esterases were found. Furthermore, other interesting enzymes able to catalyse the hydrolysis of ester compounds were obtained from compost (cutinase [59]), wheat field soil (feruloyl esterase [116]) or contaminated soil from wood-processing industries (carboxylesterases [50]).

The group of glycosyl hydrolases includes several enzymes which promote the breakdown of carbohydrates into simple sugars through the hydrolysis of specific glycosidic bonds. This kind of enzyme has been widely explored in biorefining processes for the hydrolysis of different polysaccharides from plant origin and the subsequent production of important added-value products like biofuels. Furthermore, glycosyl hydrolases are also interesting enzymes for the cosmetic (e.g., toothpaste additives), food (e.g., dairy, baking or brewing processes) and feed industries [212]. Three metagenomic studies performed with water samples from marine water and freshwater led to the identification of β-glucosidases [186,187,188]—important enzymes in the hydrolysis of short-chain oligosaccharides into glucose. As expected, enzymes belonging to the family of cellulases (e.g., endoglucanases and β-glucosidases), xylanases (e.g., endoxylanases and β-xylosidase) and amylases were widely found in soil samples. Chitinases able to catalyse the hydrolysis of chitin commonly present in the exoskeleton of some animals were also identified in metagenomic studies with soil. The natural presence of plant and animal detritus in soils justifies the existence of microorganisms with the catalytic ability for the conversion of cellulose, xylan, starch or chitin.

The family of proteases include the enzymes which promote proteolysis, i.e., the breakdown of proteins into smaller peptides or amino acids. Proteases are well-established enzymes in the food and feed industry, being used as stabilizers, meat tenderizers and additives for better digestion, or even for the improvement of brewing and baking processes. They can also be applied in leather and detergent industries, as well as in photography or therapeutic uses [213]. According to the metagenomic studies analysed here, proteolytic activity seems to be more common in soils than in water. Alkaline proteases were found in desert environments [66], forest soil [177], hot springs sediments [67] and oil-contaminated soil [111].

Oxidoreductases are a vast number of enzymes able to catalyse the transfer of electrons from one molecule (reductant) to another (oxidant). This kind of enzymes is regarded as important biocatalysts in the textile (dye bleaching), pulp and paper (bleaching), food and beverage (stabilizer) and pharmaceutical industries (synthesis of bioactive compounds), as well as in biorefining processes for the bioconversion of lignocellulose [214]. They can also have an important role in the bioremediation of industrial and municipal wastewater contaminated with organic compounds such as textile dyes, pharmaceuticals, hormones or personal care products [215]. Diverse oxidoreductases were reported in the metagenomic studies analysed here, including laccases [154,192], oxygenases [137], alcohol and aldehyde dehydrogenases [63,65], D-amino acid oxidases [136] and bilirubin oxidases [135]. Nevertheless, the majority of these enzymes were identified in soil samples.

Phosphatases are enzymes involved in the cleavage of ester bonds and the release of phosphate groups. Phytases are a type of phosphatases that act in the hydrolysis of phytic acid, a specific organic form of phosphorous. These enzymes are of great interest in human and animal nutrition by reducing the phytate content of food products and contributing to more efficient digestion and absorption of phosphorus [216]. Phytases are common and very useful biocatalysts in soils since they are considered as primary agents for dephosphorylation [217]. In fact, phytases were only found in the metagenomic studies performed with soil samples [141,159].

5. CONCLUSIONS

Metagenomics proved to be an efficient technique to explore natural resources such as soil and water. Following either a sequence- or a function-based approach, several enzymes with distinct catalytic activities were identified. In the last decade, the number of metagenomic studies performed with soil samples was considerably higher than compared with water samples. This fact can be related to some limitations reported for DNA extraction. When working with water, high volumes of samples need to be collected and filtered to allow the extraction of an appropriate quantity of DNA considered representative of the microbial communities present in the environment. Furthermore, water sources are immensely vast and, in most cases, not fully attainable without specific equipment and protection (e.g., oceans, seas and groundwater). On the other hand, as terrestrial living organisms, we found the soil as a much more accessible and near environment. Nevertheless, it is clear that the biotechnological potential of soil remains barely explored. The global distribution of the metagenomic studies carried out for enzyme discovery in the last decade noticeably shows that a vast number of promising environments are still waiting for their potential to be unravelled. The lack of economic and human resources in the African continent may justify the reporting of only two metagenomic studies, namely with compost samples from the same composting facility. However, other continents like America and Europe could also be explored more. The same conclusions are obviously taken for water samples.

The catalytic activities found in the metagenomic studies were mostly distributed in five representative groups. The origin and composition of the samples are generally connected with the biocatalytic potential of the microbial communities. Parameters like the pH and temperature of the sampling location are frequently related to the optimal conditions reported for the biocatalysts. Additionally, in most cases, it was possible to establish an association between the type of enzymes and the main constituents of the sample. Lipolytic enzymes were often found in environments containing petroleum hydrocarbons, oils or fat. Glycosyl hydrolases were mostly identified in soils since these environments are naturally composed of plant and animal debris. Thus, the microbial communities take advantage of producing these types of enzymes to convert the complex polysaccharides of plant and animal origin. Similarly, phytases are enzymes with a catalytic action that properly fits the soil environments where phosphorus assumes an important role. The presence of oxidoreductases can also be more expected in soils due to the key role they have in the conversion of lignin from plants. For proteases, no clear association was established for the metagenomic data reported.

In the last years, more accurate and advanced software to access and analyse the complex sequencing data generated in metagenomic studies have been developed. This fact greatly contributed to a better understanding of the interactions established between microbial communities and the environment by accessing important information about biosynthetic routes and taxonomic annotation. As a consequence, the motivation of the metagenomic studies slightly diverged. The studies initially focused on microbial functionality have been replaced by studies exclusively directed to whole-genome sequencing and metabolite prediction. It is expected that metagenomics consolidates in the future its status as a key technique to unravel the biotechnological potential of microbial resources, thus contributing to the search for novel and effective bioactive compounds according to the market trends.

6. CONCISE SUMMARY

Natural resources are considered a promising source of microorganisms responsible for producing biocatalysts with great relevance in several industrial areas. However, a significant fraction of the environmental microorganisms remains unknown or unexploited due to the limitations associated with their cultivation in the laboratory through classical techniques. Metagenomics has emerged as an innovative and strategic approach to explore these unculturable microorganisms through the analysis of DNA extracted from environmental samples. In this review, a detailed discussion is presented on the application of metagenomics to unravel the biotechnological potential of natural resources for the discovery of promising biocatalysts. An extensive bibliographic survey was carried out between 2010 and 2021, covering diverse metagenomic studies using soil and/or water samples from different types and locations. The review comprises, for the first time, an overview of the worldwide metagenomic studies performed in soil and water and provides a complete and global vision of the enzyme diversity associated with each specific environment.

REFERENCES

1. Arocena, J.M.; Driscoll, K.G. Natural resources of the world. In *Knowledge for Sustainable Development: An Insight into the Encyclopaedia of Life Support Systems*; UNESCO/EOLSS: Zurich, Switzerland, 2002; pp. 261–290. [Google Scholar]
2. Kattumuri, R. Sustaining natural resources in a changing environment: Evidence, policy and impact. *Contemp. Soc. Sci.* **2018**, *13*, 1–16. [Google Scholar] [CrossRef][Green Version]
3. Lampert, A. Over-exploitation of natural resources is followed by inevitable declines in economic growth and discount rate. *Nat. Commun.* **2019**, *10*, 1419. [Google Scholar] [CrossRef] [PubMed]
4. Kirsch, S. Running out? Rethinking resource depletion. *Extr. Ind. Soc.* **2020**, *7*, 838–840. [Google Scholar] [CrossRef] [PubMed]
5. Chettri, D.; Verma, A.K.; Verma, A.K. Innovations in CAZyme gene diversity and its modification for biorefinery applications. *Biotechnol. Rep.* **2020**, *28*, e00525. [Google Scholar] [CrossRef] [PubMed]
6. Gudiña, E.J.; Amorim, C.; Braga, A.; Costa, Â.; Rodrigues, J.L.; Silvério, S.; Rodrigues, L.R. Biotech green approaches to unravel the potential of residues into valuable products. In *Sustainable Green Chemical Processes and Their Allied Applications*; Inamuddin, M., Asiri, A.M., Eds.; Springer: Cham, Switzerland, 2020; pp. 97–150. [Google Scholar]
7. Berini, F.; Casciello, C.; Marcone, G.L.; Marinelli, F. Metagenomics: Novel enzymes from non-culturable microbes. *FEMS Microbiol. Lett.* **2017**, *364*, fnx211. [Google Scholar] [CrossRef] [PubMed]
8. Costa, Â.M.A.; Santos, A.O.; Sousa, J.; Rodrigues, J.L.; Gudiña, E.J.; Silvério, S.C.; Rodrigues, L.R. Improved method for the extraction of high-quality DNA from lignocellulosic compost samples for metagenomic studies. *Appl. Microbiol. Biotechnol.* **2021**, *105*, 8881–8893. [Google Scholar] [CrossRef] [PubMed]
9. Robinson, S.L.; Piel, J.; Sunagawa, S. A roadmap for metagenomic enzyme discovery. *Nat. Prod. Rep.* **2021**, *38*, 1994–2023. [Google Scholar] [CrossRef] [PubMed]
10. Batista-García, R.A.; del Rayo Sánchez-Carbente, M.; Talia, P.; Jackson, S.A.; O'Leary, N.D.; Dobson, A.D.W.; Folch-Mallol, J.L. From lignocellulosic metagenomes to lignocellulolytic genes: Trends, challenges and future prospects. *Biofuels Bioprod. Biorefining* **2016**, *10*, 864–882. [Google Scholar] [CrossRef]
11. Wang, H.; Hart, D.J.; An, Y. Functional metagenomic technologies for the discovery of novel enzymes for biomass degradation and biofuel production. *BioEnergy Res.* **2019**, *12*, 457–470. [Google Scholar] [CrossRef]

12. Escuder-Rodríguez, J.-J.; DeCastro, M.-E.; Becerra, M.; Rodríguez-Belmonte, E.; González-Siso, M.-I. Advances of functional metagenomics in harnessing thermozymes. In *Metagenomics: Perspectives, Methods, and Applications*; Academic Press: Cambridge, MA, USA, 2018; pp. 289–307. [Google Scholar]
13. Koorevaar, G.P.; Menelik, C.D. (Eds.) 1 Composition and physical properties of soils. In *Developments in Soil Science*; Elsevier: Amsterdam, The Netherlands, 1983; pp. 1–36. [Google Scholar]
14. Bhattacharyya, T.; Pal, K.D. The Soil: A natural resource. In *Soil Science: An Introduction*; Rattan, R.K., Katyal, J.C., Dwivedi, B.S., Sarkar, A.K., Tapas Bhattacharyya, J.C., Tarafdar, S.K., Eds.; ISSS: New Delhi, India, 2015; pp. 1–19. [Google Scholar]
15. Jindal, S. Microbes in soil and their Metagenomics. In *Microbial Diversity, Interventions and Scope*; Sharma, S.G., Neeta Raj Sharma, M.S., Eds.; Springer: Singapore, 2020; pp. 85–96. [Google Scholar]
16. Wang, L.; D'Odorico, P. Decomposition and mineralization. In *Reference Module in Earth Systems and Environmental Sciences*; Elsevier: Amsterdam, The Netherlands, 2013; pp. 280–285. [Google Scholar]
17. Merino, N.; Aronson, H.S.; Bojanova, D.P.; Feyhl-Buska, J.; Wong, M.L.; Zhang, S.; Giovannelli, D. Living at the extremes: Extremophiles and the limits of life in a planetary context. *Front. Microbiol.* **2019**, *10*, 780. [Google Scholar] [CrossRef] [PubMed][Green Version]
18. Mirete, S.; Morgante, V.; González-Pastor, J.E. Functional metagenomics of extreme environments. *Curr. Opin. Biotechnol.* **2016**, *38*, 143–149. [Google Scholar] [CrossRef] [PubMed]
19. Sharma, S.; Khan, F.G.; Qazi, G.N. Molecular cloning and characterization of amylase from soil metagenomic library derived from Northwestern Himalayas. *Appl. Microbiol. Biotechnol.* **2010**, *86*, 1821–1828. [Google Scholar] [CrossRef] [PubMed]
20. Liu, Y.; Lei, Y.; Zhang, X.; Gao, Y.; Xiao, Y.; Peng, H. Identification and phylogenetic characterization of a new subfamily of α-amylase enzymes from marine microorganisms. *Mar. Biotechnol.* **2012**, *14*, 253–260. [Google Scholar] [CrossRef] [PubMed]
21. Vester, J.K.; Glaring, M.A.; Stougaard, P. Discovery of novel enzymes with industrial potential from a cold and alkaline environment by a combination of functional metagenomics and culturing. *Microb. Cell Fact.* **2014**, *13*, 72. [Google Scholar] [CrossRef] [PubMed][Green Version]

22. Gupta, R.; Govil, T.; Capalash, N.; Sharma, P. Characterization of a glycoside hydrolase family 1 β-galactosidase from hot spring metagenome with transglycosylation activity. *Appl. Biochem. Biotechnol.* **2012**, *168*, 1681–1693. [Google Scholar] [CrossRef] [PubMed]

23. Zhang, X.; Li, H.; Li, C.J.; Ma, T.; Li, G.; Liu, Y.H. Metagenomic approach for the isolation of a thermostable β-galactosidase with high tolerance of galactose and glucose from soil samples of Turpan Basin. *BMC Microbiol.* **2013**, *13*, 237. [Google Scholar] [CrossRef] [PubMed][Green Version]

24. Okano, H.; Ozaki, M.; Kanaya, E.; Kim, J.J.; Angkawidjaja, C.; Koga, Y.; Kanaya, S. Structure and stability of metagenome-derived glycoside hydrolase family 12 cellulase (LC-CelA) a homolog of Cel12A from Rhodothermus marinus. *FEBS Open Bio* **2014**, *4*, 936–946. [Google Scholar] [CrossRef] [PubMed][Green Version]

25. Sae-Lee, R.; Boonmee, A. Newly derived GH43 gene from compost metagenome showing dual xylanase and cellulase activities. *Folia Microbiol.* **2014**, *59*, 409–417. [Google Scholar] [CrossRef] [PubMed]

26. Yeh, Y.-F.; Chang, S.C.-Y.; Kuo, H.-W.; Tong, C.-G.; Yu, S.-M.; Ho, T.-H.D. A metagenomic approach for the identification and cloning of an endoglucanase from rice straw compost. *Gene* **2013**, *519*, 360–366. [Google Scholar] [CrossRef]

27. Suleiman, M.; Schröder, C.; Klippel, B.; Schäfers, C.; Krüger, A.; Antranikian, G. Extremely thermoactive archaeal endoglucanase from a shallow marine hydrothermal vent from Vulcano Island. *Appl. Microbiol. Biotechnol.* **2019**, *103*, 1267–1274. [Google Scholar] [CrossRef] [PubMed]

28. Schröder, C.; Elleuche, S.; Blank, S.; Antranikian, G. Characterization of a heat-active archaeal β-glucosidase from a hydrothermal spring metagenome. *Enzym. Microb. Technol.* **2014**, *57*, 48–54. [Google Scholar] [CrossRef]

29. Cao, L.C.; Wang, Z.J.; Ren, G.H.; Kong, W.; Li, L.; Xie, W.; Liu, Y.H. Engineering a novel glucose-tolerant β-glucosidase as supplementation to enhance the hydrolysis of sugarcane bagasse at high glucose concentration. *Biotechnol. Biofuels* **2015**, *8*, 202. [Google Scholar] [CrossRef] [PubMed][Green Version]

30. Matsuzawa, T.; Yaoi, K. Screening, identification, and characterization of a novel saccharide-stimulated β-glycosidase from a soil metagenomic library. *Appl. Microbiol. Biotechnol.* **2017**, *101*, 633–646. [Google Scholar] [CrossRef] [PubMed]

31. Klippel, B.; Sahm, K.; Basner, A.; Wiebusch, S.; John, P.; Lorenz, U.; Peters, A.; Abe, F.; Takahashi, K.; Kaiser, O.; et al. Carbohydrate-active enzymes identified by metagenomic analysis of deep-sea sediment bacteria. *Extremophiles* **2014**, *18*, 853–863. [Google Scholar] [CrossRef] [PubMed]

32. Kaushal, G.; Kumar, J.; Sangwan, R.S.; Singh, S.P. Metagenomic analysis of geothermal water reservoir sites exploring carbohydrate-related thermozymes. *Int. J. Biol. Macromol.* **2018**, *119*, 882–895. [Google Scholar] [CrossRef] [PubMed]

33. Colombo, L.T.; de Oliveira, M.N.V.; Carneiro, D.G.; de Souza, R.A.; Alvim, M.C.T.; dos Santos, J.C.; da Silva, C.C.; Vidigal, P.M.P.; da Silveira, W.B.; Passos, F.M.L. Applying functional metagenomics to search for novel lignocellulosic enzymes in a microbial consortium derived from a thermophilic composting phase of sugarcane bagasse and cow manure. *Antonie Leeuwenhoek* **2016**, *109*, 1217–1233. [Google Scholar] [CrossRef] [PubMed]

34. Strazzulli, A.; Cobucci-Ponzano, B.; Iacono, R.; Giglio, R.; Maurelli, L.; Curci, N.; Schiano-di-Cola, C.; Santangelo, A.; Contursi, P.; Lombard, V.; et al. Discovery of hyperstable carbohydrate-active enzymes through metagenomics of extreme environments. *FEBS J.* **2020**, *287*, 1116–1137. [Google Scholar] [CrossRef] [PubMed]

35. Verma, D.; Kawarabayasi, Y.; Miyazaki, K.; Satyanarayana, T. Cloning, expression and characteristics of a novel alkalistable and thermostable xylanase encoding gene (Mxyl) retrieved from compost-soil metagenome. *PLoS ONE* **2013**, *8*, e52459. [Google Scholar] [CrossRef][Green Version]

36. Knapik, K.; Becerra, M.; González-Siso, M.I. Microbial diversity analysis and screening for novel xylanase enzymes from the sediment of the Lobios Hot Spring in Spain. *Sci. Rep.* **2019**, *9*, 11195. [Google Scholar] [CrossRef] [PubMed]

37. Matsuzawa, T.; Kimura, N.; Suenaga, H.; Yaoi, K. Screening, identification, and characterization of α-xylosidase from a soil metagenome. *J. Biosci. Bioeng.* **2016**, *122*, 393–399. [Google Scholar] [CrossRef] [PubMed]

38. Dai, Y.; Yang, F.; Liu, X.; Wang, H. The discovery and characterization of a novel chitinase with dual catalytic domains from a Qinghai-Tibetan Plateau wetland soil metagenome. *Int. J. Biol. Macromol.* **2021**, *188*, 482–490. [Google Scholar] [CrossRef] [PubMed]

39. Hu, X.P.; Heath, C.; Taylor, M.P.; Tuffin, M.; Cowan, D. A novel, extremely alkaliphilic and cold-active esterase from Antarctic desert soil. *Extremophiles* **2012**, *16*, 79–86. [Google Scholar] [CrossRef] [PubMed]

40. Martínez-martínez, M.; Alcaide, M.; Tchigvintsev, A.; Reva, O.; Polaina, J.; Bargiela, R.; Guazzaroni, M.-E.; Chicote, Á.; Canet, A.; Valero, F.; et al. Biochemical diversity of carboxyl esterases and lipases from Lake Arreo (Spain): A metagenomic approach. *Appl. Environ. Microbiol.* **2013**, *79*, 3553–3562. [Google Scholar] [CrossRef] [PubMed][Green Version]

41. Sudan, A.K.; Vakhlu, J. Isolation of a thioesterase gene from the metagenome of a mountain peak, Apharwat, in the northwestern Himalayas. *3 Biotech* **2013**, *3*, 19–27. [Google Scholar] [CrossRef][Green Version]

42. Okano, H.; Hong, X.; Kanaya, E.; Angkawidjaja, C.; Kanaya, S. Structural and biochemical characterization of a metagenome-derived esterase with a long N-terminal extension. *Protein Sci.* **2015**, *24*, 93–104. [Google Scholar] [CrossRef][Green Version]

43. Leis, B.; Angelov, A.; Mientus, M.; Li, H.; Pham, V.T.T.; Lauinger, B.; Bongen, P.; Pietruszka, J.; Gonçalves, L.G.; Sutherland, R. Identification of novel esterase-active enzymes from hot environments by use of the host bacterium Thermus thermophilus. *Front. Microbiol.* **2015**, *6*, 275. [Google Scholar] [CrossRef] [PubMed][Green Version]

44. Sudan, A.K.; Vakhlu, J. Isolation and in silico characterization of novel esterase gene with β-lactamase fold isolated from metagenome of north western Himalayas. *3 Biotech* **2015**, *5*, 553–559. [Google Scholar] [CrossRef][Green Version]

45. De Santi, C.; Ambrosino, L.; Tedesco, P.; de Pascale, D.; Zhai, L.; Zhou, C.; Xue, Y.; Ma, Y. Identification and characterization of a novel salt-tolerant esterase from a Tibetan glacier metagenomic library. *Biotechnol. Prog.* **2015**, *31*, 890–899. [Google Scholar] [CrossRef]

46. Zarafeta, D.; Moschidi, D.; Ladoukakis, E.; Gavrilov, S.; Chrysina, E.D.; Chatziioannou, A.; Kublanov, I.; Skretas, G.; Kolisis, F.N. Metagenomic mining for thermostable esterolytic enzymes uncovers a new family of bacterial esterases. *Sci. Rep.* **2016**, *6*, 38886. [Google Scholar] [CrossRef][Green Version]

47. Petrovskaya, L.E.; Novototskaya-Vlasova, K.A.; Spirina, E.V.; Durdenko, E.V.; Lomakina, G.Y.; Zavialova, M.G.; Nikolaev, E.N.; Rivkina, E.M. Expression and characterization of a new esterase with GCSAG motif from a permafrost metagenomic library. *FEMS Microbiol. Ecol.* **2016**, *92*, fiw046. [Google Scholar] [CrossRef] [PubMed][Green Version]

48. De Santi, C.; Altermark, B.; Pierechod, M.M.; Ambrosino, L.; De Pascale, D.; Willassen, N.P. Characterization of a cold-active and salt tolerant esterase identified by functional screening of Arctic metagenomic libraries. *BMC Biochem.* **2016**, *17*, 1–13. [Google Scholar] [CrossRef] [PubMed][Green Version]

49. Gao, W.; Wu, K.; Chen, L.; Fan, H.; Zhao, Z.; Gao, B.; Wang, H.; Wei, D. A novel esterase from a marine mud metagenomic library for biocatalytic synthesis of short-chain flavor esters Wenyuan. *Microb. Cell Fact.* **2016**, *15*, 41. [Google Scholar] [CrossRef] [PubMed][Green Version]

50. Popovic, A.; Hai, T.; Tchigvintsev, A.; Hajighasemi, M.; Nocek, B.; Khusnutdinova, A.N.; Brown, G.; Glinos, J.; Flick, R.; Skarina, T.; et al. Activity screening of environmental metagenomic libraries reveals novel carboxylesterase families. *Sci. Rep.* **2017**, *7*, 44103. [Google Scholar] [CrossRef] [PubMed]

51. Li, P.-Y.; Yao, Q.-Q.; Wang, P.; Zhang, Y.; Li, Y.; Zhang, Y.-Q.; Hao, J.; Zhou, B.-C.; Chen, X.-L.; Shi, M.; et al. A novel subfamily esterase with a homoserine transacetylase-like fold but no transferase activity. *Appl. Environ. Microbiol.* **2017**, *83*, e00131-17. [Google Scholar] [CrossRef] [PubMed][Green Version]

52. Jayanath, G.; Mohandas, S.P.; Kachiprath, B.; Solomon, S.; Sajeevan, T.P.; Bright Singh, I.S.; Philip, R. A novel solvent tolerant esterase of GDSGG motif subfamily from solar saltern through metagenomic approach: Recombinant expression and characterization. *Int. J. Biol. Macromol.* **2018**, *119*, 393–401. [Google Scholar] [CrossRef] [PubMed]

53. Hu, Y.; Fu, C.; Huang, Y.; Yin, Y.; Cheng, G.; Lei, F.; Lu, N.A.; Li, J.; Ashforth, E.J.; Zhang, L.; et al. Novel lipolytic genes from the microbial metagenomic library of the South China Sea marine sediment. *FEMS Microbiol. Ecol.* **2010**, *72*, 228–237. [Google Scholar] [CrossRef][Green Version]

54. Fu, C.; Hu, Y.; Xie, F.; Guo, H.; Ashforth, E.J.; Polyak, S.W.; Zhu, B.; Zhang, L. Molecular cloning and characterization of a new cold-active esterase from a deep-sea metagenomic library. *Appl. Microbiol. Biotechnol.* **2011**, *90*, 961–970. [Google Scholar] [CrossRef] [PubMed]

55. Peng, Q.; Zhang, X.; Shang, M.; Wang, X.; Wang, G.; Li, B.; Guan, G.; Li, Y.; Wang, Y. A novel esterase gene cloned from a metagenomic library from neritic sediments of the South China Sea. *Microb. Cell Fact.* **2011**, *10*, 95. [Google Scholar] [CrossRef] [PubMed][Green Version]

56. Yu, E.Y.; Kwon, M.A.; Lee, M.; Oh, J.Y.; Choi, J.E.; Lee, J.Y.; Song, B.K.; Hahm, D.H.; Song, J.K. Isolation and characterization of cold-active family VIII esterases from an arctic soil metagenome. *Appl. Microbiol. Biotechnol.* **2011**, *90*, 573–581. [Google Scholar] [CrossRef]

57. Fan, X.; Liu, X.; Huang, R.; Liu, Y. Identification and characterization of a novel thermostable pyrethroid-hydrolyzing enzyme isolated through metagenomic approach. *Microb. Cell Fact.* **2012**, *11*, 33. [Google Scholar] [CrossRef] [PubMed] [Green Version]

58. Jiang, X.; Xu, X.; Huo, Y.; Wu, Y.; Zhu, X.; Zhang, X.; Wu, M. Identification and characterization of novel esterases from a deep-sea sediment metagenome. *Arch. Microbiol.* **2012**, *194*, 207–214. [Google Scholar] [CrossRef] [PubMed]

59. Sulaiman, S.; Yamato, S.; Kanaya, E.; Kim, J.J.; Koga, Y.; Takano, K.; Kanaya, S. Isolation of a novel cutinase homolog with polyethylene terephthalate-degrading activity from leaf-branch compost by using a metagenomic approach. *Appl. Environ. Microbiol.* **2012**, *78*, 1556–1562. [Google Scholar] [CrossRef] [PubMed][Green Version]

60. Sahoo, R.K.; Kumar, M.; Sukla, L.B.; Subudhi, E. Bioprospecting hot spring metagenome: Lipase for the production of biodiesel. *Environ. Sci. Pollut. Res.* **2017**, *24*, 3802–3809. [Google Scholar] [CrossRef] [PubMed]

61. Sahoo, R.K.; Das, A.; Sahoo, K.; Sahu, A.; Subudhi, E. Characterization of novel metagenomic-derived lipase from Indian hot spring. *Int. Microbiol.* **2020**, *23*, 233–240. [Google Scholar] [CrossRef] [PubMed]

62. Yang, Q.; Zhang, M.; Zhang, M.; Wang, C.; Liu, Y.; Fan, X.; Li, H. Characterization of a novel, cold-adapted, and thermostable laccase-like enzyme with high tolerance for organic solvents and salt and potent dye decolorization ability, derived from a marine metagenomic library. *Front. Microbiol.* **2018**, *9*, 2998. [Google Scholar] [CrossRef] [PubMed][Green Version]

63. Chen, R.; Li, C.; Pei, X.; Wang, Q.; Yin, X.; Xie, T. Isolation an aldehyde dehydrogenase gene from Metagenomics based on semi-nest touch-down PCR. *Indian J. Microbiol.* **2014**, *54*, 74–79. [Google Scholar] [CrossRef] [PubMed] [Green Version]

64. Itoh, N.; Isotani, K.; Makino, Y.; Kato, M.; Kitayama, K.; Ishimota, T. PCR-based amplification and heterologous expression of Pseudomonas alcohol dehydrogenase genes from the soil metagenome for biocatalysis. *Enzym. Microb. Technol.* **2014**, *55*, 140–150. [Google Scholar] [CrossRef] [PubMed]

65. Itoh, N.; Kariya, S.; Kurokawa, J. Efficient PCR-based amplification of diverse alcohol dehydrogenase genes from metagenomes for improving biocatalysis: Screening of gene-specific amplicons from metagenomes. *Appl. Environ. Microbiol.* **2014**, *80*, 6280–6289. [Google Scholar] [CrossRef] [PubMed][Green Version]

66. Neveu, J.; Regeard, C.; Dubow, M.S. Isolation and characterization of two serine proteases from metagenomic libraries of the Gobi and Death Valley deserts. *Appl. Microbiol. Biotechnol.* **2011**, *91*, 635–644. [Google Scholar] [CrossRef] [PubMed]

67. Singh, R.; Chopra, C.; Gupta, V.K.; Akhlaq, B.; Verma, V.; Rasool, S. Purification and characterization of CHpro1, a thermotolerant, alkali-stable and oxidation-resisting protease of Chumathang hotspring. *Sci. Bull.* **2015**, *60*, 1252–1260. [Google Scholar] [CrossRef][Green Version]

68. Chen, W.; Zeng, Y.; Zheng, L.; Liu, W.; Lyu, Q. Discovery and characterization of a novel protease from the Antarctic soil. *Process Biochem.* **2021**, *111*, 270–277. [Google Scholar] [CrossRef]

69. Kanaya, E.; Sakabe, T.; Nguyen, N.T.; Koikeda, S.; Koga, Y.; Takano, K.; Kanaya, S. Cloning of the RNase H genes from a metagenomic DNA library: Identification of a new type 1 RNase H without a typical active-site motif. *J. Appl. Microbiol.* **2010**, *109*, 974–983. [Google Scholar] [CrossRef] [PubMed]

70. Bhat, A.; Riyaz-Ul-Hassan, S.; Srivastava, N.; Johri, S. Molecular cloning of rhodanese gene from soil metagenome of cold desert of North-West Himalayas: Sequence and structural features of the rhodanese enzyme. *3 Biotech* **2015**, *5*, 513–521. [Google Scholar] [CrossRef] [PubMed][Green Version]

71. Silva-Portela, R.C.B.; Carvalho, F.M.; Pereira, C.P.M.; De Souza-Pinto, N.C.; Modesti, M.; Fuchs, R.P.; Agnez-Lima, L.F. ExoMeg1: A new exonuclease from metagenomic library. *Sci. Rep.* **2016**, *6*, 19712. [Google Scholar] [CrossRef] [PubMed][Green Version]

72. Liu, J.; Jia, Z.; Li, S.; Li, Y.; You, Q.; Zhang, C.; Zheng, X.; Xiong, G.; Zhao, J.; Qi, C.; et al. Identification and characterization of a chitin deacetylase from a metagenomic library of deep-sea sediments of the Arctic Ocean. *Gene* **2016**, *590*, 79–84. [Google Scholar] [CrossRef]

73. Agarwal, N.; Singh, S.P. A novel trehalose synthase for the production of trehalose and trehalulose. *Microbiol. Spectr.* **2021**, *9*, e01333-21. [Google Scholar] [CrossRef] [PubMed]

74. Perfumo, A.; Banat, I.M.; Marchant, R. Going green and cold: Biosurfactants from low-temperature environments to biotechnology applications. *Trends Biotechnol.* **2018**, *36*, 277–289. [Google Scholar] [CrossRef][Green Version]

75. Morton, B.; Blackmore, G. South China Sea. *Mar. Pollut. Bull.* **2001**, *42*, 1236–1263. [Google Scholar] [CrossRef] [PubMed]

76. Pacchioni, R.G.; Carvalho, F.M.; Thompson, C.E.; Faustino, A.L.F.; Nicolini, F.; Pereira, T.S.; Silva, R.C.B.; Cantão, M.E.; Gerber, A.; Ana, T.R.; et al. Taxonomic and functional profiles of soil samples from Atlantic forest and Caatinga biomes in northeastern Brazil. *Microbiologyopen* **2014**, *3*, 299–315. [Google Scholar] [CrossRef] [PubMed]

77. DeCastro, M.E.; Rodríguez-Belmonte, E.; González-Siso, M.I. Metagenomics of thermophiles with a focus on discovery of novel thermozymes. *Front. Microbiol.* **2016**, *7*, 1521. [Google Scholar] [CrossRef] [PubMed][Green Version]

78. Sánchez, Ó.J.; Ospina, D.A.; Montoya, S. Compost supplementation with nutrients and microorganisms in composting process. *Waste Manag.* **2017**, *69*, 136–153. [Google Scholar] [CrossRef] [PubMed]

79. Madhuri, R.J.; Saraswathi, M.; Gowthami, K.; Bhargavi, M.; Divya, Y.; Deepika, V. Recent approaches in the production of novel enzymes from environmental samples by enrichment culture and metagenomic approach. In *Recent Developments in Applied Microbiology and Biochemistry*; Buddolla, V., Ed.; Elsevier: Amsterdam, The Netherlands, 2019; pp. 251–262. [Google Scholar]

80. Wang, K.; Li, G.; Yu, S.Q.; Zhang, C.T.; Liu, Y.H. A novel metagenome-derived *β*-galactosidase: Gene cloning, overexpression, purification and characterization. *Appl. Microbiol. Biotechnol.* **2010**, *88*, 155–165. [Google Scholar] [CrossRef]

81. Cheng, J.; Romantsov, T.; Engel, K.; Doxey, A.C.; Rose, D.R.; Neufeld, J.D.; Charles, T.C. Functional metagenomics reveals novel β-galactosidases not predictable from gene sequences. *PLoS ONE* **2017**, *12*, e0172545. [Google Scholar] [CrossRef] [PubMed][Green Version]

82. Allgaier, M.; Reddy, A.; Park, J.I.; Ivanova, N.; D'Haeseleer, P.; Lowry, S.; Sapra, R.; Hazen, T.C.; Simmons, B.A.; Vandergheynst, J.S.; et al. Targeted discovery of glycoside hydrolases from a switchgrass-adapted compost community. *PLoS ONE* **2010**, *5*, e8812. [Google Scholar] [CrossRef] [PubMed]

83. Alvarez, T.M.; Paiva, J.H.; Ruiz, D.M.; Cairo, J.P.L.F.; Pereira, I.O.; Paixão, D.A.A.; De Almeida, R.F.; Tonoli, C.C.C.; Ruller, R.; Santos, C.R.; et al. Structure and function of a novel cellulase 5 from sugarcane soil metagenome. *PLoS ONE* **2013**, *8*, e83635. [Google Scholar] [CrossRef][Green Version]

84. Jiang, C.; Li, S.X.; Luo, F.F.; Jin, K.; Wang, Q.; Hao, Z.Y.; Wu, L.L.; Zhao, G.C.; Ma, G.F.; Shen, P.H.; et al. Biochemical characterization of two novel β-glucosidase genes by metagenome expression cloning. *Bioresour. Technol.* **2011**, *102*, 3272–3278. [Google Scholar] [CrossRef]

85. Liu, J.; Liu, W.D.; Zhao, X.L.; Shen, W.J.; Cao, H.; Cui, Z.L. Cloning and functional characterization of a novel endo-β-1,4-glucanase gene from a soil-derived metagenomic library. *Appl. Microbiol. Biotechnol.* **2011**, *89*, 1083–1092. [Google Scholar] [CrossRef] [PubMed]

86. Xiang, L.; Li, A.; Tian, C.; Zhou, Y.; Zhang, G.; Ma, Y. Identification and characterization of a new acid-stable endoglucanase from a metagenomic library. *Protein Expr. Purif.* **2014**, *102*, 20–26. [Google Scholar] [CrossRef] [PubMed]

87. Pandey, S.; Gulati, S.; Goyal, E.; Singh, S.; Kumar, K.; Nain, L.; Saxena, A.K. Construction and screening of metagenomic library derived from soil for β-1,

4-endoglucanase gene. *Biocatal. Agric. Biotechnol.* **2016**, *5*, 186–192. [Google Scholar] [CrossRef]

88. Yang, C.; Xia, Y.; Qu, H.; Li, A.D.; Liu, R.; Wang, Y.; Zhang, T. Discovery of new cellulases from the metagenome by a metagenomics-guided strategy. *Biotechnol. Biofuels* **2016**, *9*, 138. [Google Scholar] [CrossRef] [PubMed][Green Version]
89. Meneses, C.; Silva, B.; Medeiros, B.; Serrato, R.; Johnston-Monje, D. A metagenomic advance for the cloning and characterization of a cellulase from red rice crop residues. *Molecules* **2016**, *21*, 831. [Google Scholar] [CrossRef] [Green Version]
90. Pimentel, A.C.; Ematsu, G.C.G.; Liberato, M.V.; Paixão, D.A.A.; Franco Cairo, J.P.L.; Mandelli, F.; Tramontina, R.; Gandin, C.A.; de Oliveira Neto, M.; Squina, F.M.; et al. Biochemical and biophysical properties of a metagenome-derived GH5 endoglucanase displaying an unconventional domain architecture. *Int. J. Biol. Macromol.* **2017**, *99*, 384–393. [Google Scholar] [CrossRef] [PubMed] [Green Version]
91. Lee, J.P.; Lee, H.W.; Na, H.B.; Lee, J.H.; Hong, Y.J.; Jeon, J.M.; Kwon, E.J.; Kim, S.K.; Kim, H. Characterization of truncated endo-β-1,4-glucanases from a compost metagenomic library and their saccharification potentials. *Int. J. Biol. Macromol.* **2018**, *115*, 554–562. [Google Scholar] [CrossRef]
92. Wierzbicka-Woś, A.; Henneberger, R.; Batista-García, R.A.; Martínez-Ávila, L.; Jackson, S.A.; Kennedy, J.; Dobson, A.D.W. Biochemical characterization of a novel monospecific endo-β-1,4-glucanase belonging to GH family 5 from a rhizosphere metagenomic library. *Front. Microbiol.* **2019**, *10*, 1342. [Google Scholar] [CrossRef] [PubMed]
93. Aymé, L.; Hébert, A.; Henrissat, B.; Lombard, V.; Franche, N.; Perret, S.; Jourdier, E.; Heiss-Blanquet, S. Characterization of three bacterial glycoside hydrolase family 9 endoglucanases with different modular architectures isolated from a compost metagenome. *Biochim. Biophys. Acta Gen. Subj.* **2021**, *1865*, 129848. [Google Scholar] [CrossRef] [PubMed]
94. Zhou, Y.; Wang, X.; Wei, W.; Xu, J.; Wang, W.; Xie, Z.; Zhang, Z.; Jiang, H.; Wang, Q.; Wei, C. A novel efficient β-glucanase from a paddy soil microbial metagenome with versatile activities. *Biotechnol. Biofuels* **2016**, *9*, 36. [Google Scholar] [CrossRef] [PubMed][Green Version]
95. Hjort, K.; Bergström, M.; Adesina, M.F.; Jansson, J.K.; Smalla, K.; Sjöling, S. Chitinase genes revealed and compared in bacterial isolates, DNA extracts and a metagenomic library from a phytopathogen-suppressive soil. *FEMS Microbiol. Ecol.* **2010**, *71*, 197–207. [Google Scholar] [CrossRef] [PubMed] [Green Version]

96. Hjort, K.; Presti, I.; Elväng, A.; Marinelli, F.; Sjöling, S. Bacterial chitinase with phytopathogen control capacity from suppressive soil revealed by functional metagenomics. *Appl. Microbiol. Biotechnol.* **2014**, *98*, 2819–2828. [Google Scholar] [CrossRef] [PubMed][Green Version]

97. Cretoiu, M.S.; Berini, F.; Kielak, A.M.; Marinelli, F.; van Elsas, J.D. A novel salt-tolerant chitobiosidase discovered by genetic screening of a metagenomic library derived from chitin-amended agricultural soil. *Appl. Microbiol. Biotechnol.* **2015**, *99*, 8199–8215. [Google Scholar] [CrossRef][Green Version]

98. Stöveken, J.; Singh, R.; Kolkenbrock, S.; Zakrzewski, M.; Wibberg, D.; Eikmeyer, F.G.; Pühler, A.; Schlüter, A.; Moerschbacher, B.M. Successful heterologous expression of a novel chitinase identified by sequence analyses of the metagenome from a chitin-enriched soil sample. *J. Biotechnol.* **2015**, *201*, 60–68. [Google Scholar] [CrossRef]

99. Thimoteo, S.S.; Glogauer, A.; Faoro, H.; de Souza, E.M.; Huergo, L.F.; Moerschbacher, B.M.; Pedrosa, F.O. A broad pH range and processive chitinase from a metagenome library. *Braz. J. Med. Biol. Res.* **2017**, *50*, e5658. [Google Scholar] [CrossRef] [PubMed][Green Version]

100. Delavat, F.; Phalip, V.; Forster, A.; Plewniak, F.; Lett, M.C.; Liévremont, D. Amylases without known homologues discovered in an acid mine drainage: Significance and impact. *Sci. Rep.* **2012**, *2*, 354. [Google Scholar] [CrossRef] [PubMed]

101. Soares, F.L.; Marcon, J.; Pereira e Silva, M.D.C.; Khakhum, N.; Cerdeira, L.T.; Ottoni, J.R.; Domingos, D.F.; Taketani, R.G.; De Oliveira, V.M.; Lima, A.O.D.S.; et al. A Novel Multifunctional β-N-Acetylhexosaminidase Revealed through Metagenomics of an Oil-Spilled Mangrove. *Bioengineering* **2017**, *4*, 62. [Google Scholar] [CrossRef][Green Version]

102. Jeong, Y.S.; Na, H.B.; Kim, S.K.; Kim, Y.H.; Kwon, E.J.; Kim, J.; Yun, H.D.; Lee, J.K.; Kim, H. Characterization of Xyn10J, a novel family 10 xylanase from a compost metagenomic library. *Appl. Biochem. Biotechnol.* **2012**, *166*, 1328–1339. [Google Scholar] [CrossRef] [PubMed]

103. Alvarez, T.M.; Goldbeck, R.; dos Santos, C.R.; Paixão, D.A.A.; Gonçalves, T.A.; Franco Cairo, J.P.L.; Almeida, R.F.; de Oliveira Pereira, I.; Jackson, G.; Cota, J.; et al. Development and biotechnological application of a novel endoxylanase family GH10 identified from sugarcane soil metagenome. *PLoS ONE* **2013**, *8*, e70014. [Google Scholar] [CrossRef] [PubMed]

104. Ellilä, S.; Bromann, P.; Nyyssönen, M.; Itävaara, M.; Koivula, A.; Paulin, L.; Kruus, K. Cloning of novel bacterial xylanases from lignocellulose-enriched compost metagenomic libraries. *AMB Express* **2019**, *9*, 124. [Google Scholar] [CrossRef] [PubMed][Green Version]

105. Wang, J.; Liang, J.; Li, Y.; Tian, L.; Wei, Y. Characterization of efficient xylanases from industrial-scale pulp and paper wastewater treatment microbiota. *AMB Express* **2021**, *11*, 19. [Google Scholar] [CrossRef]

106. Ndata, K.; Nevondo, W.; Cekuse, B.; van Zyl, L.J.; Trindade, M. Characterization of a highly xylose tolerant β-xylosidase isolated from high temperature horse manure compost. *BMC Biotechnol.* **2021**, *21*, 61. [Google Scholar] [CrossRef] [PubMed]

107. Kwon, E.J.; Jeong, Y.S.; Kim, Y.H.; Kim, S.K.; Na, H.B.; Kim, J.; Yun, H.D.; Kim, H. Construction of a metagenomic library from compost and screening of cellulase- and xylanase-positive clones. *J. Appl. Biol. Chem.* **2010**, *53*, 702–708. [Google Scholar] [CrossRef]

108. Gladden, J.M.; Allgaier, M.; Miller, C.S.; Hazen, T.C.; VanderGheynst, J.S.; Hugenholtz, P.; Simmons, B.A.; Singer, S.W. Glycoside hydrolase activities of thermophilic bacterial consortia adapted to switchgrass. *Appl. Environ. Microbiol.* **2011**, *77*, 5804–5812. [Google Scholar] [CrossRef] [PubMed][Green Version]

109. Dougherty, M.J.; D'haeseleer, P.; Hazen, T.C.; Simmons, B.A.; Adams, P.D.; Hadi, M.Z. Glycoside hydrolases from a targeted compost metagenome, activity-screening and functional characterization. *BMC Biotechnol.* **2012**, *12*, 38. [Google Scholar] [CrossRef] [PubMed][Green Version]

110. Kanokratana, P.; Eurwilaichitr, L.; Pootanakit, K.; Champreda, V. Identification of glycosyl hydrolases from a metagenomic library of microflora in sugarcane bagasse collection site and their cooperative action on cellulose degradation. *J. Biosci. Bioeng.* **2015**, *119*, 384–391. [Google Scholar] [CrossRef] [PubMed]

111. Berlemont, R.; Pipers, D.; Delsaute, M.; Angiono, F.; Feller, G.; Galleni, M.; Power, P. Exploring the Antarctic soil metagenome as a source of novel cold-adapted enzymes and genetic mobile elements. *Rev. Argent. Microbiol.* **2011**, *43*, 94–103. [Google Scholar] [CrossRef] [PubMed]

112. Stroobants, A.; Portetelle, D.; Vandenbol, M. New carbohydrate-active enzymes identified by screening two metagenomic libraries derived from the soil of a winter wheat field. *J. Appl. Microbiol.* **2014**, *117*, 1045–1055. [Google Scholar] [CrossRef][Green Version]

113. Liaw, R.B.; Cheng, M.P.; Wu, M.C.; Lee, C.Y. Use of metagenomic approaches to isolate lipolytic genes from activated sludge. *Bioresour. Technol.* **2010**, *101*, 8323–8329. [Google Scholar] [CrossRef] [PubMed]

114. Kim, Y.H.; Kwon, E.J.; Kim, S.K.; Jeong, Y.S.; Kim, J.; Yun, H.D.; Kim, H. Molecular cloning and characterization of a novel family VIII alkaline esterase from a compost metagenomic library. *Biochem. Biophys. Res. Commun.* **2010**, *393*, 45–49. [Google Scholar] [CrossRef] [PubMed]

115. Lee, M.H.; Hong, K.S.; Malhotra, S.; Park, J.H.; Hwang, E.C.; Choi, H.K.; Kim, Y.S.; Tao, W.; Lee, S.W. A new esterase EstD2 isolated from plant rhizosphere soil metagenome. *Appl. Microbiol. Biotechnol.* **2010**, *88*, 1125–1134. [Google Scholar] [CrossRef]

116. Sang, S.L.; Li, G.; Hu, X.P.; Liu, Y.H. Molecular cloning, overexpression and characterization of a novel feruloyl esterase from a soil metagenomic library. *J. Mol. Microbiol. Biotechnol.* **2011**, *20*, 196–203. [Google Scholar] [CrossRef] [PubMed]

117. Pereira, M.R.; Mercaldi, G.F.; Maester, T.C.; Balan, A.; De Macedo Lemos, E.G. Est16, a new esterase isolated from a metagenomic library of a microbial consortium specializing in diesel oil degradation. *PLoS ONE* **2015**, *10*, e0133723. [Google Scholar] [CrossRef]

118. Ohlhoff, C.W.; Kirby, B.M.; Van Zyl, L.; Mutepfa, D.L.R.; Casanueva, A.; Huddy, R.J.; Bauer, R.; Cowan, D.A.; Tuffin, M. An unusual feruloyl esterase belonging to family VIII esterases and displaying a broad substrate range. *J. Mol. Catal. B Enzym.* **2015**, *118*, 79–88. [Google Scholar] [CrossRef][Green Version]

119. Lee, H.W.; Jung, W.K.; Kim, Y.H.; Ryu, B.H.; Doohun Kim, T.; Kim, J.; Kim, H. Characterization of a novel alkaline family viii esterase with S-enantiomer preference from a compost metagenomic library. *J. Microbiol. Biotechnol.* **2016**, *26*, 315–325. [Google Scholar] [CrossRef]

120. Dukunde, A.; Schneider, D.; Lu, M.; Brady, S.; Daniel, R. A novel, versatile family IV carboxylesterase exhibits high stability and activity in a broad pH spectrum. *Biotechnol. Lett.* **2017**, *39*, 577–587. [Google Scholar] [CrossRef] [PubMed]

121. Lu, M.; Dukunde, A.; Daniel, R. Biochemical profiles of two thermostable and organic solvent–tolerant esterases derived from a compost metagenome. *Appl. Microbiol. Biotechnol.* **2019**, *103*, 3421–3437. [Google Scholar] [CrossRef] [PubMed]

122. Yao, J.; Gui, L.; Yin, S. A novel esterase from a soil metagenomic library displaying a broad substrate range. *AMB Express* **2021**, *11*, 38. [Google Scholar] [CrossRef] [PubMed]

123. Yan, Z.; Ding, L.; Zou, D.; Wang, L.; Tan, Y.; Guo, S.; Zhang, Y.; Xin, Z. Identification and characterization of a novel carboxylesterase EstQ7 from a soil metagenomic library. *Arch. Microbiol.* **2021**, *203*, 4113–4125. [Google Scholar] [CrossRef] [PubMed]

124. Yan, Z.; Ding, L.; Zou, D.; Qiu, J.; Shao, Y.; Sun, S.; Li, L.; Xin, Z. Characterization of a novel carboxylesterase with catalytic activity toward di(2-ethylhexyl) phthalate from a soil metagenomic library. *Sci. Total Environ.* **2021**, *785*, 147260. [Google Scholar] [CrossRef]

125. Park, J.E.; Jeong, G.S.; Lee, H.W.; Kim, H. Biochemical characterization of a family IV esterase with R-form enantioselectivity from a compost metagenomic library. *Appl. Biol. Chem.* **2021**, *64*, 81. [Google Scholar] [CrossRef]

126. Park, J.E.; Jeong, G.S.; Lee, H.W.; Kim, H. Molecular characterization of novel family iv and viii esterases from a compost metagenomic library. *Microorganisms* **2021**, *9*, 1614. [Google Scholar] [CrossRef] [PubMed]

127. Park, J.E.; Jeong, G.S.; Lee, H.W.; Kim, S.K.; Kim, J.; Kim, H. Characterization of a novel family iv esterase containing a predicted czco domain and a family v esterase with broad substrate specificity from an oil-polluted mud flat metagenomic library. *Appl. Sci.* **2021**, *11*, 5905. [Google Scholar] [CrossRef]

128. O'Mahony, M.M.; Henneberger, R.; Selvin, J.; Kennedy, J.; Doohan, F.; Marchesi, J.R.; Dobson, A.D.W. Inhibition of the growth of Bacillus subtilis DSM10 by a newly discovered antibacterial protein from the soil metagenome. *Bioengineered* **2015**, *6*, 89–98. [Google Scholar] [CrossRef] [Green Version]

129. Glogauer, A.; Martini, V.P.; Faoro, H.; Couto, G.H.; Müller-Santos, M.; Monteiro, R.A.; Mitchell, D.A.; de Souza, E.M.; Pedrosa, F.O.; Krieger, N. Identification and characterization of a new true lipase isolated through metagenomic approach. *Microb. Cell Fact.* **2011**, *10*, 54. [Google Scholar] [CrossRef] [PubMed] [Green Version]

130. Zheng, J.; Liu, C.; Liu, L.; Jin, Q. Characterisation of a thermo-alkali-stable lipase from oil-contaminated soil using a metagenomic approach. *Syst. Appl. Microbiol.* **2013**, *36*, 197–204. [Google Scholar] [CrossRef] [PubMed]

131. Kim, H.J.; Jeong, Y.S.; Jung, W.K.; Kim, S.K.; Lee, H.W.; Kahng, H.Y.; Kim, J.; Kim, H. Characterization of novel family IV esterase and family I.3 lipase from an oil-polluted mud flat metagenome. *Mol. Biotechnol.* **2015**, *57*, 781–792. [Google Scholar] [CrossRef]

132. Yao, J.; Fan, X.J.; Lu, Y.; Liu, Y.H. Isolation and characterization of a novel tannase from a metagenomic library. *J. Agric. Food Chem.* **2011**, *59*, 3812–3818. [Google Scholar] [CrossRef]

133. Nacke, H.; Will, C.; Herzog, S.; Nowka, B.; Engelhaupt, M.; Daniel, R. Identification of novel lipolytic genes and gene families by screening of metagenomic libraries derived from soil samples of the German Biodiversity Exploratories. *FEMS Microbiol. Ecol.* **2011**, *78*, 188–201. [Google Scholar] [CrossRef] [PubMed][Green Version]

134. Stroobants, A.; Martin, R.; Roosens, L.; Portetelle, D.; Vandenbol, M. New lipolytic enzymes identified by screening two metagenomic libraries derived from the soil of a winter wheat field. *Biotechnol. Agron. Soc. Environ.* **2015**, *19*, 125–131. [Google Scholar]

135. Kimura, N.; Kamagata, Y. A thermostable bilirubin-oxidizing enzyme from activated sludge isolated by a metagenomic approach. *Microbes Environ.* **2016**, *31*, 435–441. [Google Scholar] [CrossRef] [PubMed][Green Version]

136. Ou, Q.; Liu, Y.; Deng, J.; Chen, G.; Yang, Y.; Shen, P.; Wu, B.; Jiang, C. A novel d-amino acid oxidase from a contaminated agricultural soil metagenome and its characterization. *Antonie Leeuwenhoek Int. J. Gen. Mol. Microbiol.* **2015**, *107*, 1615–1623. [Google Scholar] [CrossRef] [PubMed]

137. Kimura, N.; Sakai, K.; Nakamura, K. Isolation and characterization of a 4-nitrotoluene-oxidizing enzyme from activated sludge by a metagenomic approach. *Microbes Environ.* **2010**, *25*, 133–139. [Google Scholar] [CrossRef] [PubMed][Green Version]

138. Singh, A.; Singh Chauhan, N.; Thulasiram, H.V.; Taneja, V.; Sharma, R. Identification of two flavin monooxygenases from an effluent treatment plant sludge metagenomic library. *Bioresour. Technol.* **2010**, *101*, 8481–8484. [Google Scholar] [CrossRef] [PubMed]

139. Chemerys, A.; Pelletier, E.; Cruaud, C.; Martin, F.; Violet, F.; Jouanneaua, Y. Characterization of novel polycyclic aromatic hydrocarbon dioxygenases from the bacterial metagenomic DNA of a contaminated soil. *Appl. Environ. Microbiol.* **2014**, *80*, 6591–6600. [Google Scholar] [CrossRef] [PubMed][Green Version]

140. Lu, Y.; Yu, Y.; Zhou, R.; Sun, W.; Dai, C.; Wan, P.; Zhang, L.; Hao, D.; Ren, H. Cloning and characterisation of a novel 2,4-dichlorophenol hydroxylase from a metagenomic library derived from polychlorinated biphenyl-contaminated soil. *Biotechnol. Lett.* **2011**, *33*, 1159–1167. [Google Scholar] [CrossRef] [PubMed]

141. Tan, H.; Mooij, M.J.; Barret, M.; Hegarty, P.M.; Harrington, C.; Dobson, A.D.W.; O'Gara, F. Identification of novel phytase genes from an agricultural soil-derived metagenome. *J. Microbiol. Biotechnol.* **2014**, *24*, 113–118. [Google Scholar] [CrossRef] [PubMed][Green Version]

142. Kim, N.H.; Park, J.H.; Chung, E.; So, H.A.; Lee, M.H.; Kim, J.C.; Hwang, E.C.; Lee, S.W. Characterization of a soil metagenome-derived gene encoding wax ester synthase. *J. Microbiol. Biotechnol.* **2016**, *26*, 248–254. [Google Scholar] [CrossRef] [PubMed]

143. Jiang, L.; Lin, M.; Zhang, Y.; Li, Y.; Xu, X.; Li, S.; Huang, H. Identification and characterization of a novel trehalose synthase gene derived from saline-alkali soil metagenomes. *PLoS ONE* **2013**, *8*, e77437. [Google Scholar] [CrossRef] [PubMed][Green Version]

144. Wang, S.D.; Guo, G.S.; Li, L.; Cao, L.C.; Tong, L.; Ren, G.H.; Liu, Y.H. Identification and characterization of an unusual glycosyltransferase-like enzyme with β-galactosidase activity from a soil metagenomic library. *Enzym. Microb. Technol.* **2014**, *57*, 26–35. [Google Scholar] [CrossRef] [PubMed]

145. Colin, P.Y.; Kintses, B.; Gielen, F.; Miton, C.M.; Fischer, G.; Mohamed, M.F.; Hyvönen, M.; Morgavi, D.P.; Janssen, D.B.; Hollfelder, F. Ultrahigh-throughput discovery of promiscuous enzymes by picodroplet functional metagenomics. *Nat. Commun.* **2015**, *6*, 10008. [Google Scholar] [CrossRef] [PubMed]

146. Ajibade, F.O.; Adelodun, B.; Lasisi, K.H.; Fadare, O.O.; Ajibade, T.F.; Nwogwu, N.A.; Sulaymon, I.D.; Ugya, A.Y.; Wang, H.C.; Wang, A. Chapter 25—Environmental pollution and their socioeconomic impacts. In *Microbe Mediated Remediation of Environmental Contaminants*; Kumar, A., Singh, V.K., Singh, P., Mishra, V.K., Eds.; Woodhead Publishing: Cambridge, UK, 2021; pp. 321–354. ISBN 9780128211991. [Google Scholar]

147. Dindar, E.; Topaç Şağban, F.O.; Başkaya, H.S. Variations of soil enzyme activities in petroleum-hydrocarbon contaminated soil. *Int. Biodeterior. Biodegrad.* **2015**, *105*, 268–275. [Google Scholar] [CrossRef]

148. Waldhauser, F.; Schaff, D.; Richards, P.G.; Kim, W.-Y. Lop Nor Revisited: Underground nuclear explosion locations, 1976–1996, from double-difference analysis of regional and teleseismic data. *Bull. Seismol. Soc. Am.* **2004**, *94*, 1879–1889. [Google Scholar] [CrossRef][Green Version]

149. Van Leeuwen, J.P.; Djukic, I.; Bloem, J.; Lehtinen, T.; Hemerik, L. Effects of land use on soil microbial biomass, activity and community structure at different soil depths in the Danube floodplain. *Eur. J. Soil Biol.* **2017**, *79*, 14–20. [Google Scholar] [CrossRef]

150. Souza, R.C.; Hungria, M.; Cantão, M.E.; Vasconcelos, A.T.R.; Nogueira, M.A.; Vicente, V.A. Metagenomic analysis reveals microbial functional redundancies and specificities in a soil under different tillage and crop-management regimes. *Appl. Soil Ecol.* **2015**, *86*, 106–112. [Google Scholar] [CrossRef]

151. Siles-Castellano, A.B.; López, M.J.; López-González, J.A.; Suárez-Estrella, F.; Jurado, M.M.; Estrella-González, M.J.; Moreno, J. Comparative analysis of phytotoxicity and compost quality in industrial composting facilities processing different organic wastes. *J. Clean. Prod.* **2020**, *252*, 119820. [Google Scholar] [CrossRef]

152. Bergmann, J.C.; Costa, O.Y.A.; Gladden, J.M.; Singer, S.; Heins, R.; D'haeseleer, P.; Simmons, B.A.; Quirino, B.F. Discovery of two novel β-glucosidases from

an Amazon soil metagenomic library. *FEMS Microbiol. Lett.* **2014**, *351*, 147–155. [Google Scholar] [CrossRef] [PubMed][Green Version]

153. Couto, G.H.; Glogauer, A.; Faoro, H.; Chubatsu, L.S.; Souza, E.M.; Pedrosa, F.O. Isolation of a novel lipase from a metagenomic library derived from mangrove sediment from the south Brazilian coast. *Genet. Mol. Res.* **2010**, *9*, 514–523. [Google Scholar] [CrossRef]
154. Ye, M.; Li, G.; Liang, W.Q.; Liu, Y.H. Molecular cloning and characterization of a novel metagenome-derived multicopper oxidase with alkaline laccase activity and highly soluble expression. *Appl. Microbiol. Biotechnol.* **2010**, *87*, 1023–1031. [Google Scholar] [CrossRef] [PubMed]
155. Li, G.; Jiang, Y.; Fan, X.J.; Liu, Y.H. Molecular cloning and characterization of a novel β-glucosidase with high hydrolyzing ability for soybean isoflavone glycosides and glucose-tolerance from soil metagenomic library. *Bioresour. Technol.* **2012**, *123*, 15–22. [Google Scholar] [CrossRef]
156. Mai, Z.; Su, H.; Yang, J.; Huang, S.; Zhang, S. Cloning and characterization of a novel GH44 family endoglucanase from mangrove soil metagenomic library. *Biotechnol. Lett.* **2014**, *36*, 1701–1709. [Google Scholar] [CrossRef] [PubMed]
157. Mai, Z.; Su, H.; Zhang, S. Isolation and characterization of a glycosyl hydrolase family 16 β-agarase from a mangrove soil metagenomic library. *Int. J. Mol. Sci.* **2016**, *17*, 1360. [Google Scholar] [CrossRef] [PubMed][Green Version]
158. Bunterngsook, B.; Kanokratana, P.; Thongaram, T.; Tanapongpipat, S.; Uengwetwanit, T.; Rachdawong, S.; Vichitsoonthonkul, T.; Eurwilaichitr, L. Identification and characterization of lipolytic enzymes from a peat-swamp forest soil metagenome. *Biosci. Biotechnol. Biochem.* **2010**, *74*, 1848–1854. [Google Scholar] [CrossRef][Green Version]
159. Tan, H.; Wu, X.; Xie, L.; Huang, Z.; Peng, W.; Gan, B. A novel phytase derived from an acidic peat-soil microbiome showing high stability under acidic plus pepsin conditions. *J. Mol. Microbiol. Biotechnol.* **2016**, *26*, 291–301. [Google Scholar] [CrossRef] [PubMed]
160. Berlemont, R.; Spee, O.; Delsaute, M.; Lara, Y.; Schuldes, J.; Simon, C.; Power, P.; Daniel, R.; Galleni, M. Novel organic solvent-tolerant esterase isolated by metagenomics: Insights into the lipase/esterase classification. *Rev. Argent. Microbiol.* **2013**, *45*, 3–12. [Google Scholar] [PubMed]
161. Gu, X.; Wang, S.; Wang, S.; Zhao, L.X.; Cao, M.; Feng, Z. Identification and characterization of two novel esterases from a metagenomic library. *Food Sci. Technol. Res.* **2015**, *21*, 649–657. [Google Scholar] [CrossRef]
162. Gomes-Pepe, E.S.; Sierra, E.G.M.; Pereira, M.R.; Castellane, T.C.L.; De Lemos, E.G.M. Bg10: A novel metagenomics alcohol-tolerant and glucose-stimulated

gh1 β-glucosidase suitable for lactose-free milk preparation. *PLoS ONE* **2016**, *11*, e0167932. [Google Scholar] [CrossRef] [PubMed]

163. Tao, W.; Lee, M.H.; Yoon, M.Y.; Kim, J.C.; Malhotra, S.; Wu, J.; Hwang, E.C.; Lee, S.W. Characterization of two metagenome-derived esterases that reactivate chloramphenicol by counteracting chloramphenicol acetyltransferase. *J. Microbiol. Biotechnol.* **2011**, *21*, 1203–1210. [Google Scholar] [CrossRef][Green Version]

164. Li, Z.; Li, X.; Liu, T.; Chen, S.; Liu, H.; Wang, H.; Li, K.; Song, Y.; Luo, X.; Zhao, J.; et al. The critical roles of exposed surface residues for the thermostability and halotolerance of a novel GH11 xylanase from the metagenomic library of a saline-alkaline soil. *Int. J. Biol. Macromol.* **2019**, *133*, 316–323. [Google Scholar] [CrossRef] [PubMed]

165. Istvan, P.; Souza, A.A.; Garay, A.V.; dos Santos, D.F.K.; de Oliveira, G.M.; Santana, R.H.; Lopes, F.A.C.; de Freitas, S.M.; Barbosa, J.A.R.G.; Krüger, R.H. Structural and functional characterization of a novel lipolytic enzyme from a Brazilian Cerrado soil metagenomic library. *Biotechnol. Lett.* **2018**, *40*, 1395–1406. [Google Scholar] [CrossRef]

166. dos Santos, D.F.K.; Istvan, P.; Noronha, E.F.; Quirino, B.F.; Krüger, R.H. New dioxygenase from metagenomic library from Brazilian soil: Insights into antibiotic resistance and bioremediation. *Biotechnol. Lett.* **2015**, *37*, 1809–1817. [Google Scholar] [CrossRef] [PubMed]

167. Zhao, X.; Liu, L.; Deng, Z.; Liu, S.; Yun, J.; Xiao, X.; Li, H. Screening, cloning, enzymatic properties of a novel thermostable cellulase enzyme, and its potential application on water hyacinth utilization. *Int. Microbiol.* **2021**, *24*, 337–349. [Google Scholar] [CrossRef]

168. Erich, S.; Kuschel, B.; Schwarz, T.; Ewert, J.; Böhmer, N.; Niehaus, F.; Eck, J.; Lutz-Wahl, S.; Stressler, T.; Fischer, L. Novel high-performance metagenome β-galactosidases for lactose hydrolysis in the dairy industry. *J. Biotechnol.* **2015**, *210*, 27–37. [Google Scholar] [CrossRef][Green Version]

169. Lezyk, M.; Jers, C.; Kjaerulff, L.; Gotfredsen, C.H.; Mikkelsen, M.D.; Mikkelsen, J.D. Novel α-L-fucosidases from a soil metagenome for production of fucosylated human milk oligosaccharides. *PLoS ONE* **2016**, *11*, e0147438. [Google Scholar] [CrossRef][Green Version]

170. Kim, S.Y.; Oh, D.B.; Kwon, O. Characterization of a lichenase isolated from soil metagenome. *J. Microbiol. Biotechnol.* **2014**, *24*, 1699–1706. [Google Scholar] [CrossRef] [PubMed][Green Version]

171. Sathya, T.A.; Jacob, A.M.; Khan, M. Cloning and molecular modelling of pectin degrading glycosyl hydrolase of family 28 from soil metagenomic library. *Mol. Biol. Rep.* **2014**, *41*, 2645–2656. [Google Scholar] [CrossRef] [PubMed]

172. Mori, T.; Kamei, I.; Hirai, H.; Kondo, R. Identification of novel glycosyl hydrolases with cellulolytic activity against crystalline cellulose from metagenomic libraries constructed from bacterial enrichment cultures. *SpringerPlus* **2014**, *3*, 365. [Google Scholar] [CrossRef][Green Version]

173. Chai, S.; Zhang, X.; Jia, Z.; Xu, X.; Zhang, Y.; Wang, S.; Feng, Z. Identification and characterization of a novel bifunctional cellulase/hemicellulase from a soil metagenomic library. *Appl. Microbiol. Biotechnol.* **2020**, *104*, 7563–7572. [Google Scholar] [CrossRef] [PubMed]

174. Faoro, H.; Glogauer, A.; Couto, G.H.; de Souza, E.M.; Rigo, L.U.; Cruz, L.M.; Monteiro, R.A.; de Oliveira Pedrosa, F. Characterization of a new Acidobacteria-derived moderately thermostable lipase from a Brazilian Atlantic Forest soil metagenome. *FEMS Microbiol. Ecol.* **2012**, *81*, 386–394. [Google Scholar] [CrossRef] [PubMed][Green Version]

175. Lim, H.K.; Han, Y.J.; Hahm, M.S.; Park, S.Y.; Hwang, I.T. Isolation and characterization of a novel triolein selective lipase from soil environmental genes. *Microbiol. Biotechnol. Lett.* **2021**, *48*, 480–490. [Google Scholar] [CrossRef]

176. Jin, P.; Pei, X.; Du, P.; Yin, X.; Xiong, X.; Wu, H.; Zhou, X.; Wang, Q. Overexpression and characterization of a new organic solvent-tolerant esterase derived from soil metagenomic DNA. *Bioresour. Technol.* **2012**, *116*, 234–240. [Google Scholar] [CrossRef]

177. Biver, S.; Vandenbol, M. Characterization of three new carboxylic ester hydrolases isolated by functional screening of a forest soil metagenomic library. *J. Ind. Microbiol. Biotechnol.* **2013**, *40*, 191–200. [Google Scholar] [CrossRef]

178. Jeon, J.H.; Lee, H.S.; Lee, J.H.; Koo, B.S.; Lee, C.M.; Lee, S.H.; Kang, S.G.; Lee, J.H. A novel family VIII carboxylesterase hydrolysing third- and fourth-generation cephalosporins. *SpringerPlus* **2016**, *5*, 525. [Google Scholar] [CrossRef][Green Version]

179. Park, J.M.; Won, S.M.; Kang, C.H.; Park, S.; Yoon, J.H. Characterization of a novel carboxylesterase belonging to family VIII hydrolyzing β-lactam antibiotics from a compost metagenomic library. *Int. J. Biol. Macromol.* **2020**, *164*, 4650–4661. [Google Scholar] [CrossRef]

180. Nagayama, H.; Sugawara, T.; Endo, R.; Ono, A.; Kato, H.; Ohtsubo, Y.; Nagata, Y.; Tsuda, M. Isolation of oxygenase genes for indigo-forming activity from an artificially polluted soil metagenome by functional screening using Pseudomonas putida strains as hosts. *Appl. Microbiol. Biotechnol.* **2015**, *99*, 4453–4470. [Google Scholar] [CrossRef] [PubMed]

181. Biver, S.; Portetelle, D.; Vandenbol, M. Characterization of a new oxidant-stable serine protease isolated by functional metagenomics. *SpringerPlus* **2013**, *2*, 410. [Google Scholar] [CrossRef] [PubMed][Green Version]

182. Stephens, G.L.; Slingo, J.M.; Rignot, E.; Reager, J.T.; Hakuba, M.Z.; Durack, P.J.; Worden, J.; Rocca, R. Earth's water reservoirs in a changing climate. *Proc. R. Soc. A Math. Phys. Eng. Sci.* **2020**, *476*, 20190458. [Google Scholar] [CrossRef] [Green Version]

183. Parages, M.L.; Gutiérrez-Barranquero, J.A.; Reen, F.J.; Dobson, A.D.W.; O'Gara, F. Integrated (Meta) genomic and synthetic biology approaches to develop new biocatalysts. *Mar. Drugs* **2016**, *14*, 62. [Google Scholar] [CrossRef]

184. Jiang, T.; Sun, S.; Chen, Y.; Qian, Y.; Guo, J.; Dai, R.; An, D. Microbial diversity characteristics and the influence of environmental factors in a large drinking-water source. *Sci. Total Environ.* **2021**, *769*, 144698. [Google Scholar] [CrossRef]

185. Kamble, P.; Vavilala, S.L. Discovering novel enzymes from marine ecosystems: A metagenomic approach. *Bot. Mar.* **2018**, *61*, 161–175. [Google Scholar] [CrossRef]

186. Fang, W.; Liu, J.; Hong, Y.; Peng, H.; Zhang, X.; Sun, B.; Xiao, Y. Cloning and characterization of a β -glucosidase from marine microbial metagenome with excellent glucose tolerance. *J. Microbiol. Biotechnol.* **2010**, *20*, 1351–1358. [Google Scholar] [CrossRef]

187. Wierzbicka-Woś, A.; Bartasun, P.; Cieśliński, H.; Kur, J. Cloning and characterization of a novel cold-active glycoside hydrolase family 1 enzyme with β-glucosidase, β-fucosidase and β-galactosidase activities. *BMC Biotechnol.* **2013**, *13*, 22. [Google Scholar] [CrossRef][Green Version]

188. Toyama, D.; de Morais, M.A.B.; Ramos, F.C.; Zanphorlin, L.M.; Tonoli, C.C.C.; Balula, A.F.; de Miranda, F.P.; Almeida, V.M.; Marana, S.R.; Ruller, R.; et al. A novel β-glucosidase isolated from the microbial metagenome of Lake Poraquê (Amazon, Brazil). *Biochim. Biophys. Acta Prot. Proteom.* **2018**, *1866*, 569–579. [Google Scholar] [CrossRef] [PubMed]

189. Mohamed, Y.M.; Ghazy, M.A.; Sayed, A.; Ouf, A.; El-Dorry, H.; Siam, R. Isolation and characterization of a heavy metal-resistant, thermophilic esterase from a Red Sea Brine Pool. *Sci. Rep.* **2013**, *3*, 3358. [Google Scholar] [CrossRef] [PubMed][Green Version]

190. Fang, Z.; Li, J.; Wang, Q.; Fang, W.; Peng, H.; Zhang, X.; Xiao, Y. A novel esterase from a marine metagenomic library exhibiting salt tolerance ability. *J. Microbiol. Biotechnol.* **2014**, *24*, 771–780. [Google Scholar] [CrossRef] [PubMed]

191. López-López, O.; Knapik, K.; Cerdán, M.E.; González-Siso, M.I. Metagenomics of an alkaline hot spring in Galicia (Spain): Microbial diversity analysis and

screening for novel lipolytic enzymes. *Front. Microbiol.* **2015**, *6*, 1291. [Google Scholar] [CrossRef][Green Version]

192. Fang, Z.; Li, T.; Wang, Q.; Zhang, X.; Peng, H.; Fang, W.; Hong, Y.; Ge, H.; Xiao, Y. A bacterial laccase from marine microbial metagenome exhibiting chloride tolerance and dye decolorization ability. *Appl. Microbiol. Biotechnol.* **2011**, *89*, 1103–1110. [Google Scholar] [CrossRef] [PubMed]
193. Sayed, A.; Ghazy, M.A.; Ferreira, A.J.S.; Setubal, J.C.; Chambergo, F.S.; Ouf, A.; Adel, M.; Dawe, A.S.; Archer, J.A.C.; Bajic, V.B.; et al. A novel mercuric reductase from the unique deep brine environment of Atlantis II in the Red Sea. *J. Biol. Chem.* **2014**, *289*, 1675–1687. [Google Scholar] [CrossRef] [PubMed][Green Version]
194. Badiea, E.A.; Sayed, A.A.; Maged, M.; Fouad, W.M.; Said, M.M.; Esmat, A.Y. A novel thermostable and halophilic thioredoxin reductase from the Red Sea Atlantis II hot brine pool. *PLoS ONE* **2019**, *14*, e0217565. [Google Scholar] [CrossRef] [PubMed]
195. Witte, B.; John, D.; Wawrik, B.; Paul, J.H.; Dayan, D.; Robert Tabita, F. Functional prokaryotic rubisCO from an oceanic metagenomic library. *Appl. Environ. Microbiol.* **2010**, *76*, 2997–3003. [Google Scholar] [CrossRef] [PubMed][Green Version]
196. Jiang, C.; Wu, L.L.; Zhao, G.C.; Shen, P.H.; Jin, K.; Hao, Z.Y.; Li, S.X.; Ma, G.F.; Luo, F.F.; Hu, G.Q.; et al. Identification and characterization of a novel fumarase gene by metagenome expression cloning from marine microorganisms. *Microb. Cell Fact.* **2010**, *9*, 91. [Google Scholar] [CrossRef][Green Version]
197. Sonbol, S.A.; Ferreira, A.J.S.; Siam, R. Red Sea Atlantis II brine pool nitrilase with unique thermostability profile and heavy metal tolerance. *BMC Biotechnol.* **2016**, *16*, 14. [Google Scholar] [CrossRef][Green Version]
198. Elbehery, A.H.A.; Leak, D.J.; Siam, R. Novel thermostable antibiotic resistance enzymes from the Atlantis II Deep Red Sea brine pool. *Microb. Biotechnol.* **2017**, *10*, 189–202. [Google Scholar] [CrossRef]
199. Sobat, M.; Asad, S.; Kabiri, M.; Mehrshad, M. Metagenomic discovery and functional validation of L-asparaginases with anti-leukemic effect from the Caspian Sea. *iScience* **2021**, *24*, 101973. [Google Scholar] [CrossRef]
200. Shu, W.S.; Huang, L.N. Microbial diversity in extreme environments. *Nat. Rev. Microbiol.* **2021**, *20*, 219–235. [Google Scholar] [CrossRef] [PubMed]
201. Poli, A.; Finore, I.; Romano, I.; Gioiello, A.; Lama, L.; Nicolaus, B. Microbial diversity in extreme marine habitats and their biomolecules. *Microorganisms* **2017**, *5*, 25. [Google Scholar] [CrossRef][Green Version]

202. Ojaveer, H.; Jaanus, A.; Mackenzie, B.R.; Martin, G.; Olenin, S.; Radziejewska, T.; Telesh, I.; Zettler, M.L.; Zaiko, A. Status of biodiversity in the Baltic Sea. *PLoS ONE* **2010**, *5*, e12467. [Google Scholar] [CrossRef] [PubMed][Green Version]

203. Sass, A.M.; Sass, H.; Coolen, M.J.L.; Cypionka, H.; Overmann, J. Microbial communities in the chemocline of a hypersaline deep-sea basin (Urania Basin, Mediterranean Sea). *Appl. Environ. Microbiol.* **2001**, *67*, 5392–5402. [Google Scholar] [CrossRef][Green Version]

204. Borah, P.; Kumar, M.; Devi, P. Types of inorganic pollutants: Metals/metalloids, acids, and organic forms. In *Inorganic Pollutants in Water*; Devi, P., Singh, P., Kansal, S.K., Eds.; Elsevier: Amsterdam, The Netherlands, 2020; pp. 17–31. ISBN 9780128189658. [Google Scholar]

205. Ouyang, L.M.; Liu, J.Y.; Qiao, M.; Xu, J.H. Isolation and biochemical characterization of two novel metagenome-derived esterases. *Appl. Biochem. Biotechnol.* **2013**, *169*, 15–28. [Google Scholar] [CrossRef] [PubMed]

206. Singh, D.N.; Gupta, A.; Singh, V.S.; Mishra, R.; Kateriya, S.; Tripathi1, A.K. Identification and characterization of a novel phosphodiesterase from the metagenome of an Indian coalbed. *PLoS ONE* **2015**, *10*, e0118075. [Google Scholar] [CrossRef] [PubMed]

207. Yan, W.; Li, F.; Wang, L.; Zhu, Y.; Dong, Z.; Bai, L. Discovery and characterizaton of a novel lipase with transesterification activity from hot spring metagenomic library. *Biotechnol. Rep.* **2017**, *14*, 27–33. [Google Scholar] [CrossRef]

208. Kotik, M.; Vanacek, P.; Kunka, A.; Prokop, Z.; Damborsky, J. Metagenome-derived haloalkane dehalogenases with novel catalytic properties. *Appl. Microbiol. Biotechnol.* **2017**, *101*, 6385–6397. [Google Scholar] [CrossRef]

209. Apolinar, M.M.; Peña, Y.J.; Pérez-rueda, E.; Canto-canché, B.B.; Santos-briones, C.D.L.; Connor-sánchez, A.O. Identi fi cation and in silico characterization of two novel genes encoding peptidases S8 found by functional screening in a metagenomic library of Yucatán underground water. *Gene* **2016**, *593*, 154–161. [Google Scholar] [CrossRef]

210. Aviv-Reuven, S.; Rosenfeld, A. Publication patterns' changes due to the COVID-19 pandemic: A longitudinal and short-term scientometric analysis. *Scientometrics* **2021**, *126*, 6761–6784. [Google Scholar] [CrossRef]

211. Lai, O.M.; Lee, Y.Y.; Phuah, E.T.; Akoh, C.C. Lipase/esterase: Properties and industrial applications. In *Encyclopedia of Food Chemistry*; Melton, L., Shahidi, P.V.F., Eds.; Elsevier: Amsterdam, The Netherlands, 2018; pp. 158–167. ISBN 9780128140451. [Google Scholar]

212. Linares-Pasten, J.; Andersson, M.; Karlsson, E. Thermostable glycoside hydrolases in biorefinery technologies. *Curr. Biotechnol.* **2014**, *3*, 26–44. [Google Scholar] [CrossRef]

213. Liu, X.; Kokare, C. Microbial enzymes of use in industry. In *Biotechnology of Microbial Enzymes*; Brahmachari, G., Ed.; Elsevier: Amsterdam, The Netherlands, 2017; pp. 267–298. [**Google Scholar**]

214. Martínez, A.T.; Ruiz-Dueñas, F.J.; Camarero, S.; Serrano, A.; Linde, D.; Lund, H.; Vind, J.; Tovborg, M.; Herold-Majumdar, O.M.; Hofrichter, M.; et al. Oxidoreductases on their way to industrial biotransformations. *Biotechnol. Adv.* **2017**, *35*, 815–831. [**Google Scholar**] [**CrossRef**] [**PubMed**][**Green Version**]

215. Alneyadi, A.H.; Rauf, M.A.; Ashraf, S.S. Oxidoreductases for the remediation of organic pollutants in water—A critical review. *Crit. Rev. Biotechnol.* **2018**, *38*, 971–988. [**Google Scholar**] [**CrossRef**] [**PubMed**]

216. Jatuwong, K.; Suwannarach, N.; Kumla, J.; Penkhrue, W.; Kakumyan, P.; Lumyong, S. Bioprocess for production, characteristics, and biotechnological applications of fungal phytases. *Front. Microbiol.* **2020**, *11*, 188. [**Google Scholar**] [**CrossRef**] [**PubMed**]

217. Berry, D.F.; Shang, C.; Zelazny, L.W. Measurement of phytase activity in soil using a chromophoric tethered phytic acid probe. *Soil Biol. Biochem.* **2009**, *41*, 192–200. [**Google Scholar**] [**CrossRef**]

CHAPTER-13

CONTINUOUS SYNTHESIS OF DHA AND EPA ETHYL ESTERS IS ACHIEVED THROUGH LIPASE-CATALYZED TRANSESTERIFICATION WITHIN AN ULTRASONIC PACKED-BED BIOREACTOR

1. INTRODUCTION: AN OVERVIEW

In the pursuit of sustainable and efficient methods for the production of valuable omega-3 polyunsaturated fatty acids (PUFAs) - specifically docosahexaenoic acid (DHA) and eicosapentaenoic acid (EPA) ethyl esters - continuous processes have emerged as pivotal strategies. This study focuses on a novel approach utilizing lipase-catalyzed transesterification within an ultrasonic packed-bed bioreactor, representing a cutting-edge synergy of enzymatic catalysis and advanced reactor design.

DHA and EPA, renowned for their numerous health benefits, are traditionally sourced from marine organisms. However, the increasing demand for these PUFAs, coupled with concerns about overfishing and environmental sustainability, has prompted the exploration of alternative production methods. Enzymatic transesterification, facilitated by lipases, offers a bio-friendly and selective route for the synthesis of DHA and EPA ethyl esters. The integration of ultrasonic technology into a packed-bed bioreactor introduces an innovative dimension, enhancing mass transfer and enzyme-substrate interactions.

Continuous production is a key element in addressing scalability and economic viability in industrial processes. By employing a packed-bed bioreactor configuration with ultrasonic assistance, this study aims to optimize the transesterification process, enabling a continuous and efficient generation of DHA and EPA ethyl esters. The incorporation of ultrasonic energy not only enhances reaction kinetics but also contributes to the uniform distribution of reactants within the bioreactor, thereby improving overall process efficiency.

At present, the global annual production of fish oil is around 1,000,000 tons, and the global market of ω-3 polyunsaturated fatty acids was valued at USD 2.49 billion in 2019 [1]. As people's consumption levels improve, the fish oil market continues to grow, and the amount of fish oil used in health care products is also rapidly expanding. DHA and EPA are well-known active ingredients in fish oil. Fish oils vary in DHA and EPA levels, depending on the source [2], so their value varies greatly. As nutritional supplements, the total content of DHA and EPA is generally greater than 30%. As a pharmaceutical grade, the total DHA and EPA content must be over 60% [3]. However, the DHA and EPA content in fish oil is typically below 30%. Several methods have been used for concentrating ω-3 PUFA to meet the specifications for health-promoting food or medicine, including urea complexation [4], molecular distillation [5], supercritical fluid method [6], reverse phase HPLC [7], solvent fractionation [8], and enzyme method [9].

The concentrated or purified fish oil products rich in DHA and EPA can be divided into three types according to their chemical structure: (1) glyceride; (2) fatty acid ethyl ester; (3) free fatty acid (FFA). The glyceride type is commonly sold as a health-promoting food. For pharmaceutical grade products, high purity is required and usually present as ethyl esters sold on the market, such as Lovaza® and Vascepa® [10,11]. To increase the content of DHA and EPA in the fish oil, the concentrated FFA with a high amount of DHA and EPA can be used as raw material to synthesize into the form of glyceride or ester through lipase catalyzed esterification, i.e., the transesterification of phophatidylcholine with PUFA in hexane [12], acidolysis of ethyl acetate (EA) with DHA+ EPA concentrate in hexane [13], esterification of tuna–FFA with various alcohol [14,15], esterification of sardine–FFA with glycerol [16].

Lipase-catalyzed esterification was performed in the packed-bed bioreactor [17,18]. The packed-bed bioreactor consists of simple equipment, a pump and a column packed with immobilized lipase. The reagents are pumped into a long tubular column to react directly with the immobilized lipase, resulting in high reaction rates and shortened reaction times to achieve a higher conversion [19,20]. Due to the advantages of high efficiency, lower energy consumption, minimum reaction volume and convenient operation, the packed-bed bioreactor is often used in the industry to maximize the efficiency for continuous operation of enzyme reaction [21]. In addition, ultrasound improves the mass transfer rate [22] and delivers activation energy to trigger reactions, which is a useful tool to accelerate enzyme reactions in different systems [23]. Ultrasonic-assisted enzymatic reactions have been widely investigated and reported [24,25]. Ultrasonication accelerated lipase-catalyzed acetylation of resveratrol and their potential kinetic response models were widely reported [26]. Lipase-catalyzed transesterification and esterification reactions were reported that could proceed with a two-substrate reaction model, ping-pong bi–bi [27] and order bi–bi mechanism [13,28]. In contrast, there are limited data available on the packed-bed bioreactors for continuous synthesis of DHA and EPA ethyl esters. In particular, DHA and EPA are long carbon chain FFA; when the enzyme acts on the substrate, it should be in a specific position to synthesize the DHA and EPA ethyl esters, making the reaction longer.

A solvent-free system means that the reactant also plays the role of solvent in the reaction. Esters synthesized using solvent-free systems have several advantages, such as high substrate concentrations, high conversion and less reaction time [29]. The ethyl acetate (EA) has been used in solvent-free system for the lipase-catalyzed synthesis of many esters. Lipase-catalyzed transesterification of geraniol to geranyl acetate at a mole ratio of 1:7 of geraniol to ethyl acetate, obtaining 83% conversion [30]. A high substrate concentration of 2-phenethyl alcohol (493.4 mM) was reacted with EA in a solvent-free system to produce 2-phenylethyl acetate, and a conversion of 97.64% was achieved at reaction time of 3.07 h [31]. In a previous study, we used lipase-catalyzed transesterification of EA with DHA and EPA concentrate in n-hexane to obtain in DHA and EPA ethyl ester [13]. Therefore, a solvent-free system employing the reactant EA as the solvent for the transesterification reaction to synthesize DHA and EPA ethyl esters was explored. It should accelerate the reaction to the direction in which the product is synthesized.

In this study, a packed-bed bioreactor was developed for the continuous synthesis of DHA and EPA ethyl esters using a solvent-free system. For understanding the relationship between conversion, flow rate and substrate concentration, we applied the central composite design and response surface methodology (RSM). More specifically, the mass transfer effects on the performance of the packed bed bioreactor with ultrasonication were also studied. Besides, we developed a kinetic model to evaluate ultrasonication's effects on the lipase-catalyzed synthesis of DHA and EPA ethyl esters. Finally, the stability of the packed-bed bioreactor was evaluated under long-term operation.

2. RESULTS AND DISCUSSION

2.1. Effect of Ultrasonication on DHA and EPA Ethyl Ester Synthesized Using Lipase Packed-Bed Bioreactor

Figure 1 shows a diagram of the apparatus. The column packing with immobilized lipase was placed in a temperature-controlled ultrasonic bath. The DHA + EPA concentrate was well mixed with EA in a feeding flask. The experimental design is based on the three-level–two-factor central composite design as shown in Table 1. Initially, the lipase-catalyzed synthesis of DHA + EPA ethyl ester was carried out at a reaction temperature of 60°C with a flow rate of 1 to 5 mL min^{-1}, without ultrasonication or with ultrasonication. Zanwar and Pangarkar have reported that the mass transfer coefficient was increased with acoustic power [32]. Therefore, the ultrasonic bath was set at 100% output power. The substrate mixture was fed through the column using a reciprocating piston pump. The HPLC analysis of reactants and products eluting from the packed-bed bioreactor is shown in Figure 2a,b, respectively. After lipase catalysis, the DHA + EPA concentrate was mostly converted into DHA + EPA ethyl ester, as shown in Table 1. The highest conversion was 98.84% in treatment 1 (DHA + EPA 100 mM, flow rate 1 mL min^{-1}), and the lowest conversion was 86.61% in treatment 10 (DHA + EPA 500 mM, flow rate 5 mL min^{-1}) for the packed-bed bioreactor. In the contrast, the ultrasonic packed-bed bioreactor had the highest conversion of 99.21% in treatment 1 and the lowest conversion of 93.03% in treatment 10. As a result, the ultrasonication-improved synthetic conversion was obtained for all treatments in Table 1. These results showed that ultrasonication can effectively enhance the synthesis of DHA + EPA ethyl ester in the packed-bed bioreactor and can obtain a higher conversion when it operates at higher substrate concentration and higher flow rates.

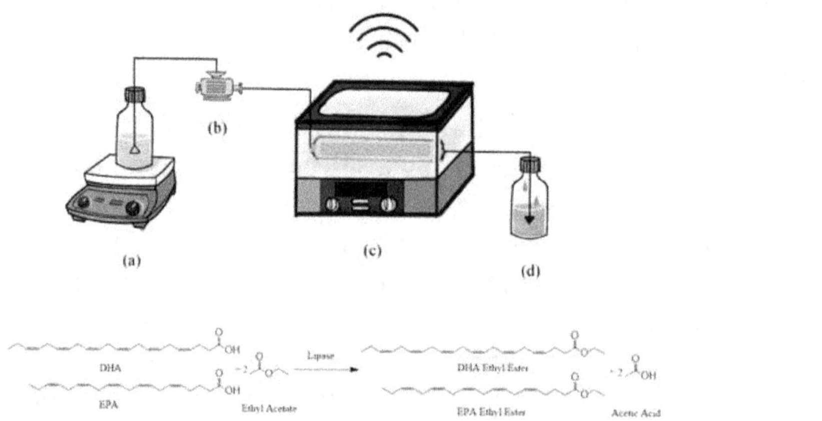

Figure 1. **Diagram of the ultrasonic packed-bed bioreactor: (a) DHA + EPA concentrate in the EA; (b) pump; (c) temperature-controlled ultrasonic bath; (d) product.**

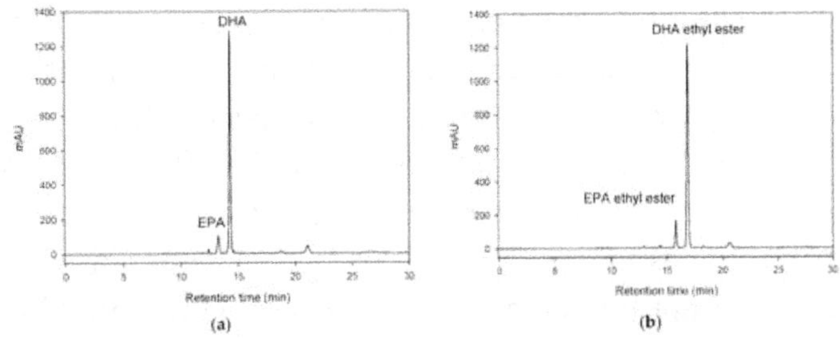

Figure 2. HPLC analysis of (a) the reaction mixture and (b) product eluting from the lipase packed-bed bioreactor.

Table 1. Central composite design and experimental conversion of DHA + EPA ethyl ester from the packed-bed bioreactor and ultrasonic packed-bed bioreactor.

Treatment No.	Factor		Conversion (%)	
	x_1 Flow Rate (mL min^{-1})	x_2 DHA + EPA (mM)	Packed-Bed Bioreactor	Ultrasonic Packed-Bed Bioreactor
1	-1 [a] (1)	-1 (100)	98.84 ± 0.05	99.21 ± 0.00
2	-1 (1)	0 (300)	97.01 ± 1.30	97.61 ± 0.03
3	-1 (1)	1 (500)	95.49 ± 0.42	97.34 ± 0.10
4	0 (3)	-1 (100)	96.59 ± 0.91	98.82 ± 0.01
5	0 (3)	0 (300)	95.48 ± 0.94	96.34 ± 0.19
6	0 (3)	0 (300)	95.13 ± 0.14	96.00 ± 0.05
7	0 (3)	1 (500)	91.79 ± 1.33	95.69 ± 0.32
8	1 (5)	-1 (100)	95.51 ± 0.35	96.74 ± 0.61
9	1 (5)	0 (300)	92.41 ± 1.03	94.50 ± 0.66
10	1 (5)	1 (500)	86.61 ± 1.28	93.03 ± 1.55

[a] The values -1, 0 and 1 are coded levels.

The second-order polynomial Equations (1) and (2) were obtained from the Design-Expert software by fitting the data in **Table 1**.

$Y_1\,(\%) = 97.985136 + 0.088092x_1 + 0.008523x_2 - 0.074570x_1x_1 - 0.003473x_2x_1 - 0.000020509x_2x_2$ (1)

$Y_2\,(\%) = 100.683977 + 0.172551x_1 - 0.015220x_2 - 0.108444x_1x_1 - 0.001154x_2x_1 + 0.000019055x_2x_2$ (2)

where Y_1 is the conversion of DHA + EPA ethyl ester synthesized using a packed-bed bioreactor; Y_2 is the conversion of DHA + EPA ethyl ester synthesized using an ultrasonic packed-bed bioreactor; x_1 is the concentration of DHA + EPA (100–500 mM), x_2 is the flow rate (1–5 mL min^{-1}).

According to the results of the analysis of variance (ANOVA) data in **Table 2**, the polynomial model can adequately describe the actual relationship between response and significant variables, as indicated by a significant total model for $p < 0.05$, and a satisfactory coefficient of determination ($R^2 = 0.99$). Moreover, the ANOVA results indicate that the linear term and interaction term had a significant influence ($p < 0.05$) on responses for both models. The quadratic terms had less influence ($p > 0.05$), except for the quadratic terms of Equation (2) ($p < 0.05$).

Table 2. Analysis of variance (ANOVA) analysis for continuous lipase-catalyzed synthesis of DHA + EPA ethyl ester by packed-bed bioreactor and ultrasonic packed-bed bioreactor.

Factor [a]	Packed-Bed Bioreactor (Y_1)			Ultrasonic Packed-Bed Bioreactor (Y_2)		
	Degree of Freedom	Sum of Squares	Prob > F	Degree of Freedom	Sum of Squares	Prob > F
Model	5	105.27	0.0009 *	5	31.35	0.0004 *
Linear term						
x_1	1	47.12	0.0004 *	1	16.31	0.0001 *
x_2	1	48.40	0.0004 *	1	12.61	0.0002 *
Quadratic						
x_{11}	1	0.21	0.5053	1	0.44	0.0735
x_{22}	1	1.57	0.1147	1	1.36	0.0133 *
Interaction						
x_2x_1	1	7.72	0.0112 *	1	0.85	0.0284 *
	$R^2 = 0.99$			$R^2 = 0.99$		

[a] Independent variable x_1: flow rate, x_2: DHA + EPA concentration. * Significant at *p*-Value less than 0.05.

The response surface plots of the packed-bed bioreactor show the effect of the DHA + EPA concentration and flow rate on the conversion as shown in **Figure 3**a. The conversion was only ~87% when the flow rate was 5 mL min^{-1} and the DHA + EPA concentration was 500 mM. However, the conversion was ~98% when the flow rate was decreased to 1 mL min^{-1} and the DHA + EPA concentration was decreased to 100 mM. In contrast, the conversions of DHA + EPA ethyl ester for the ultrasonic packed-bed bioreactor (**Figure 3**b) were ~93% at a flow rate of 5 mL min^{-1} and DHA + EPA concentration of 500 mM and ~99% at a flow rate of 1 mL min^{-1} and DHA + EPA concentration of 100 mM, the conversion was higher than that of the packed-bed bioreactor at the same flow rate and DHA + EPA concentration. Therefore, ultrasonication can significantly improve the efficiency for the synthesis of DHA/EPA ethyl ester, allowing the packed-bed bioreactor to operate at higher concentrations and flow rates. Stavarache et al. used base-catalyzed synthesis of biodiesel with low-frequency ultrasonic waves (28 and 40 kHz); the time required for ultrasonic synthesis was shorter [33]. Ji et al. studied soybean oil transesterification using alkali as a catalyst to produce biodiesel with ultrasonication; the highest yield of 100% can be obtained within the reaction time of 10–30 min [34]. Liu et al. compared the effect of lipase-catalyzed hydrolysis of soy oil in a shaking bath and in an ultrasonic bath. The overall hydrolysis reaction rate in the ultrasonic bath was above 2-fold than that in the shaking bath [35]. Yu et al. conducted the transesterification of soybean oil and methanol with Novozym® 435 to synthesize biodiesel and discussed the effect of vibration speed and ultrasonication. The authors found that enzymes work at a vibration speed of 50 rpm with ultrasonication; a yield of 96% can be obtained in 4 h [36]. Patchimpet et al. reported that ultrasonic irradiation coupled with stirring enhanced the transesterification with the highest yield of 97.59% [37]. Our experimental results may be shown in the same way as the results of the above literature. It is deduced that ultrasonication can assist the enzymatic reaction, which increases the reaction rate and decreases the reaction time.

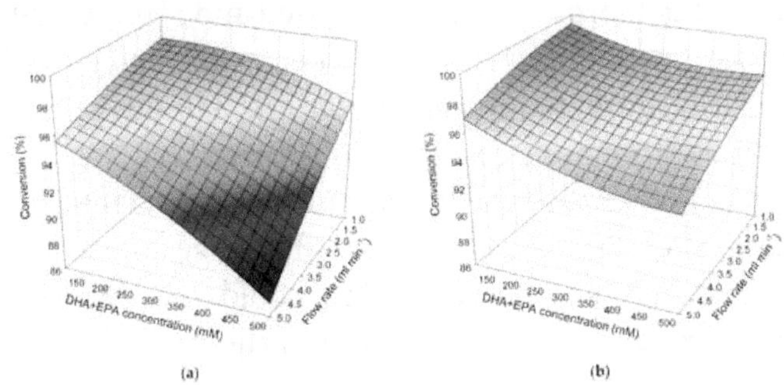

Figure 3. Response surface plots show the mutual effects of DHA + EPA concentration and flow rate on the conversion for (a) packed-bed bioreactor and (b) ultrasonic packed-bed bioreactor.

2.2. Mass Transfer Kinetic Model of an Ultrasonic Packed-Bed Reactor

It is possible to determine a mass transfer limited model of a packed-bed bioreactor based on the mass balance of DHA + EPA in the column and the mass transfer rate of DHA + EPA from the bulk fluid to the surface of the immobilized enzyme. The mass transfer limited model is as follows:

$$\frac{VdS}{AxdZ} = -km \cdot am(S - Si) \quad (3)$$

where V is the substrate flow rate (mL min^{-1}), Ax is the cross-sectional area of column (cm^2), dS/dZ is the concentration gradient along the column length (mM cm^{-1}), Z is the column length (cm), km is the external mass transfer coefficient (cm min^{-1}), am is area of mass transfer (cm^2 cm^{-3}), S is DHA + EPA concentration in the bulk liquid (mM), and Si is DHA + EPA concentration at the external surface of the immobilized enzyme (mM) under mass transfer condition.

Equation (3) uses the boundary conditions as $Z = 0$; $S = S_0$;$Z = h$; $S = Si$ to obtain Equation (4).

$$ln(\frac{S_0 - Si}{S - Si}) = \frac{km \cdot am \cdot Ax \cdot Z}{V} \quad (4)$$

The km can be obtained by plotting $ln(S_0 - Si/S - Si)$ versus $amAxZ/V$. The experimental data obtained from response surface plots were then used to plot the graphs for the mass transfer limited model. The value of Si was obtained by fitting the data to Equation (4), which gives a high value of R^2. **Figure 4** is an example to show a plot $ln(S_0 - Si/S - Si)$ versus $amAxZ/V$ for the mass transfer limited condition at a DHA + EPA concentration of 500 mM. In this way, the best fit of Si value gives the highest value of R^2 for packed-bed bioreactor and ultrasonic packed-bed bioreactor listed in **Table 3**. In both bioreactors, km decreased and Si increased with increasing substrate concentration. Similar results have been reported by Todero et al. in the case of the enzymatic synthesis of isoamyl butyrate under optimal experimental conditions [38]. In addition, our results also showed that the ultrasonic packed-bed bioreactor had a higher external mass transfer coefficient km and a lower Si as compared to the packed-bed bioreactor. At this stage, the mass transfer model experiments confirmed that ultrasonication could indeed assist the action of enzymes, which was reflected in the increase in the reaction rate and the higher conversion.

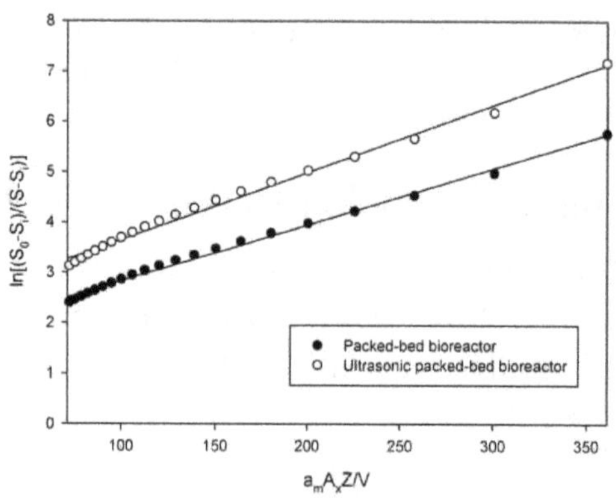

Figure 4. The mass transfer limited model of • packed-bed bioreactor and ○ ultrasonic packed-bed bioreactor operated at a DHA + EPA concentration of 500 mM and different flow rates.

Table 3. ***km*** **and** ***Si*** **obtained from Equation (5) for packed-bed bioreactor and ultrasonic packed-bed bioreactor operated at a different substrate concentration.**

DHA + EPA Conc. (mM)	Packed-Bed Reactor			Ultrasonic Packed-Bed Reactor		
	k_m (cm min^{-1})	S_i (mM)	R^2	k_m (cm min^{-1})	S_i (mM)	R^2
100	0.0138	1.66	0.99	0.0163	0.68	0.99
300	0.0123	6.60	0.99	0.0140	7.20	0.99
500	0.0112	21.50	0.99	0.0133	13.00	0.99

2.3. Continuous Synthesis of DHA + EPA Ethyl Ester by Packed-Bed Bioreactor with Long-Term Operation

In the batch reactor, the by-product, acetic acid, is produced continuously and accumulates in the reactor. It has been reported that the high acid concentration may cause enzyme inhibition [39,40]. Unlike batch reactors, by-products continually flow out of the packed-bed bioreactor, and there is no accumulation of by-products in the reactor. Therefore, the reusability and productivity of enzymes are better when used in continuous production than that in batch production [41]. In this way, the experiments in this section were to test the long-term stability of the packed-bed bioreactor for the continuous synthesis of DHA + EPA ethyl ester. Because the long-term operation required a large amount of DHA + EPA concentrate, the long-term operation was carried out at a low flow rate and low DHA + EPA concentration. In the packed-bed bioreactor or ultrasonic packed-bed bioreactor, the bioreactors were in continuous operation for 5 days at a DHA + EPA concentration of 100 mM, a flow rate of 1 mL min^{-1} and a temperature of 60 °C. The results are shown in Figure 5. After continuous operation for 5 days, the conversion remained at 98% without a downtrend. The reason for no difference in conversion between ultrasonic packed-bed and packed-bed bioreactors is that the experiments were operated at a low flow rate and low DHA + EPA concentration. The conversion is proportional to the residence time of the substrate in the packed-bed bioreactor and is inversely proportional to the substrate concentration. A lower flow rate indicated that the substrate had sufficient residence time to fully convert to product, so the benefit of ultrasonication cannot be seen. However, operating under conditions of high flow rate (5 mL min^{-1}) and high substrate concentration (500 mM), as shown in Figure 3a,b, due to insufficient residence time, the conversion of the packed-bed bioreactor was 86.61%, while the conversion of the packed-bed bioreactor was 93.03%. Under the conditions of high substrate concentration and high flow rate, the ultrasound showed a significant positive impact on reaction conversion. Zenevicz et al. [42] found that the enzymatic production of ethyl esters in continuous mode coupled with an ultrasound bath increased conversion from 87% to 95%, which is consistent with our results. Furthermore, the long-term operation showed the packed-bed bioreactor was very stable. This result confirmed Novozym® 435 is suitable for the packed-bed bioreactor for continuous production of DHA + EPA ethyl ester. The packed-bed bioreactor is a continuous operation system, which is superior to batch production in terms of enzyme utilization and production cost and has good feasibility for future application in production capacity or industrial mass production. It should be kept in mind that mass production required the operating conditions of flow rate and substrate concentration as high as possible. Our results confirmed that the use of ultrasound improved the conversion effectively when production of DHA + EPA ethyl ester operated at a higher flow rate or substrate concentration. In addition, extending the length of the column to increase the residence time of the substrate, or connecting the columns in series, can also achieve the effect of improving the conversion [43,44].

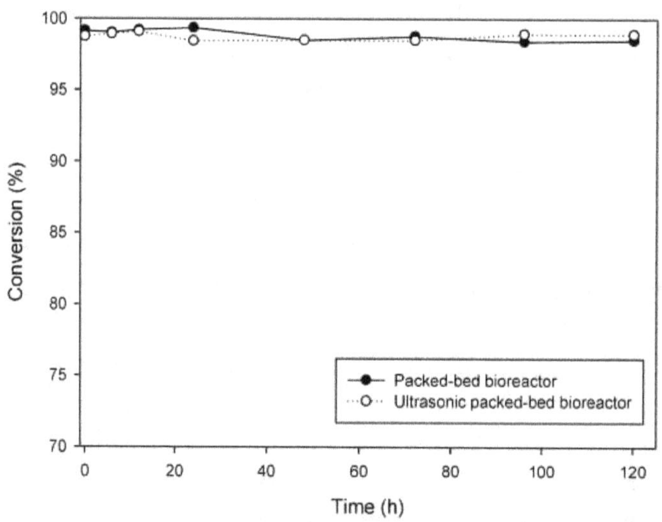

Figure 5. Long-term operation of • packed-bed bioreactor and ○ ultrasonic packed-bed bioreactor for continuous synthesis of DHA + EPA ethyl ester.

2.4. Recovery of DHA/EPA Ethyl Ester

An amount of 100 mL of the product produced by the ultrasonic packed-bed bioreactor using a DHA + EPA concentration of 100 mM after 5 days of operation as described in **Section 2.3** was collected for recovery of DHA + EPA ethyl ester. The vacuum rotary evaporator was used to remove ethyl acetate and acetic acid at 80°C; a 3.15 g of DHA + EPA ethyl ester was obtained with a yield of 96%, and a conversion of 98%. Next, the ultrasonic packed-bed bioreactor was operated at 60°C with increasing DHA + EPA concentration to 500 mm and flow rate to 5 mL min^{-1}. A 15.9 g of DHA + EPA ethyl ester was obtained, with a yield of 97% and a conversion of 93%. It can be seen from the recovery results that when the conversion is high enough, the purified product can be easily obtained.

2.5. Evaluate the Ultrasonication Effect by the Kinetic Model in the Batch Reaction

In the packed-bed bioreactor, the substrate was reacted in the presence of excess enzymes. In order to further examine the effect of ultrasonication, we performed enzyme kinetic studies in the batch reaction with trace amount of enzyme. The lipase-catalyzed synthesis of ester is a two-substrates reaction. The kinetic mechanism of the dual substrate reaction is more complicated as compared with one substrate reaction. Generally, three mechanisms can be used to describe the multi-substrate enzyme-catalyzed reaction: the ping-pong Bi–Bi mechanism, the ordered mechanism and the random order mechanism. In the first mechanism, the first substrate is combined with the enzyme to form a substituted enzyme intermediate and release the first product. As the second substrate interacts with the substituted enzyme intermediate, the second product is produced, and the native enzyme is regenerated. The latter two mechanisms are before the release of the product; the enzyme must be fully combined with the substrate to react, and then the product will appear. Our previous studies have shown that lipase-catalyzed transesterification can be represented by an ordered mechanism [28]. The kinetic model is as follows:

$$v = \frac{Vmax[A][B]}{K_{dA}K_{mB} + K_{mB}[A] + K_{mA}[B] + [A][B]} \quad (5)$$

where v is the initial reaction rate, *Vmax* is the maximum initial reaction rate, [A] is the initial concentration of DHA + EPA, [B] is the initial concentration of EA, and K_{mA} and K_{mB} are the Michaelis constants for DHA + EPA and EA, respectively. K_{dA} is the dissociation constant of the DHA + EPA–lipase complex.

Equation (5) can be arranged into the following equation by combining the parameters:

$$v = \frac{VmaxK_1[A]}{K_2 + [A]} \quad (6)$$

where $K_1 = \frac{[B]}{K_{mB}+[B]}$, $K_2 = \frac{K_{dA}K_{mB}+K_{mA}[B]}{K_{mB}+[B]}$, let $V \; max = VmaxK_1$, obtain Equation (7) that is similar to Michaelis–Menten equation.

$$v = \frac{V'max[A]}{K_2 + [A]} \quad (7)$$

Therefore, the kinetic model of the reaction is only related to the concentration of DHA + EPA. This particular kinetic model can be used to analyze lipase-catalyzed reactions in solvent-free systems, such as esterification of formic acid [45], synthesis of amyl levulinate [46], synthesis of monoglyceryl phenolic acids [47], synthesis of ethyl valerate [48], synthesis of geranyl acetate [49]. Since the solvent concentration is saturated and much greater than the substrate concentration, the solvent concentration can be regarded as a constant. Thus, the kinetic parameters can be obtained by using the Lineweaver–Burk plot method [50].

Lipase-catalyzed synthesis of DHA + EPA ethyl esters was performed using different DHA + EPA concentrations (50–400 mM) in EA. The reactions were carried out in a shaking bath at 150 rpm or in an ultrasonic bath at 37 Hz, respectively. According to Equation (7), plot the reciprocal initial reaction rate ($1/v$) versus the reciprocal substrate concentrations ($1/[A]$) (i.e., Lineweaver–Burk plot) is shown in **Figure 6**. An acceptable value of the determination coefficient ($R^2 = 0.98$) confirmed the fitness of the kinetic Equation (7). The Lineweaver–Burk plot showed both curves were linear and no upward curve at high substrate concentrations (lower $1/[A]$), indicating that the substrate inhibition did not occur in this solvent-free system at high concentrations of DHA + EPA. Moreover, the kinetic parameters can be obtained from the slope and intercept in **Figure 6**. The values of the apparent Michaelis constant (K_2), apparent maximum initial reaction rate ($V'max$) and specificity constant ($V'max/K_2$) were 960.02 mM, 11.76 mM min^{-1} and 0.012 min^{-1}, respectively, for shaking bath, and 715.98 mM, 76.92 mM min^{-1} and 0.107 min^{-1} for ultrasonic bath. The $V'max$ of the lipase-catalyzed synthesis of DHA/EPA ethyl ester in the ultrasonic bath increased about 6.54 times, while the K_2 decreased. K_2 represents the affinity of enzymes and substrates; the smaller K_2 means the greater affinity of the enzyme and the substrate. The ultrasonic bath showed lower K_2 indicating that the ultrasonication increased the substrate affinity toward immobilized lipase. $V'max/K_2$ can be used to express the specificity constant, which reflects both affinity and catalytic ability [28,51,52]. The ultrasonic bath showed 8.9 times higher $V'max/K_2$ value than that of the shaking bath, indicating that ultrasonication highly enhanced the efficiency of lipase-catalyzed transesterification. The higher reaction rate increased by ultrasonication can shorten the reaction time. Lipase-catalyzed synthesis of cetyl oleate with ultrasonication has been reported to reduce reaction time by about 75% [53]. Ultrasonication-assisted lipase-catalyzed synthesis of cinnamyl acetate also found that the reaction time for the maximum conversion was reduced from 60 min to 20 min [54].

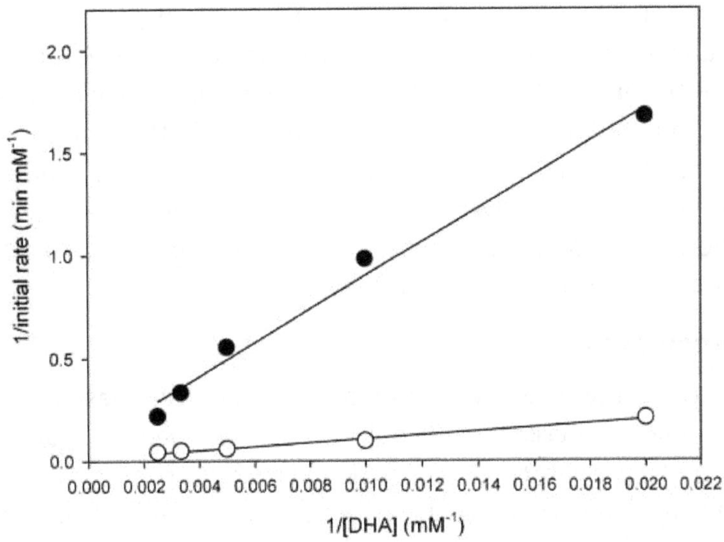

Figure 6. Lineweaver–Burk plot of reciprocal initial reaction rate versus reciprocal concentrations of DHA + EPA concentrate for lipase-catalyzed transesterification in • shaking bath and ○ ultrasonic bath.

3. MATERIALS AND METHODS

3.1. Materials

Immobilized lipase (Novozym® 435, 10,000 PLU U/g; propyl laurate units) was purchased from Novo Nordisk Bioindustrials Inc. (Copenhagen, Denmark). As described previously [8,13], DHA + EPA concentrate with an average molecular weight of 312 was produced from cobia liver oil by fractionating fatty acids salts using acetone. There was 54.4% DHA, 16.8% EPA, 7.0% docosadienoic acid, 14.2% oleic acid, 0.8% linoleic acid, 0.5% linolenic acid, 3.1% palmitoleic acid, 2.7% palmitic acid and 0.5% myristic acid in DHA + EPA concentrate. Ethyl acetate (99.8%) was produced by Merck (Darmstadt, Germany). *cis*-4,7,10,13,16,19-Docosahexaenoic acid and *cis*-5,8,11,14,17-eicosapentaenoic acid used for analysis were purchased from Acros (Fair Lawn, NJ, USA) and TCI Co., LTD. (Tokyo, Japan), respectively. All chemicals and reagents used in this study were of analytical grade, unless otherwise noted.

3.2. Continuous Synthesis of DHA and EPA Ethyl Ester in Packed-Bed Bioreactor

The apparatus of the packed-bed bioreactor is shown in **Figure 2**. The packed-bed bioreactor was implemented in a stainless-steel column at 60°C. The stainless-steel column was 25 cm in length with an inner diameter of 0.25 cm and packed with 1.3 g of Novozym® 435. The upper and lower end of the column was layered with a 2 μm filter. The flow rates were controlled by a Hitachi Pump (L-2130, Hitachi, Tokyo, Japan). The ultrasonic packed-bed bioreactor was carried out in the ultrasonic bath and operated at 37 kHz with 100% output. A feeding flask containing various concentrations of DHA + EPA (100–500 mM) in EA was thoroughly mixed before reaction. The reaction mixture in the feeding flask was pumped continuously into the column under designed flow rates (1–5 mL min^{-1}). The product was collected at the end of the column. After each experiment, ethyl acetate was used to flush the column at a flow rate of 1 mL min^{-1} for 30 min.

3.3. Experimental Design

In this study, the three-level-two-factor central composite design was employed, and the experiments were carried out in a packed-bed reactor or ultrasonic packed-bed bioreactor at 60°C. The experimental variables were DHA + EPA concentration (100–500 mM), mixture flow rate (1–5 mL min^{-1}), as shown in **Table 1**. The experimental data (**Table 1**) were analyzed by Design-Expert software 8.0 (StatEase Inc., Minneapolis, MN, USA) to fit the following second-order polynomial equation:

$$Y = \beta_0 + \beta_1 X_1 + \beta_2 X_2 + \beta_{12} X_1 X_2 + \beta_{11} X_1^2 + \beta_{22} X_2^2$$

where *Y* is the response (conversion of DHA+EPA ethyl ester; %), β_0 is the constant term; β_1 and β_2 are coefficients of the linear effects, β_{11} and β_{22} are coefficients of quadratic effects and β_{12} is coefficients of interaction effects; X_1 and X_2 are the uncoded independent variables.

3.4. Measure Initial Rate of DHA and EPA Ethyl Ester Synthesized in the Batch Reaction

Various amounts of DHA + EPA concentrate were added with EA to a total volume of 3 mL to make the solution concentration of 50–400 mM. Then, 20 mg immobilized lipase Novozym® 435 was added to 3 mL of the reaction mixture. The reactions were performed at 60°C for 5 min and 20 min with or without ultrasonication, respectively. The ultrasonic bath (Elmasonic P 70 H, Elma, Siegen, Germany) was operated at 37 kHz with 100% output power. A liquid sample was withdrawn from the reaction mixture to determine the production of DHA + EPA ethyl ester using HPLC after the reaction. Initial reaction rates are expressed as produced DHA + EPA ethyl ester mM per min (mM min^{-1}). The $V'max$ and K_2, were obtained from the Lineweaver–Burk plot.

3.5. Analytical Methods

Inertsil ODS-3 column (5 μM, 250 mm × 4.6 mm) and HPLC system, consisting of a Hitachi L-2130 HPLC pump and a Hitachi L-2420 UV/VIS detector (Hitachi, Tokyo, Japan), were used to analyze the samples. Deionized water and methanol containing

0.1% acetic acid were used for eluting the sample. From 80% to 100% methanol, gradient elution was carried out for 10 min, followed by 20 min at 100% methanol. Flow rate and wavelength were both set to 1.0 mL min^{-1} and 303 nm, respectively. The conversion (%) was calculated from the peak areas of the substrate (DHA + EPA) and product (DHA + EPA ethyl ester).

4. CONCLUSIONS

Continuous synthesis of DHA + EPA ethyl ester using a lipase packed-bed bioreactor was successfully studied. The benefits of ultrasonication in increasing the reaction rate were successfully demonstrated in both kinetic and mass transfer models. A three-level–two-factor central composite design and RSM were employed for the experimental design and data analysis. A model for the DHA + EPA ethyl ester synthesis was established. With ultrasonication, the highest conversion of 99% was obtained with ultrasonication to be a flow rate of 1 mL min^{-1}, and substrate concentration of 100 mM. The mass transfer rates could be increased by ultrasonication. Therefore, when the substrate concentration increased to 500 mM and the flow rate increased to 5 mL min^{-1}, the conversion rate still is maintained at 93%. The ultrasonic packed-bed bioreactor used for continuous synthesis of DHA + EPA ethyl ester was quite stable under long-term operation, and no reduction in conversion was observed. The high conversion makes the product easy to recover. Therefore, an ultrasonic packed-bed bioreactor combined with a solvent-free system has the potential to be utilized in industrial applications for the continuous production of DHA + EPA ethyl ester. In addition, we also successfully developed an apparent Michaelis–Menten equation that can be used to compare the efficiency of lipase-catalyzed reactions under different reactor operations.

5. CONCISE SUMMARY

Ethyl esters of omega-3 fatty acids are active pharmaceutical ingredients used for the reduction in triglycerides in the treatment of hyperlipidemia. Herein, an ultrasonic packed-bed bioreactor was developed for continuous production of docosahexaenoic acid (DHA) and eicosapentaenoic acid (EPA) ethyl esters from DHA+EPA concentrate and ethyl acetate (EA) using an immobilized lipase, Novozym® 435, as a biocatalyst. A three-level–two-factor central composite design combined with a response surface methodology (RSM) was employed to evaluate the packed-bed bioreactor with or without ultrasonication on the conversion of DHA + EPA ethyl ester. The highest conversion of 99% was achieved with ultrasonication at the condition of 1 mL min^{-1} flow rate and 100 mM DHA + EPA concentration. Our results also showed that the ultrasonic packed-bed bioreactor has a higher external mass transfer coefficient and a lower external substrate concentration on the surface of the immobilized enzyme. The effect of ultrasound was also demonstrated by a kinetic model in the batch reaction that the specificity constant ($V'max/K_2$) in the ultrasonic bath was 8.9 times higher than that of the shaking bath, indicating the ultrasonication increased the affinity between enzymes and substrates and, therefore, increasing reaction rate. An experiment performed under the highest conversion conditions showed that the enzyme in the bioreactor remained stable at least for 5 days and maintained a 98% conversion.

REFERENCES

1. Oliver, L.; Dietrich, T.; Marañón, I.; Villarán, M.C.; Barrio, R.J. Producing omega-3 polyunsaturated fatty acids: A Review of Sustainable Sources and Future Trends for the EPA and DHA Market. *Resources* **2020**, *9*, 148. [Google Scholar] [CrossRef]
2. Mohanty, B.P.; Ganguly, S.; Mahanty, A.; Sankar, T.V.; Anandan, R.; Chakraborty, K.; Paul, B.N.; Sarma, D.; Syama Dayal, J.; Venkateshwarlu, G.; et al. DHA and EPA Content and Fatty Acid Profile of 39 Food Fishes from India. *BioMed Res. Int.* **2016**, *2016*, 4027437. [Google Scholar] [CrossRef] [PubMed][Green Version]
3. Innes, J.K.; Calder, P.C. Marine omega-3 (N-3) fatty acids for cardiovascular health: An Update for 2020. *Int. J. Mol. Sci.* **2020**, *21*, 1362. [Google Scholar] [CrossRef] [PubMed][Green Version]
4. Dovale-Rosabal, G.; Rodríguez, A.; Contreras, E.; Ortiz-Viedma, J.; Muñoz, M.; Trigo, M.; Aubourg, S.P.; Espinosa, A. Concentration of EPA and DHA from refined salmon oil by optimizing the urea–fatty acid adduction reaction conditions using response surface methodology. *Molecules* **2019**, *24*, 1642. [Google Scholar] [CrossRef][Green Version]
5. Lin, W.; Wu, F.; Yue, L.; Du, Q.; Tian, L.; Wang, Z. Combination of urea complexation and molecular distillation to purify DHA and EPA from sardine oil ethyl esters. *J. Am. Oil Chem. Soc.* **2014**, *91*, 687–695. [Google Scholar] [CrossRef]
6. Montañés, F.; Tallon, S. Supercritical fluid chromatography as a technique to fractionate high-valued compounds from lipids. *Separations* **2018**, *5*, 38. [Google Scholar] [CrossRef][Green Version]
7. Oh, C.-E.; Kim, G.-J.; Park, S.-J.; Choi, S.; Park, M.-J.; Lee, O.-M.; Seo, J.-W.; Son, H.-J. Purification of high purity docosahexaenoic acid from Schizochytrium sp. SH103 using preparative-scale HPLC. *Appl. Biol. Chem.* **2020**, *63*, 1–8. [Google Scholar] [CrossRef]
8. Kuo, C.-H.; Huang, C.-Y.; Chen, J.-W.; Wang, H.-M.D.; Shieh, C.-J. Concentration of Docosahexaenoic and Eicosapentaenoic Acid from Cobia Liver Oil by

Acetone Fractionation of Fatty Acid Salts. *Appl. Biochem. Biotechnol.* **2020**, *192*, 517–529. [Google Scholar] [CrossRef]

9. Kralovec, J.A.; Zhang, S.; Zhang, W.; Barrow, C.J. A review of the progress in enzymatic concentration and microencapsulation of omega-3 rich oil from fish and microbial sources. *Food Chem.* **2012**, *131*, 639–644. [Google Scholar] [CrossRef]
10. Chiesa, G.; Busnelli, M.; Manzini, S.; Parolini, C. Nutraceuticals and bioactive components from fish for dyslipidemia and cardiovascular risk reduction. *Mar. Drugs* **2016**, *14*, 113. [Google Scholar] [CrossRef][Green Version]
11. Katanaev, V.L.; Di Falco, S.; Khotimchenko, Y. The anticancer drug discovery potential of marine invertebrates from Russian Pacific. *Mar. Drugs* **2019**, *17*, 474. [Google Scholar] [CrossRef] [PubMed][Green Version]
12. Chojnacka, A.; Gładkowski, W.; Grudniewska, A. Lipase-catalyzed transesterification of egg-yolk phophatidylcholine with concentrate of n-3 polyunsaturated fatty acids from cod liver oil. *Molecules* **2017**, *22*, 1771. [Google Scholar] [CrossRef][Green Version]
13. Kuo, C.-H.; Huang, C.-Y.; Lee, C.-L.; Kuo, W.-C.; Hsieh, S.-L.; Shieh, C.-J. Synthesis of DHA/EPA ethyl esters via lipase-catalyzed acidolysis using Novozym® 435: A kinetic study. *Catalysts* **2020**, *10*, 565. [Google Scholar] [CrossRef]
14. Shimada, Y.; Sugihara, A.; Nakano, H.; Kuramoto, T.; Nagao, T.; Gemba, M.; Tominaga, Y. Purification of docosahexaenoic acid by selective esterification of fatty acids from tuna oil with Rhizopus delemar lipase. *J. Am. Oil Chem. Soc.* **1997**, *74*, 97–101. [Google Scholar] [CrossRef]
15. Shimada, Y.; Watanabe, Y.; Sugihara, A.; Baba, T.; Ooguri, T.; Moriyama, S.; Terai, T.; Tominaga, Y. Ethyl esterification of docosahexaenoic acid in an organic solvent-free system with immobilized Candida antarctica lipase. *J. Biosci. Bioeng.* **2001**, *92*, 19–23. [Google Scholar] [CrossRef]
16. Bispo, P.; Batista, I.; Bernardino, R.J.; Bandarra, N.M. Preparation of triacylglycerols rich in omega-3 fatty acids from sardine oil using a Rhizomucor miehei lipase: Focus in the EPA/DHA Ratio. *Appl. Biochem. Biotechnol.* **2014**, *172*, 1866–1881. [Google Scholar] [CrossRef]
17. Chen, H.C.; Kuo, C.H.; Twu, Y.K.; Chen, J.H.; Chang, C.M.J.; Liu, Y.C.; Shieh, C.J. A continuous ultrasound-assisted packed-bed bioreactor for the lipase-catalyzed synthesis of caffeic acid phenethyl ester. *J. Chem. Technol. Biotechnol.* **2011**, *86*, 1289–1294. [Google Scholar] [CrossRef]
18. Kuo, C.-H.; Shieh, C.-J. Biocatalytic Process Optimization. *Catalysts* **2020**, *10*, 1303. [Google Scholar] [CrossRef]

19. Huang, S.-M.; Huang, H.-Y.; Chen, Y.-M.; Kuo, C.-H.; Shieh, C.-J. Continuous production of 2-phenylethyl acetate in a solvent-free system using a packed-bed reactor with Novozym® 435. *Catalysts* **2020**, *10*, 714. [Google Scholar] [CrossRef]

20. Zhou, N.; Shen, L.; Dong, Z.; Shen, J.; Du, L.; Luo, X. Enzymatic synthesis of thioesters from thiols and vinyl esters in a continuous-flow microreactor. *Catalysts* **2018**, *8*, 249. [Google Scholar] [CrossRef][Green Version]

21. Lindeque, R.M.; Woodley, J.M. Reactor selection for effective continuous biocatalytic production of pharmaceuticals. *Catalysts* **2019**, *9*, 262. [Google Scholar] [CrossRef][Green Version]

22. Yang, K.-R.; Tsai, M.-F.; Shieh, C.-J.; Arakawa, O.; Dong, C.-D.; Huang, C.-Y.; Kuo, C.-H. Ultrasonic-Assisted Extraction and Structural Characterization of Chondroitin Sulfate Derived from Jumbo Squid Cartilage. *Foods* **2021**, *10*, 2363. [Google Scholar] [CrossRef] [PubMed]

23. Lin, J.-A.; Kuo, C.-H.; Chen, B.-Y.; Li, Y.; Liu, Y.-C.; Chen, J.-H.; Shieh, C.-J. A novel enzyme-assisted ultrasonic approach for highly efficient extraction of resveratrol from Polygonum cuspidatum. *Ultrason. Sonochem.* **2016**, *32*, 258–264. [Google Scholar] [CrossRef] [PubMed]

24. Murillo, G.; He, Y.; Yan, Y.; Sun, J.; Bartocci, P.; Ali, S.S.; Fantozzi, F. Scaled-up biodiesel synthesis from Chinese Tallow Kernel oil catalyzed by Burkholderia cepacia lipase through ultrasonic assisted technology: A non-edible and alternative source of bio energy. *Ultrason. Sonochem.* **2019**, *58*, 104658. [Google Scholar] [CrossRef] [PubMed]

25. Alves, J.S.; Garcia-Galan, C.; Schein, M.F.; Silva, A.M.; Barbosa, O.; Ayub, M.A.; Fernandez-Lafuente, R.; Rodrigues, R.C. Combined effects of ultrasound and immobilization protocol on butyl acetate synthesis catalyzed by CALB. *Molecules* **2014**, *19*, 9562–9576. [Google Scholar] [CrossRef][Green Version]

26. Kuo, C.-H.; Hsiao, F.-W.; Chen, J.-H.; Hsieh, C.-W.; Liu, Y.-C.; Shieh, C.-J. Kinetic aspects of ultrasound-accelerated lipase catalyzed acetylation and optimal synthesis of 4′-acetoxyresveratrol. *Ultrason. Sonochem.* **2013**, *20*, 546–552. [Google Scholar] [CrossRef]

27. Onoja, E.; Chandren, S.; Razak, F.I.A.; Wahab, R.A. Enzymatic synthesis of butyl butyrate by Candida rugosa lipase supported on magnetized-nanosilica from oil palm leaves: Process Optimization, Kinetic and Thermodynamic Study. *J. Taiwan Inst. Chem. Eng.* **2018**, *91*, 105–118. [Google Scholar] [CrossRef]

28. Kuo, C.-H.; Chen, G.-J.; Chen, C.-I.; Liu, Y.-C.; Shieh, C.-J. Kinetics and optimization of lipase-catalyzed synthesis of rose fragrance 2-phenylethyl acetate through transesterification. *Process Biochem.* **2014**, *49*, 437–444. [Google Scholar] [CrossRef]

29. Sousa, R.R.; Silva, A.; Fernandez-Lafuente, R.; Ferreira-Leitão, V.S. Solvent-free esterifications mediated by immobilized lipases: A Review from Thermodynamic and Kinetic Perspectives. *Catal Sci. Technol.* **2021**, *11*, 5696–5711. [Google Scholar] [CrossRef]

30. Bhavsar, K.V.; Yadav, G.D. Synthesis of geranyl acetate by transesterification of geraniol with ethyl acetate over Candida antarctica lipase as catalyst in solvent-free system. *Flavour Fragr. J.* **2019**, *34*, 288–293. [Google Scholar] [CrossRef]

31. Kuo, C.-H.; Liu, T.-A.; Chen, J.-H.; Chang, C.-M.J.; Shieh, C.-J. Response surface methodology and artificial neural network optimized synthesis of enzymatic 2-phenylethyl acetate in a solvent-free system. *Biocatal. Agric. Biotechnol.* **2014**, *3*, 1–6. [Google Scholar] [CrossRef]

32. Zanwar, S.; Pangarkar, V. Solid-liquid mass transfer in packed beds: Enhancement Due to Ultrasound. *Chem. Eng. Commun.* **1988**, *68*, 133–142. [Google Scholar] [CrossRef]

33. Stavarache, C.; Vinatoru, M.; Nishimura, R.; Maeda, Y. Fatty acids methyl esters from vegetable oil by means of ultrasonic energy. *Ultrason. Sonochem.* **2005**, *12*, 367–372. [Google Scholar] [CrossRef]

34. Ji, J.; Wang, J.; Li, Y.; Yu, Y.; Xu, Z. Preparation of biodiesel with the help of ultrasonic and hydrodynamic cavitation. *Ultrasonics* **2006**, *44*, e411–e414. [Google Scholar] [CrossRef]

35. Liu, Y.; Jin, Q.; Shan, L.; Liu, Y.; Shen, W.; Wang, X. The effect of ultrasound on lipase-catalyzed hydrolysis of soy oil in solvent-free system. *Ultrason. Sonochem.* **2008**, *15*, 402–407. [Google Scholar] [CrossRef]

36. Yu, D.; Tian, L.; Wu, H.; Wang, S.; Wang, Y.; Ma, D.; Fang, X. Ultrasonic irradiation with vibration for biodiesel production from soybean oil by Novozym 435. *Process Biochem.* **2010**, *45*, 519–525. [Google Scholar] [CrossRef]

37. Patchimpet, J.; Zhang, Y.; Simpson, B.K.; Rui, X.; Sangkharak, K.; Eiad-ua, A.; Klomklao, S. Ultrasonic enhancement of lipase-catalyzed transesterification for biodiesel production from used cooking oil. *Biomass Convers. Biorefinery* **2021**, 1–10. [Google Scholar] [CrossRef]

38. Todero, L.M.; Bassi, J.J.; Lage, F.A.; Corradini, M.C.C.; Barboza, J.; Hirata, D.B.; Mendes, A.A. Enzymatic synthesis of isoamyl butyrate catalyzed by immobilized lipase on poly-methacrylate particles: Optimization, Reusability

and Mass Transfer Studies. *Bioprocess Biosyst. Eng.* **2015**, *38*, 1601–1613. [Google Scholar] [CrossRef]

39. Cavallaro, V.; Tonetto, G.; Ferreira, M.L. Optimization of the enzymatic synthesis of pentyl oleate with lipase immobilized onto novel structured support. *Fermentation* **2019**, *5*, 48. [Google Scholar] [CrossRef][Green Version]
40. Claon, P.A.; Akoh, C.C. Effect of reaction parameters on SP435 lipase-catalyzed synthesis of citronellyl acetate in organic solvent. *Enzym. Microb. Technol.* **1994**, *16*, 835–838. [Google Scholar] [CrossRef]
41. Kuo, C.; Chen, C.; Chiang, B. Process characteristics of hydrolysis of chitosan in a continuous enzymatic membrane reactor. *J. Food Sci.* **2004**, *69*, 332–337. [Google Scholar] [CrossRef]
42. Zenevicz, M.C.P.; Jacques, A.; Silva, M.J.A.; Furigo, A.; Oliveira, V.; de Oliveira, D. Study of a reactor model for enzymatic reactions in continuous mode coupled to an ultrasound bath for esters production. *Bioprocess. Biosyst. Eng.* **2018**, *41*, 1589–1597. [Google Scholar] [CrossRef] [PubMed]
43. Tran, D.-T.; Chen, C.-L.; Chang, J.-S. Continuous biodiesel conversion via enzymatic transesterification catalyzed by immobilized Burkholderia lipase in a packed-bed bioreactor. *Appl. Energy* **2016**, *168*, 340–350. [Google Scholar] [CrossRef]
44. Wang, X.; Liu, X.; Zhao, C.; Ding, Y.; Xu, P. Biodiesel production in packed-bed reactors using lipase–nanoparticle biocomposite. *Bioresour. Technol.* **2011**, *102*, 6352–6355. [Google Scholar] [CrossRef]
45. Aljawish, A.; Heuson, E.; Bigan, M.; Froidevaux, R. Lipase catalyzed esterification of formic acid in solvent and solvent-free systems. *Biocatal. Agric. Biotechnol.* **2019**, *20*, 101221. [Google Scholar] [CrossRef]
46. Jaiswal, K.S.; Rathod, V.K. Green synthesis of amyl levulinate using lipase in the solvent free system: Optimization, Mechanism and Thermodynamics Studies. *Catal. Today* **2021**, *375*, 120–131. [Google Scholar] [CrossRef]
47. Xu, C.; Zhang, H.; Shi, J.; Zheng, M.; Xiang, X.; Huang, F.; Xiao, J. Ultrasound irradiation promoted enzymatic alcoholysis for synthesis of monoglyceryl phenolic acids in a solvent-free system. *Ultrason. Sonochem.* **2018**, *41*, 120–126. [Google Scholar] [CrossRef]
48. Bayramoğlu, G.; Hazer, B.; Altıntaş, B.; Arıca, M.Y. Covalent immobilization of lipase onto amine functionalized polypropylene membrane and its application in green apple flavor (ethyl valerate) synthesis. *Process Biochem.* **2011**, *46*, 372–378. [Google Scholar] [CrossRef]

49. Mahapatra, P.; Kumari, A.; Kumar, G.V.; Banerjee, R.; Nag, A. Kinetics of solvent-free geranyl acetate synthesis by Rhizopus oligosporus NRRL 5905 lipase immobilized on to cross-linked silica. *Biocatal. Biotransform.* **2009**, *27*, 124–130. [Google Scholar] [CrossRef]

50. Rodriguez, J.-M.G.; Hux, N.P.; Philips, S.J.; Towns, M.H. Michaelis–Menten graphs, Lineweaver–Burk plots, and reaction schemes: Investigating Introductory Biochemistry Students' Conceptions of Representations in Enzyme Kinetics. *J. Chem. Educ.* **2019**, *96*, 1833–1845. [Google Scholar] [CrossRef]

51. Kuo, C.-H.; Shieh, C.-J.; Huang, S.-M.; Wang, H.-M.D.; Huang, C.-Y. The effect of extrusion puffing on the physicochemical properties of brown rice used for saccharification and Chinese rice wine fermentation. *Food Hydrocoll.* **2019**, *94*, 363–370. [Google Scholar] [CrossRef]

52. Bi, Y.; Wang, Z.; Tian, Y.; Fan, H.; Huang, S.; Lu, Y.; Jin, Z. Highly efficient regioselective decanoylation of hyperoside using nanobiocatalyst of Fe_3O_4@PDA-thermomyces lanuginosus lipase: Insights of Kinetics and Stability Evaluation. *Front. Bioeng. Biotechnol.* **2020**, *8*, 485. [Google Scholar] [CrossRef] [PubMed]

53. Khan, N.R.; Jadhav, S.V.; Rathod, V.K. Lipase catalysed synthesis of cetyl oleate using ultrasound: Optimisation and Kinetic Studies. *Ultrason. Sonochem.* **2015**, *27*, 522–529. [Google Scholar] [CrossRef]

54. Tomke, P.D.; Rathod, V.K. Ultrasound assisted lipase catalyzed synthesis of cinnamyl acetate via transesterification reaction in a solvent free medium. *Ultrason. Sonochem.* **2015**, *27*, 241–246. [Google Scholar] [CrossRef]

CHAPTER-14

DESIGNING A TRYPSIN VARIANT INDEPENDENT OF CALCIUM THROUGH RATIONAL MEANS

1. INTRODUCTION: AN OVERVIEW

In the pursuit of enhancing the versatility and applicability of trypsin, a pivotal enzyme widely employed in various biotechnological and industrial processes, the focus has shifted towards the rational design of calcium-independent trypsin variants. Traditionally, trypsin activity has been closely associated with calcium ions, necessitating their presence for optimal enzymatic function. However, the development of a calcium-independent trypsin variant holds significant promise for overcoming limitations associated with calcium dependency. This endeavour involves a meticulous and rational approach to molecular design, with the aim of creating a trypsin variant capable of exhibiting robust enzymatic activity in the absence of calcium ions. This innovative approach not only broadens the scope of trypsin applications but also contributes to the fundamental understanding of enzyme structure-function relationships, paving the way for advancements in enzymology and biocatalysis.

In nature, the serine protease trypsin (EC 3.4.21.4) is involved in the digestive hydrolysis of proteins into peptides in many vertebrates and has been used widely in various biotechnological processes. The enzyme cleaves peptide bonds on the carboxyl side of lysine and arginine. Catalytic activity is mediated by three highly conserved residues corresponding to the catalytic triad H57, D102, and S195 (chymotrypsinogen numbering) [1,2]. Like many other proteases, trypsin is synthesized as an inactive zymogen and converted into the active enzyme by limited proteolytical cleavage between Arg15 and Ile16 [3]. This cleavage induces conformational changes in the enzyme by forming a salt bridge between Ile16 and Asp194. This interaction stabilizes the substrate-binding site and the oxyanion hole, solidifying the transition state negative charge of the scissile peptide bond carbonyl oxygen [4,5]. The stability of the active state of trypsin is commonly mediated by three disulfide bonds (C42-C58, C168-C182, and C191-C220), which mainly contribute to the high catalytic activity, substrate specificity as well as catalytic robustness in non-physiological reaction conditions [6]. Furthermore, the enzyme has no need for cofactors, which is crucial for many applications.

Nevertheless, trypsin is highly dependent on Ca^{2+}-ions [7]. There are two calcium-binding sites within the enzyme. The first one is localized in the zymogenic peptide and necessary for activation processes via enterokinase. The other calcium-binding site is localized in the calcium-binding loop (CBL) and is necessary for the enzyme's activity. The absence of calcium ions leads to a considerable increase in autodigestion [8]. Furthermore, a temperature-dependent trypsin activation by calcium ions can be observed [7]. This Ca^{2+}-induced thermal stability is also described for other proteases like thermolysin or subtilisins [9,10].

One of the most striking uses of trypsin is the detachment of adherent cells from culture flasks in animal cell cultures. Since exogenous calcium causes cell differentiation and inhibition of proliferation, trypsin is used with EDTA for this application [11]. Since trypsin activity and stability are Ca^{2+} dependent, the use of EDTA makes it necessary to increase the amount of enzyme, which could alter the physiology, protein expression, and metabolism of cultured cells [12].

Besides cell culture, trypsin is also the most used enzyme in proteomics. Its cleavage carboxyterminal of Arg and Lys results in a positive charge at the peptide C-terminus, being advantageous for MS analysis [13]. Furthermore, some MS applications would profit from a Ca^{2+}-free buffer system due to the prevention of insoluble salts resulting in a possible decrease of ion signals due to buffer-induced ionization suppression [14].

The insertion of artificial cysteines forming cross-linkage within the protein of interest to increase stability is valuable in protein chemistry [15,16,17]. In the case of subtilisin, the deletion of the calcium-binding loop leads to a drastic stability reduction, which can be restored by inserting a disulfide bridge within the enzyme [18]. This work demonstrates that reduced stability in subtilisin can be compensated by inserting a disulfide bridge. In the present work we were able to show that in trypsin the calcium binding site can even be directly replaced by a disulfide bridge, and thus a more stable, calcium-independent trypsin variant was generated.

2. RESULTS AND DISCUSSION

2.1. Biosynthesis and Titration of Free Thiols

In the CBL of anionic rat trypsin II (Tn), the Ca^{2+}-ion is coordinated by electrostatic interactions with the side chain of Glu70 and Glu80 and the backbone carbonyl oxygens of Asn72 and Val75 (**Figure 1**b) [**19**]. The distance between Cα atoms of Glu70 and Glu80 is 5.6 Å and fits ideally to the length of a disulfide bond. This fact, and a relatively high conservation score of six (determined by the bioinformatics analysis with Consurf, **Figure 1**a), encouraged us to create a trypsin variant with an additional disulfide bridge at this position [**20**]. Therefore, a calcium-independent, stabilized apo-trypsin (aTn) bearing the amino acid substitutions E70C and E80C was generated by site-directed mutagenesis, expressed as inactive zymogen in *Saccharomyces cerevisiae,* and finally purified and activated by enterokinase. To prove the concept, wildtype Tn (wt-Tn) and a single cysteine variant (mC-Tn) bearing a single E70C mutation were chosen as controls. Both enzyme variants were generated similarly to aTn.

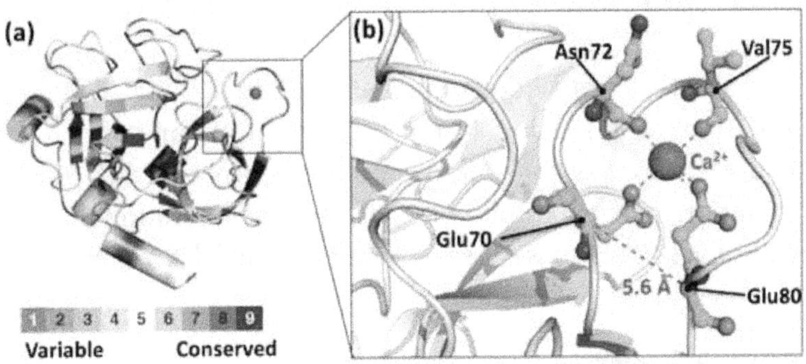

Figure 1. Level of conservation and localization of the CBL of trypsin. (a) Illustration of the degree of conservation of trypsin. Highly conserved residues are shown in red and variable residues are shown in blue. (b) Depiction of the CBL of anionic rat trypsin II. The coordinating residues Glu70, Asn72, Val75, and Glu80 are shown in green and the stick form. Additionally, electrostatic interactions, as well as Ca^{2+}, are shown in magenta. For analysis and the depiction, the pdb-file 1ITZ was used.

After biosynthesis and preparation, Tn species were investigated for the presence of free thiol functionalities, to indirectly determine the formation of disulfide bonds. Therefore, free thiols were titrated with Ellmann's reagent (DTNB: 5,5'-dithio-bis-(2-nitrobenzoic acid)), which enables detection of reduced cysteine side chains by converting DTNB to 2-nitro-5-thiobenzoate, which can be quantified spectrophotometrically [**21**]. In the case of wt-Tn, no absorption signal at 412 nm, correlating to thionitrobenzoic acid products, was detected. This result implies that all 12 cysteines are disulfide-bridged (**Table 1**). In contrast, 65% of mC-Tn have a free sulfhydryl moiety, indicating that the artificial cysteine is not involved in forming an intra-molecular disulfide bridge. It can be assumed that the remaining 35% are involved in the formation of intermolecular disulfide bonds, resulting in the formation of dimers. This assumption is also confirmed by non-reducing SDS-PAGE. The aTn variant shows a proportion of free cysteines of only 4.1%, indicating that almost every one of the 14 cysteine side chains is involved in disulfide bond formation. Thus, we have generated a trypsin species containing seven disulfide bridges.

Table 1. Overview of the catalytic parameters for Bz-Arg-AMC turnover and proportion of free thiol (SH_{free}) species for Tn variants.

aTn	wt-Tn	mC-Tn	
K_M (μM) [a]	56.7 ± 2.8	70.2 ± 3.7	54.9 ± 4.8
k_{cat} (s^{-1}) [a]	0.11	0.15	0.07
k_{cat}/K_M (M^{-1} s^{-1}) [a]	1900 ± 200	2100 ± 200	1300 ±200
SH_{free} (%) [b]	4.1	0	65

[a] Kinetic parameters were determined at 30 °C in 100 mM HEPES (pH 7.8), 100 mM NaCl, and 10 mM $CaCl_2$ using 16 nM of Tn variants and varying Bz-Arg-AMC from 5 to 200 μM. Hydrolysis was monitored via fluorescence (λ_{ex}=381 nm, λ_{em}=455 nm) [b] Determination of free thiols was carried out by following the procedure described by Ellman [**21**]. Errors represent the standard deviation of three technical replicates.

2.2. Enzymatic Activity

Enzymatic activity of wt-Tn, mC-Tn, and aTn was determined for the substrate Bz-Arg-AMC (benzoyl-arginyl-7-amido-4-methylcoumarin. Enzymatic cleavage of the carboxamide bond between Arg and AMC releases 7-amino-4-methylcoumarin, which results in an increased fluorescence signal (λ_{ex} = 381 nm, λ_{em} = 455 nm). According to the Michaelis-Menten-kinetics, trypsin species were characterized by analyzing the kinetic parameters of the hydrolysis, namely k_{cat}, K_M, and k_{cat}/K_M. The v/S-plots are depicted, and the corresponding kinetic parameters are summarized in **Table 1**. As a result, the presence of artificial cysteine residues in the CBL does not influence the enzymatic activity of trypsin. Furthermore, K_M and k_{cat} are in the same order of magnitude for all Tn variants. However, the K_M value for wt-Tn is slightly higher (K_M = 70 μM) than the cysteine variants mC-Tn and aTn (K_M = 55-57 μM). On the other hand, the k_{cat} of wt-Tn is marginally higher (k_{cat} = 0.15 s^{-1}) than for aTn (k_{cat} = 0.11 s^{-1}). aTn also shows a slightly higher k_{cat} value than the single cysteine variant mC-Tn (k_{cat} = 0.07 s^{-1}). In general, all kinetic constants determined in this work are in the same order of magnitude as described in the literature for bovine trypsin (Tn(bov)) and the aforementioned Bz-Arg-AMC substrate [22].

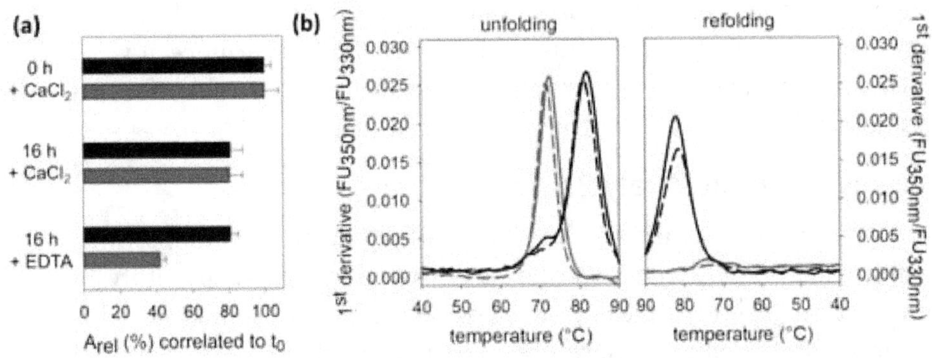

Figure 2. Evaluation of the contribution of the E70/80C disulfide bridge in aTn to the calcium dependence of enzymatic activity and the thermal stability (a) The bar chart shows the influence on the reduction of the relative activity A_{rel} in the presence and absence (EDTA) of Ca^{2+} after 16 h in comparison to the starting activity at 0 h (t_0). The activities were related to the respective starting point (A) of either aTn (shown in black) or wt-Tn (shown in blue). (b) The micro-thermophoresis method depicts the thermal denaturation of aTn (shown in black) and wt-Tn (shown in blue). A_{rel} was determined at 30°C in 100 mM HEPES (pH 7.8), 100 mM NaCl, and either 10 mM $CaCl_2$ or 0.2 mM EDTA using 16 nM of Tn variants and 200 µM Bz-Arg-AMC. Unfolding and folding was monitored using 20 µM trypsin, 10 mM $CaCl_2$ or 0.2 mM EDTA in 20 mM MES, 150 mM NaCl, pH 5.5. The temperature was increased from 40 to 90°C at a ramp rate of 1°C/min. Errors represent the standard deviation of three technical replicates.

Determination of Ca^{2+}-dependency on the activity and stability of wt-Tn and aTn was carried out after 16 h incubation in the presence and absence of Ca^{2+} and subsequent activity measurements (Figure 2a). Reaction mixtures without Ca^{2+} were additionally treated with EDTA to remove remaining calcium ions by chelation. The incubation in the presence of Ca^{2+} leads to an expectable reduction of activity ($A_{rel, wt\text{-}Tn}$ = 80%; $A_{rel, aTn}$ = 81%), resulting from autolytic processes. On the contrary, in the absence of Ca^{2+} (which was supported by the addition of EDTA) the activity of aTn (A_{rel} = 81%) is nearly unchanged (Figure 2a). At the same time, the activity of wt-Tn is drastically reduced by 41%. This effect is explained by an increased autolysis rate due to the missing Ca^{2+} in the case of wt-Tn, indicating that aTn benefits from the artificial disulfide bond [23]. In addition, the disulfide bridge also contributes to increased thermal stability. Corresponding measurements were carried out with the appropriate trypsinogen variants and at a lowered pH to prevent autolytic events, which would result in fragmentation of the trypsins during thermal denaturation. Thermal unfolding and refolding of wt-Tn and aTn was measured by real-time simultaneous monitoring of internal tryptophane fluorescence at 330 nm and 350 nm differential scanning fluorimetry. Melting curves are depicted in **Figure 2**b. Melting temperature increases from 72.4°C in the case of wt-Tn to 81.7°C in the case of aTn in the presence of Ca^{2+}. A similar stabilizing effect is also observed in the presence of EDTA (wt-Tn = 71.6°C, aTn = 81.1°C). Furthermore, it was found that the aTn does not seem to denature completely even at 90°C. Therefore, refolding of aTn is detectable, which is not the case with wt-Tn.

2.3. X-ray Structure Analysis

Anionic rat trypsin sometimes can be challenging to crystallize. One reason is the high proteolytic activity of the enzyme, which leads to autolysis and thus fragmentation [**23**]. In addition, rat trypsin variants are sometimes characterized by poor crystallizability. Therefore, as in previous studies, the sequence-like (sequence similarity 86%, for more detail see **Figure S6**) and structurally identical bovine trypsin was used for crystallization [**24**]. In addition, all variants were produced as inactive S195A variants to prevent autolysis as well as optional folding influences by inhibitors. As a positive side effect, this strategy allowed us to investigate the transferability of the motif between different trypsin species.

After biosynthesis, purification, and refolding of bovine trypsin variants, circular dichroism spectroscopy was done to prove correct protein folding, as activity measurements with the S195A mutation were not possible (**Figure S7**). In addition, thermodynamic denaturation was examined, showing a 4.2°C increase in stability comparing wt-Tn(bov) S195A (67.6°C) to aTn(bov) S195A (71.8°C). This result is similar to the stabilization effects of rat trypsin variants, although not quite as pronounced.

As a result of crystallization, the structure of aTn(bov) S195A was solved. The resolution was 1.40 Å, with a completeness of the structure of 98%. Residual 2% correspond to the region from residue 73 to 79, which could not be resolved (**Figure 2**a). This region corresponds to the original CBL. The increased flexibility can be attributed to the missing Ca^{2+}-induced coordination of the residues Asn72 and Val75. Despite this result, there is still increased structural integrity created by a new formed disulfide bridge between Cys70 and Cys80, proven by the electron density map (FoFc and 2FoFc). The correct positioning of the disulfide bond is also verified in the overlay structure of aTn(bov) S195A and wt-Tn(bov).

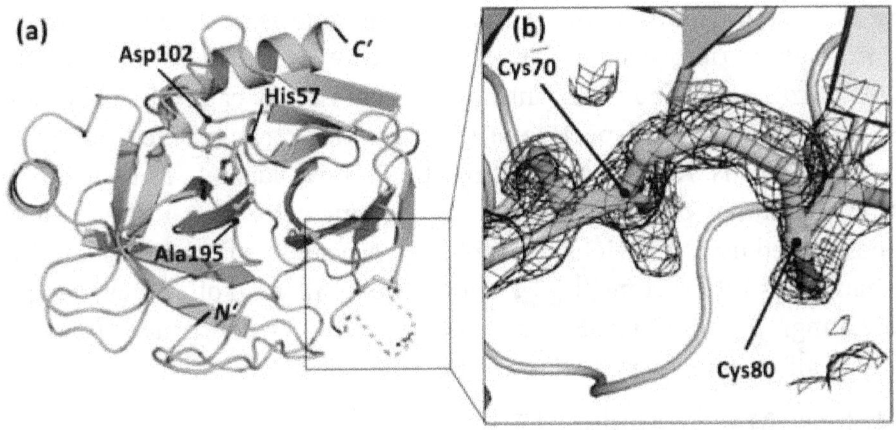

Figure 3. Crystal structure of aTn(bov) S195A. (a) The whole structure is shown as cartoon. For better orientation, the catalytic center consisting of His57, Asp102, and Ala195 are marked in the green sticks. The unresolved residues 73 to 79 are shown as dashed lines. (b) Depicted is the introduced disulfide bridge. Both cysteine residues are shown as green sticks. In addition, the electron density (FoFc and the 2FoFc) maps are indicated as black mesh (pdb-ID 8ADT).

2.4. Conclusions

Thus, the crystal structure supports the presence of the expected additional disulfide bridge in trypsin and confirms the activity and stability measurements. Furthermore, we demonstrated that a disulfide bridge between residue 70 and 80 is formed in both rat and bovine trypsin. In both cases, this is beneficial for the stability and Ca^{2+} independence of the enzyme. This feature emphasizes a potential universal stabilization strategy for the conformation and catalytic activity of Ca^{2+}-dependent trypsins in numerous applications such as mass spectrometry and biocatalysis.

3. MATERIALS AND METHODS

3.1. Construction, Biosynthesis, and Purification of wt-Trypsin and Trypsin Variants

All trypsin variants were generated by site-directed mutagenesis using *Pfu* DNA polymerase (Thermo Scientific, Waltham, MA, USA) and either a pST-vector (Tn) or a pET-vector (Tn(bov)) as described previously [25,26]. All mutations were introduced using pairs of complementary primers. The sequence of all generated constructs was confirmed by DNA sequencing (LGC Genomics, Berlin, Germany). The gene constructs of Tn-variants were subsequently cloned into pYT-expression-vector and transformed into *Saccharomyces cerevisiae* DLM 101a cells. Biosynthesis, purification of zymogenic Tn, and subsequent activation using enterokinase

(Roche Diagnostics, Mannheim, Germany) were done as described before [25]. For preventing autolytic processes, Tn variants were stored in 10 mM HCl at -20°C. For crystallization, Tn(bov) variants were expressed as catalytic inactive species (S195A). Therefore, the respective pET-vectors were transformed into *Escherichia coli* BL21 (DE3) cells. After accumulating the Tn(bov) variants as inclusion bodies, isolation, refolding, purification, and subsequent activation using enterokinase were performed using previously established procedures [27]. The overall yield of Tn variants was 1.5 to 2.1 $mg/l_{culture}$, while the overall yield of Tn(bov) variants was 0.8 to 2.3 $mg/l_{culture}$. SDS-PAGE and mass spectrometry confirmed the purity and identity of all protein variants (**Figure S11**).

3.2. Activity Measurement of Tn

The activity of Tn and Tn variants was determined while monitoring the hydrolysis of benzoyl-L-arginine-7-amido-4-methyl coumarin (Bz-Arg-AMC, Bachem, Bubendorf, Switzerland) at an F-310 fluorescence spectrometer (Hitachi, Tokio, Japan) [28]. In detail, 16 nM of the corresponding Tn was dissolved in 100 mM HEPES (pH 7.8), 10 mM $CaCl_2$, 100 mM NaCl, and 5-200 µM Bz-Arg-AMC (dissolved in *N,N*-dimethylformamide). Increasing fluorescence was monitored for 5 min at 20°C using an excitation wavelength of 381 nm and an emission wavelength of 455 nm. The catalytic properties k_{cat} and K_M were determined with the Michealis-Menten-regression from the v/[S]-regression curves using Origin8.1 (OriginLab Corporation, Northampton, MA, USA) [29]. To determine Ca^{2+}-dependency, wt-Tn and aTn were incubated in 100 mM HEPES (pH 7.8), 100 mM NaCl containing either 10 mM $CaCl_2$ or 10 mM EDTA. After 16 h of incubation, measurements using 16 nM of the corresponding Tn and 200 µM of Bz-Arg-AMC for 5 min at 20°C were performed.

3.3. Determination of Free Sulfhydryl Groups

Ellman's protocol was used to determine the number of accessible sulfhydryl functionalities within all protein variants to verify the correct formation of the disulfide bonds [21,30]. For calibration, a concentration of cysteamine (5–50 µM) was used. For measurement, 65 µM of the particular trypsin variant was dissolved in 100 mM Tris/HCl (pH 7.5), and 0.1 mM of DTNB (dissolved in dimethyl sulfoxide) was added. After incubation for 5 min at room temperature, absorbance was measured at 412 nm with a NOVOstar plate reader (BMGLabtech, Ortenberg, Germany).

3.4. Determination of Thermostability

Real-time simultaneous monitoring of the internal tryptophane fluorescence at 330 nm and 350 nm during thermal unfolding and refolding of wt-Tn and aTn was measured on a Prometheus NT.48 instrument (Nanotemper, Munich, Germany) with an excitation wavelength of 280 nm [31]. Capillaries were filled with 10 µL of a suspension containing 20 μM of trypsin in presence of $CaCl_2$ (c = 10 mM) and EDTA (c = 0.2 mM), respectively (in 20 mM MES, 150 mM NaCl pH 5.5). The temperature was increased from 40 to 90°C at a ramp rate of 1°C/min, with one fluorescence measurement per 0.027°C. The ratio of the recorded emission intensities (Em_{350nm}/Em_{330nm}) was plotted as a function of the temperature. The fluorescence intensity ratio and first derivative were calculated with the manufacturer's software (PR. ThermControl).

3.5. Crystallization

The protein solution of aTn(bov) S195A was concentrated to a final concentration of 10 mg/mL in 50 mM HEPES/NaOH (pH 7.8), 100 mM NaCl, and 10 mM $CaCl_{2,}$ and crystallized by hanging drop vapor diffusion at 20°C. Equal amounts of the protein solution and precipitant solution (0.2 M KNO_3, pH 6.5, 22% (*w*/*v*) PEG 3350) were mixed and incubated at 20°C. After approximately 14 days, the growth of trigonal crystal was observable.

Diffraction images of a single aTn(bov) S195A crystal were collected using a copper rotating-anode source (Cu K_α radiation (λ = 1.5418 Å), RA Micromax 007, Rigaku Europe, Neu-Isenburg, Germany) and a CCD detector (Saturn 944+, Rigaku Europe, Neu-Isenburg, Germany). Oscillation images were integrated, merged, and scaled using XDS to a resolution of 1.439 Å [**32**]. All datasets were processed with the HKL2000 suite, and structures were solved using Phaser's molecular replacement method using PDB coordinate file 1MTS as a search model [**33**,**34**,**35**]. Coot and REFMAC5 were used for model building and refinement, respectively [**36**,**37**]. PROCHECK analyzed structure quality [**38**]. All molecular images were generated by Pymol (Schrödinger, New York, NY, USA).

4. CONCISE SUMMARY

Trypsin is a long-known serine protease widely used in biochemical, analytical, biotechnological, or biocatalytic applications. The high biotechnological potential is based on its high catalytic activity, substrate specificity, and catalytic robustness in non-physiological reaction conditions. The latter is mainly due to its stable protein fold, to which six intramolecular disulfide bridges make a significant contribution. Although trypsin does not depend on cofactors, it essentially requires the binding of calcium ions to its calcium-binding site to obtain complete enzymatic activity and stability. This behaviour is inevitably associated with a limitation of the enzyme's applicability. To make trypsin intrinsically calcium-independent, we removed the native calcium-binding site and replaced it with another disulfide bridge. The resulting stabilized apo-trypsin (aTn) retains full catalytic activity as proven by enzyme kinetics. Studies using Ellmann's reagent further prove that the two inserted cysteines at positions Glu70 and Glu80 are in their oxidized state, creating the desired functional disulfide bond. Furthermore, aTn is independent of calcium ions, possesses increased thermal and functional stability, and significantly reduced autolysis compared to wildtype trypsin. Finally, we confirmed our experimental data by solving the X-ray crystal structure of aTn.

REFERENCES

1. Neurath, H.; Walsh, K.A. Role of proteolytic enzymes in biological regulation (a review). *Proc. Natl. Acad. Sci. USA* **1976**, *73*, 3825–3832. [Google Scholar] [CrossRef] [PubMed]
2. Schechter, I.; Berger, A. On the size of the active site in proteases. I. Papain. *Biochem. Biophys. Res. Commun.* **1967**, *27*, 157–162. [Google Scholar] [CrossRef]
3. Huber, R.; Bode, W. Structural Basis of the Activation, Action and Inhibition of Trypsin. *H S Z Physiol. Chem.* **1979**, *360*, 489. [Google Scholar] [CrossRef]
4. Fehlhammer, H.; Bode, W.; Huber, R. Crystal-Structure of Bovine Trypsinogen at 1.8 a Resolution.2. Crystallographic Refinement, Refined Crystal-Structure and Comparison with Bovine Trypsin. *J. Mol. Biol.* **1977**, *111*, 415–438. [Google Scholar] [CrossRef]
5. Bode, W.; Fehlhammer, H.; Huber, R. Crystal-Structure of Bovine Trypsinogen at 1.8 a Resolution.1. Data-Collection, Application of Patterson Search Techniques and Preliminary Structural Interpretation. *J. Mol. Biol.* **1976**, *106*, 325–335. [Google Scholar] [CrossRef]
6. Varallyay, E.; Lengyel, Z.; Graf, L.; Szilagyi, L. The role of disulfide bond C191-C220 in trypsin and chymotrypsin. *Biochem. Biophys. Res. Commun.* **1997**, *230*, 592–596. [Google Scholar] [CrossRef]
7. Sipos, T.; Merkel, J.R. An effect of calcium ions on the activity, heat stability, and structure of trypsin. *Biochemistry* **1970**, *9*, 2766–2775. [Google Scholar] [CrossRef]
8. Gabel, D.; Kasche, V. Autolysis of Beta-Trypsin-Influence of Calcium-Ions and Heat. *Acta Chem. Scand.* **1973**, *27*, 1971–1981. [Google Scholar] [CrossRef]
9. Dahlquist, F.W.; Long, J.W.; Bigbee, W.L. Role of Calcium in the thermal stability of thermolysin. *Biochemistry* **1976**, *15*, 1103–1111. [Google Scholar] [CrossRef]
10. Smith, C.A.; Toogood, H.S.; Baker, H.M.; Daniel, R.M.; Baker, E.N. Calcium-mediated thermostability in the subtilisin superfamily: The crystal structure of Bacillus Ak.1 protease at 1.8 angstrom resolution. *J. Mol. Biol.* **1999**, *294*, 1027–1040. [Google Scholar] [CrossRef]

11. Fujisaki, H.; Futaki, S.; Yamada, M.; Sekiguchi, K.; Hayashi, T.; Ikejima, T.; Hattori, S. Respective optimal calcium concentrations for proliferation on type I collagen fibrils in two keratinocyte line cells, HaCaT and FEPE1L-8. *Regen. Ther.* **2018**, *8*, 73–79. [Google Scholar] [CrossRef] [PubMed]

12. Huang, H.L.; Hsing, H.W.; Lai, T.C.; Chen, Y.W.; Lee, T.R.; Chan, H.T.; Lyu, P.C.; Wu, C.L.; Lu, Y.C.; Lin, S.T.; et al. Trypsin-induced proteome alteration during cell subculture in mammalian cells. *J. Biomed. Sci.* **2010**, *17*, 36. [Google Scholar] [CrossRef] [PubMed]

13. Huang, Y.; Triscari, J.M.; Tseng, G.C.; Pasa-Tolic, L.; Lipton, M.S.; Smith, R.D.; Wysocki, V.H. Statistical characterization of the charge state and residue dependence of low-energy CID peptide dissociation patterns. *Anal. Chem.* **2005**, *77*, 5800–5813. [Google Scholar] [CrossRef]

14. Donato, P.; Cacciola, F.; Tranchida, P.Q.; Dugo, P.; Mondello, L. Mass spectrometry detection in comprehensive liquid chromatography: Basic concepts, instrumental aspects, applications and trends. *Mass Spectrom. Rev.* **2012**, *31*, 523–559. [Google Scholar] [CrossRef]

15. Mitchinson, C.; Wells, J.A. Protein engineering of disulfide bonds in subtilisin BPN'. *Biochemistry* **1989**, *28*, 4807–4815. [Google Scholar] [CrossRef] [PubMed]

16. Neubacher, S.; Saya, J.M.; Amore, A.; Grossmann, T.N. In Situ Cyclization of Proteins (INCYPRO): Cross-Link Derivatization Modulates Protein Stability. *J. Org. Chem.* **2020**, *85*, 1476–1483. [Google Scholar] [CrossRef] [PubMed]

17. Then, J.; Wei, R.; Oeser, T.; Gerdts, A.; Schmidt, J.; Barth, M.; Zimmermann, W. A disulfide bridge in the calcium binding site of a polyester hydrolase increases its thermal stability and activity against polyethylene terephthalate. *FEBS Open Bio* **2016**, *6*, 425–432. [Google Scholar] [CrossRef] [PubMed]

18. Strausberg, S.L.; Alexander, P.A.; Gallagher, D.T.; Gilliland, G.L.; Barnett, B.L.; Bryan, P.N. Directed evolution of a subtilisin with calcium-independent stability. *Biotechnology* **1995**, *13*, 669–673. [Google Scholar] [CrossRef]

19. Pasternak, A.; Ringe, D.; Hedstrom, L. Comparison of anionic and cationic trypsinogens: The anionic activation domain is more flexible in solution and differs in its mode of BPTI binding in the crystal structure. *Protein Sci.* **1999**, *8*, 253–258. [Google Scholar] [CrossRef]

20. Ben Chorin, A.; Masrati, G.; Kessel, A.; Narunsky, A.; Sprinzak, J.; Lahav, S.; Ashkenazy, H.; Ben-Tal, N. ConSurf-DB: An accessible repository for the evolutionary conservation patterns of the majority of PDB proteins. *Protein Sci.* **2020**, *29*, 258–267. [Google Scholar] [CrossRef][Green Version]

21. Ellman, G.L. Tissue Sulfhydryl Groups. *Arch. Biochem. Biophys.* **1959**, *82*, 70–77. [Google Scholar] [CrossRef]

22. Grahn, S.; Ullmann, D.; Jakubke, H. Design and synthesis of fluorogenic trypsin peptide substrates based on resonance energy transfer. *Anal. Biochem.* **1998**, *265*, 225–231. [Google Scholar] [CrossRef] [PubMed]

23. Nord, F.F.; Bier, M.; Terminiello, L. On the Mechanism of Enzyme Action.61. The Self Digestion of Trypsin, Calcium-Trypsin and Acetyltrypsin. *Arch. Biochem. Biophys.* **1956**, *65*, 120–131. [Google Scholar] [CrossRef]

24. Perona, J.J.; Tsu, C.A.; Craik, C.S.; Fletterick, R.J. Crystal structures of rat anionic trypsin complexed with the protein inhibitors APPI and BPTI. *J. Mol. Biol.* **1993**, *230*, 919–933. [Google Scholar] [CrossRef] [PubMed]

25. Hedstrom, L.; Szilagyi, L.; Rutter, W.J. Converting trypsin to chymotrypsin: The role of surface loops. *Science* **1992**, *255*, 1249–1253. [Google Scholar] [CrossRef]

26. Lundberg, K.S.; Shoemaker, D.D.; Adams, M.W.W.; Short, J.M.; Sorge, J.A.; Mathur, E.J. High-Fidelity Amplification Using a Thermostable DNA-Polymerase Isolated from Pyrococcus-Furiosus. *Gene* **1991**, *108*, 1–6. [Google Scholar] [CrossRef]

27. Liebscher, S.; Schopfel, M.; Aumuller, T.; Sharkhuukhen, A.; Pech, A.; Hoss, E.; Parthier, C.; Jahreis, G.; Stubbs, M.T.; Bordusa, F. N-terminal protein modification by substrate-activated reverse proteolysis. *Angew Chem. Int. Ed. Engl.* **2014**, *53*, 3024–3028. [Google Scholar] [CrossRef]

28. Lee, W.S.; Park, C.H.; Byun, S.M. Streptomyces griseus trypsin is stabilized against autolysis by the cooperation of a salt bridge and cation-pi interaction. *J. Biochem.* **2004**, *135*, 93–99. [Google Scholar] [CrossRef]

29. Orsi, B.A.; Tipton, K.F. Kinetic analysis of progress curves. *Methods Enzymol.* **1979**, *63*, 159–183. [Google Scholar] [CrossRef]

30. Butterworth, P.H.; Baum, H.; Porter, J.W. A modification of the Ellman procedure for the estimation of protein sulfhydryl groups. *Arch. Biochem. Biophys.* **1967**, *118*, 716–723. [Google Scholar] [CrossRef]

31. Magnusson, A.O.; Szekrenyi, A.; Joosten, H.J.; Finnigan, J.; Charnock, S.; Fessner, W.D. nanoDSF as screening tool for enzyme libraries and biotechnology development. *FEBS J.* **2019**, *286*, 184–204. [Google Scholar] [CrossRef] [PubMed]

32. Kabsch, W. XDS. *Acta Crystallogr. D Biol. Crystallogr.* **2010**, *66*, 15–132. [Google Scholar] [CrossRef] [PubMed]

33. Otwinowski, Z.; Minor, W. Processing of X-ray Diffraction Data Collected in Oscillation Mode. *Methods Enzymol.* **1997**, *276*, 307–326. [Google Scholar] [CrossRef] [PubMed]

34. McCoy, A.J.; Grosse-Kunstleve, R.W.; Adams, P.D.; Winn, M.D.; Storoni, L.C.; Read, R.J. Phaser crystallographic software. *J. Appl. Crystallogr.* **2007**, *40*, 658–674. [Google Scholar] [CrossRef]

35. Stubbs, M.T.; Huber, R.; Bode, W. Crystal structures of factor Xa specific inhibitors in complex with trypsin: Structural grounds for inhibition of factor Xa and selectivity against thrombin. *FEBS Lett.* **1995**, *375*, 103–107. [Google Scholar] [CrossRef]

36. Emsley, P.; Lohkamp, B.; Scott, W.G.; Cowtan, K. Features and development of Coot. *Acta Crystallogr. D Biol. Crystallogr.* **2010**, *66*, 486–501. [Google Scholar] [CrossRef] [PubMed]

37. Murshudov, G.N.; Skubak, P.; Lebedev, A.A.; Pannu, N.S.; Steiner, R.A.; Nicholls, R.A.; Winn, M.D.; Long, F.; Vagin, A.A. REFMAC5 for the refinement of macromolecular crystal structures. *Acta Crystallogr. D Biol. Crystallogr.* **2011**, *67*, 355–367. [Google Scholar] [CrossRef]

38. Laskowski, R.A.; Rullmannn, J.A.; MacArthur, M.W.; Kaptein, R.; Thornton, J.M. AQUA and PROCHECK-NMR: Programs for checking the quality of protein structures solved by NMR. *J. Biomol. NMR* **1996**, *8*, 477–486. [Google Scholar] [CrossRef]

39. Corpet, F. Multiple sequence alignment with hierarchical clustering. *Nucleic Acids Res.* **1988**, *16*, 10881–10890. [Google Scholar] [CrossRef]

40. Robert, X.; Gouet, P. Deciphering key features in protein structures with the new ENDscript server. *Nucleic Acids Res.* **2014**, *42*, W320. [Google Scholar] [CrossRef][Green Version]

41. Sreerama, N.; Venyaminov, S.Y.; Woody, R.W. Estimation of protein secondary structure from circular dichroism spectra: Inclusion of denatured proteins with native proteins in the analysis. *Anal Biochem.* **2000**, *287*, 243–251. [Google Scholar] [CrossRef] [PubMed]

42. McGrath,M.E.;Haymore,B.L.;Summers,N.L.;Craik,C.S.;Fletterick,R.J.Structure of an engineered, metal-actuated switch in trypsin. *Biochemistry* **1993**, *32*, 1914–1919. [Google Scholar] [CrossRef] [PubMed]

43. Laemmli, U.K. Cleavage of structural proteins during the assembly of the head of bacteriophage T4. *Nature* **1970**, *227*, 680–685. [Google Scholar] [CrossRef] [PubMed]

44. Neuhoff, V.; Stamm, R.; Eibl, H. Clear Background and Highly Sensitive Protein Staining with Coomassie Blue Dyes in Polyacrylamide Gels-a Systematic Analysis. *Electrophoresis* **1985**, *6*, 427–448. [Google Scholar] [CrossRef]

INDEX

www.ingramcontent.com/pod-product-compliance
Lightning Source LLC
LaVergne TN
LVHW062112200926
848484LV00007B/1649